The Colour Image Processing Handbook

The Colour Image Processing Handbook

Edited by
S. J. Sangwine and R. E. N. Horne
Department of Engineering
University of Reading
UK

CHAPMAN & HALL
London · Weinheim · New York · Tokyo · Melbourne · Madras

Published by Chapman & Hall, an imprint of Thomson Science, 2–6 Boundary Row, London SE1 8HN, UK

Thomson Science, 2–6 Boundary Row, London SE1 8HN, UK

Thomson Science, 115 Fifth Avenue, New York, NY 10003, USA

Thomson Science, Suite 750, 400 Market Street, Philadelphia, PA 19106, USA

Thomson Science, Pappelallee 3, 69469 Weinheim, Germany

First edition 1998

© 1998 Chapman & Hall

Thomson Science is a division of International Thomson Publishing

Printed in Great Britain by The University Press, Cambridge

ISBN   0 412 80620 7

11798165

'Munsell', 'Judge' and 'ColorChecker' are registered trademarks of Gretag-Macbeth.
(http://www.gretagmacbeth.com/)
The names 'NCS' and 'Natural Color System' are used with permission of NCS - Scandinavian Colour Institute.
'ColorSync' is a registered trademark of Apple Computer Inc.
(http://www.apple.com)
'Calibrator' is a registered trademark of Barco Graphics.
(http://www.barco.com/)
'Fujichrome' is a registered trademark of Fujiphoto Film USA Inc.
(http://www.fujifilm.com/)
'Kodachrome Gold' and 'Ektachrome' are registered trademarks of Eastman Kodak Company.
(http://www.kodak.com/)
'Genesis' is a trademark of Matrox Electronic Systems Ltd.
(http://www.matrox.com/)
'Colour Dimensions' is a trademark of Imperial Chemical Industries.
'TekColor' and 'TekHVC' are trademarks of Tektronix Inc.
'Pentium' is a trademark of Intel Corporation.
'X Windows' is a trademark of Open Group.
(http://www.opengroup.org/tech/desktop/X/)
'Unix' is a trademark of AT&T.

Contents

List of contributors

Maciej Bartkowiak is finishing his PhD in the Institute of Electronics and Telecommunications at Poznan University of Technology, Poland. His published work to date is in the fields of colour image processing and image data compression.

Dr Mary L. Comer is currently working in the Corporate Innovation and Research Department at Thomson Consumer Electronics, Indianapolis, Indiana, USA, where she is involved in the development of digital video products.

Dr Christine Connolly formerly of the School of Engineering at the University of Huddersfield, is a Director of Colour Valid Ltd., Bradford, West Yorkshire, UK. Her current work is in the area of non-contact colour inspection for a range of industrial applications.

Dr Edward J. Delp is Professor of Electrical Engineering in the School of Electrical Engineering at Purdue University, West Lafayette, Indiana, USA. His research interests include image and video compression, medical imaging, parallel processing, multimedia systems, ill-posed inverse problems in computational vision, nonlinear filtering using mathematical morphology, communication and information theory. He has consulted for various companies and government agencies in the areas of signal and image processing, robot vision, pattern recognition, and secure communications and has published and presented over 170 papers.

Dr Marek Domański is Professor of Image Processing and Multidimensional Systems in the Institute of Electronics and Telecommunications at Poznan University of Technology, Poland. His research activities include multidimensional digital filters (linear and nonlinear), passive digital systems, image and video compression, subband coding, image enhancement and restoration. He was a visiting scientist at the University of Ruhr, Bochum, Germany in 1986–1987 and 1990–1991. He has published and presented about 100 papers.

Dr John M. Gauch is an Associate Professor in the Department of Electrical Engineering and Computer Science at the University of Kansas, USA, where he has developed a comprehensive system of image processing libraries to support teaching and computer vision research programmes.

Robin E. N. Horne has been a Senior Scientific Officer in the Department of Engineering at the University of Reading, UK since 1981. His technical interests are in video engineering and image processing and he has lectured on these subjects at undergraduate and postgraduate level since 1983.

Professor Bryan F. Jones of the Department of Computer Studies at the University of Glamorgan, UK, has worked since 1985 on medical imaging. His particular current field of interest is the measurement and analysis of wounds.

Dr M. Ronnier Luo is a Professor of Colour Science in the Colour and Imaging Institute at the University of Derby, UK. He has more than a decade of experience in applying the theories of colour science to industrial problems and is a member of numerous technical committees of the CIE. In addition, he is currently the chairman of the UK's Colour Measurement Committee (CMC) and was awarded the prestigious Bartleson Award for his contribution to colour science.

Dr William McIlhagga has an MSc in mathematics and a PhD in vision science from the Australian National University. He is currently working on models for detection and discrimination of visual stimuli, and MRI investigations of visual operations at The Psychological Laboratory, Copenhagen University, Denmark.

Ján Morovic is currently working towards a PhD in the Colour and Imaging Institute at the University of Derby, UK. He is developing a universal gamut mapping algorithm.

Dr Henryk Palus is a Lecturer in the Department of Automatic Control at the Silesian University of Technology, Gliwice, Poland. He has an extensive research background in the areas of shape analysis, sensor systems, robot vision systems, neural networks, stereovision systems, knowledge-based vision systems and colour computer vision. He was awarded the DAAD Scholarship in 1992 (University of Magdeburg and Technical College Ulm) and a British Council Scholarship in 1995 (The University of Reading). Dr Palus is an Advisory Editor of Machine Graphics and Vision and charter-member of the Polish Association of Image Processing (IAPR member).

Peter Plassmann is a Research Fellow in the Department of Computer Studies at the University of Glamorgan, UK. He is an electrical engineer with research interests in the fields of medical computing and image processing, with emphasis on thermal and 3D imaging of wounds.

Dr Konstantinos N. Plataniotis of the Department of Math, Physics & Computer Science, School of Computer Science, Ryerson Polytechnic University, Toronto, Canada, specializes in colour image processing, multimedia, content-based retrieval, adaptive systems and fuzzy logic.

Dr Peter A. Rhodes is a Research Fellow in the Colour and Imaging Institute at the University of Derby, UK. His research interests include computer-mediated colour fidelity and communication for which he received his PhD in 1995 from Loughborough University, UK.

Dr Stephen J. Sangwine is an electronics engineer and has been a Lecturer in the Department of Engineering at The University of Reading, UK since 1985. His technical interests include digital circuits and fault diagnosis, a field in which he obtained his PhD in 1991, but since 1992, his research has been in colour image processing with a particular interest in frequency domain techniques.

Amy L. Thornton is a design engineer working on both hardware and software for FFT processing. Following on from postgraduate research for a PhD at The University of Reading, UK, her area of interest is in processing data in the frequency domain.

Professor Anastasios N. Venetsanopoulos is in the Department of Electrical and Computer Engineering at the University of Toronto, Canada. His research is in the fields of multimedia, colour image processing, communications, nonlinear and multidimensional signal processing, fuzzy logic and neural networks.

Preface

This book is aimed at those using colour image processing or researching new applications or techniques of colour image processing. It has been clear for some time that there is a need for a text dedicated to colour. We foresee a great increase in the use of colour over the coming years, both in research and in industrial and commercial applications. We are sure this book will prove a useful reference text on the subject for practicing engineers and scientists, for researchers, and for students at doctoral and, perhaps masters, level. It is not intended as an introductory text on image processing, rather it assumes that the reader is already familiar with basic image processing concepts such as image representation in digital form, linear and non-linear filtering, transforms, edge detection and segmentation, and so on, and has some experience with using, at the least, monochrome equipment. There are many books covering these topics and some of them are referenced in the text, where appropriate.

The book covers a restricted, but nevertheless, a very important, subset of image processing concerned with **natural colour** (that is colour as perceived by the human visual system). This is an important field because it shares much technology and basic theory with colour television and video equipment, the market for which is worldwide and very large; and with the growing field of multimedia, including the use of colour images on the Internet.

The book is structured into four parts. The first deals with colour principles and is aimed at readers who have very little prior knowledge of colour and colour science. Chapter 2 outlines the theory of human colour vision, because this is what defines **colour**. Chapter 3 gives an overview of the field of colour science relevant to image processing, and Chapter 4 describes various possible mathematical spaces or coordinate systems in which colour may be represented, including the equations and coefficients required to convert colour pixel values between different colour spaces. The second part of the book covers colour video technology (Chapter 5), colour image sources and image acquisition (Chapter 6 and Chapter 7). This part is intended to convey the need for care in acquiring images if the results of subsequent processing are to have the best chance of success. As these chapters show, there is much more to colour image acquisition than in the case of monochrome. The third part of the book (Chapters 9 to 13) is its core, and covers the main fields of processing so far established, some of them quite mature, others still in the early stages of development. The fourth, and final, part covers some specific applications of colour image processing, including the use of colour in the textile and graphics arts industries, and some industrial and medical applications of colour measurement and analysis.

No book is ever free from errors, and we would welcome comments and corrections, for use in what we hope will be the next edition.

S. J. SANGWINE

Reading and
October 1997 R. E. N. HORNE

Acknowledgements

Thanks are due to Paul Whitlock for help with converting some of the figures to Encapsulated Postscript and for checking and correcting many of the entries in the Bibliography.

The whole book was gathered electronically over the Internet, with the exception of the work of one contributor who became temporarily, we hope, disconnected from the Internet and who submitted material by post on a floppy disk. Without this global infrastructure, our work as editors would have been much harder, and we are grateful to The University of Reading for providing high-quality facilities, chiefly email.

The book was typeset using LaTeX 2_ε software. The implementation used was MiKTeX version 1.07 running under Windows '95. Camera-ready copy was produced in 600 dpi Postscript using dvips version 5.66. GSview version 2.1 proved invaluable in correcting and checking the diagrams and the layout of the text. Our thanks are due to the many authors and contributors who have made their software freely available.

1

The present state and the future of colour image processing

Colour image processing, the subject of this book, is concerned with the manipulation of digital colour images on a computer, and is a branch of the wider field of digital image processing, which is itself a branch of the more general field of digital signal processing. Digital colour images occur very widely in the modern world, indeed few contemporary images are not digitally processed. All images on the Internet are in digital form; most images seen in the press have been in digital form at some point between reality and paper; and images seen on television have often been generated, processed and transmitted digitally. Images may be observed simply for the immediate information they convey, or they may be processed to extract useful information, such as the position, size or orientation of an object; or detail, such as a vehicle licence number.

It is obvious that a colour image has the potential to convey more information than a monochrome or black and white image, and that some information in an image cannot easily be extracted from a monochrome representation. Some colours, for example, will reproduce as almost the same shade of grey in a monochrome image, whereas, seen in colour, the two colours will be readily separable by the human eye and/or by computer processing. Television and video can be regarded simply as sequences of images, so that digital colour video processing can be seen as an extension of digital colour image processing. Of course, processing of video in real time (that is processing each **frame** or image in less time than the interval between frames) is a challenging application of digital signal processing. Some processing operations can be performed by commercially available hardware, including relatively straightforward geometrical manipulations and other effects which are seen regularly on television.

Digital image processing is a well-established discipline with a large body of theory and many practical applications in fields as diverse as industrial inspection, medicine, publishing and multimedia. Many applications require no more than monochrome images, indeed in some applications (X-ray and ultrasonic imaging, for example) the images are inherently monochromatic and the question of colour does not arise. However, in many other applications, colour could be used with advantage and yet is not. Examples include agricultural quality control (e.g. fruit sorting), surveillance of vehicles or crowds, and industrial handling. Until recently, the extensive exploitation of colour image information has been prevented partly by cost and partly by the relatively immature development of appropriate processing algorithms.

The history of research work in colour goes back over 20 years to the work of Tenenbaum, Kender and others, whose works are cited in this book. They worked with colour images at a time when even a modest-sized image required far more computer memory than was available in many minicomputers of the time (a typical 1970s minicomputer had 128 kB of RAM, whereas the cheapest personal computer (PC) today has at least 8 MB, a factor of 64 greater). In the last few years it has become possible to store very large numbers of colour images on a personal computer and to process those images quite easily. Indeed, processing power and memory are no longer serious problems for most researchers, and in many applications. Much work remains to be done on colour processing algorithms. Later chapters in this book show that in some areas there are already good techniques available for practical applications. In other areas, the practical techniques are emerging and the chapters in this book survey the current state of development.

The last decade has seen a global convergence in the technologies of computers, telecommunications and broadcasting. Historically, colour image processing has been based on the technology of colour television which was developed in the 1950s. Colour video cameras, which were, until the recent advent of digital still cameras, the main means of capturing natural scenes digitally, are based on colour television technology. This is made clear in Chapter 4 and in Chapter 5. The convergence of many technologies can only strengthen the link between colour image processing, colour television, multimedia and digital video, and the cost of image acquisition equipment based on common standards of colour representation is likely to fall further. This means that applications of colour image processing will usually be based on the standard trichromatic nature of these technologies, which is itself based on the human colour visual system discussed in Chapter 2. If the images resulting from digital processing are to be interpreted by the human eye, as many are, this could not be otherwise. But in many industrial applications the images need not be seen by the human eye: the computer extracts some information, such as the position and orientation of an object, that is used to control a machine, and there is no reason to use a visual system based on human vision. Some researchers have thought along these lines, and it is possible, using specialized equipment, to capture multi-spectral images which are not based on the human trichromatic vision system. This has been common practice for decades in astronomy and remote sensing, and much theory has been developed in multi-spectral imaging which could be applied to natural colour imaging and processing.

Some of the topics presented in this book are at an early stage of development. Morphology, discussed in Chapter 11 could be developed further, and this is only one example of a field where the use of colour has great potential. There are, almost certainly, other fields where existing techniques from monochrome image processing could be generalized to handle colour. Equally, there are fields where colour processing will have no monochrome analogue. A clear example is the emerging field of chromatically sensitive linear filters which is mentioned at the end of Chapter 12 but which has not yet advanced far enough to be reported fully in this book.

FURTHER READING

Research in image processing is reported in many conferences, but few are specifically devoted to colour, or have sessions devoted to colour. A notable exception is the series of workshops organized in Germany, the most recent of which (at the time of writing) was held in Erlangen in September 1997. These workshops have been held in German, which unfortunately excludes many researchers. Two major conferences of note are those organized by the IEE (*International Conference on Image Processing and its Applications*) held every two years, usually in July, and by the IEEE (*International Conference on Image Processing*) held annually, usually in September.

Industrial and commercial applications are presented in the magazines *Advanced Imaging* published in the USA, and *Image Processing* published in the UK.

PART ONE

COLOUR

2

Colour vision

William McIlhagga

2.1   WHAT IS COLOUR?

Everything appears coloured. One cannot look at grass, for example, without seeing green, or look at the sky without seeing blue. The pervasiveness of colour, and its impact on our senses, makes it seem as if it is an aspect of reality, but of course it is not. Neither the light reflected from grass, nor the light scattered from the sky, has colour; both are just collections of photons. The colours are created in our minds as a response to the light. Since the final sensation of colour is not at all similar to a physical description of light in terms of photons and wavelengths, there must obviously be a great deal of interpretation occurring. Fortunately, this interpretation is mostly indepen-dent of higher-order processes, such as object recognition, consciousness or emotion, so a comprehensive theory of colour is possible.

To progress, it helps to have a definition of colour. Since colour is how a human observer interprets light, any definition has to involve an observer. Colour can't be defined in physical terms, since it is not a physical property, but once a behavioural definition is in place, it can be used to deduce the correlations between colour sensation and the physical properties of the light. The following definition is simple and appealing: take a colour sample, for example a paint square. Place the sample under a variable intensity light. When the intensity of the light is varied, the colour of the sample stays the same. Colour must therefore be something which is invariant with respect to light intensity.

2.1.1   Theoretical background

It is worth beginning by asking what measurements of light have the intensity-invariant property. Light, either from a radiant source or reflected from a surface, can be described by a power distribution $P(\lambda)$ which gives the power (in watts, say) of the light around wavelength λ. An obvious intensity-invariant measure of $P(\lambda)$ is the normalized distribution $P_N(\lambda)$ given by dividing $P(\lambda)$ by the total power:

$$P_N(\lambda) = \frac{P(\lambda)}{\int P(\lambda)d\lambda}$$

When $P(\lambda)$ changes by a multiplicative factor (i.e. a change in intensity) $P_N(\lambda)$ remains unchanged. Clearly, any sum or ratio of functionals based on P_N must also be intensity-invariant. Considering only linear functionals, any measure of the following form

$$\prod_i \left(\int A_i(\lambda)P(\lambda)d\lambda \right)^{n_i} \quad \text{for constants } n_i \text{ satisfying } \sum_i n_i = 0$$

where the functions $A_i(\lambda)$ are distinct, is also intensity invariant. The logarithm:

$$\sum_i n_i \log \left(\int A_i(\lambda)P(\lambda)d\lambda \right) \tag{2.1}$$

is likewise. Although important, intensity invariance cannot be the only criterion to apply to colour vision. A colour description should also be able to distinguish between two power distributions that differ by something other than a change in intensity. The ability to distinguish such distributions depends on the number of invariants that are computed. At least one is necessary, but more are better. The normalized power $P_N(\lambda)$ is an infinite number of invariants, one for each wavelength. As we shall see later, human colour vision uses only two invariants.

2.1.2   Human colour perception

How closely does human colour perception match the theoretical outline presented above? Given that the theoretical outline is directly inspired by the intensity invariance of human colour vision, it would be surprising if it were very different. Indeed, parallels between the colour descriptor defined in

equation (2.1) and the human colour system are quite direct. The computation of a descriptor like equation (2.1) can be broken into three stages:

- the integration of $P(\lambda)$ with $A_i(\lambda)$;

- a logarithmic transform; and

- a summation weighted by coefficients n_i.

Each of these stages has a corresponding stage in the human colour pathway
(described in the next section). The integration step is implemented in the
retina, where the light is absorbed by photoreceptors (section 2.3). An approximate logarithmic transform is implemented by the response nonlinearity that occurs at the photoreceptors (section 2.3.2), and the opponent process
channels (section 2.4) are weighted summations of cone responses, which are
implemented in the cortex.

2.2 THE VISUAL PATHWAY

Colour perception occurs in stages. Each stage has a different location in the
brain and visual apparatus. The pipeline of stages is called the **visual pathway** and is represented in Fig. 2.1. Conventionally, the pathway is divided
into three parts: the retina; the lateral geniculate nucleus; and the cortex.
The cortical stage of visual processing is itself composed of many stages, but
much less is known about them, and it has only recently been possible to map
human visual areas and relate them to their function. Most of the information
about the visual pathways comes from investigations using the macaque monkey, which as far as we know has a similar visual system to humans. For a
short but comprehensive review of visual pathways, see van Essen, Anderson,
and Felleman (1992). For introductory material, see Shepherd (1988).

The first stage in the visual pathway is the retina. Light is focused by the
cornea and lens onto the retina, a thin layer of tissue at the back of the eye.
(The retina is in fact an outgrowth of the brain attached to the back of the
eyeball.) The retina is composed of a number of layers. Specialized cells in
one layer of the retina, called **photoreceptors**, absorb the light and convert it
into a neural signal. These neural signals are collected and further processed
by cells in other layers of the retina before being passed to the brain along
a nerve bundle, or cable, called the optic nerve. The layering of the retina
is oddly engineered. The photoreceptors lie in the outer layer, furthest from
the light. The other cells in the retina lie in layers on top of them, partially

Right hemisphere   Left hemisphere

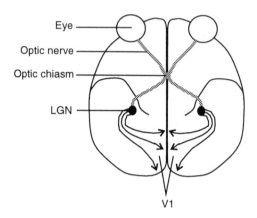

Fig. 2.1 The visual pathway in humans. The eyes are connected to the LGN by the optic nerves. Signals from the LGN pass to visual area 1 (V1). The diagram shows the brain as viewed from below.

obscuring that light, although in the fovea, where visual acuity is greatest, the inner layers of the retina have all but disappeared to give the photoreceptors an unobstructed view. The optic nerve connects to the cells in the innermost layer, called the retinal ganglion cells, so it must punch through the outer (photoreceptor) layer to escape the eye; and this leaves a small spot on the eye which has no photoreceptors - the blind spot. The cells in the inner retinal layer can be divided into two classes, called M (for magnocellular, or large-bodied) and P (for parvocellular, or small-bodied) cells. Only the P cells carry a colour signal. The optic nerve passes from the eye to a pair of small midbrain structures called the lateral geniculate nuclei (LGN). These do not seem to process the nerve signals in any significant way. From the LGN, the visual signal passes to V1 (Visual area 1, also called area 17) which lies in the cleft between the two brain hemispheres at the back of the brain. V1 is laid out like a map of the retina: each point in V1 processes signals from a corresponding point on the retina. Each hemisphere of the brain carries half of V1. The left brain holds the part of V1 that deals with the right visual field, and the right brain holds the part that deals with the left. Within V1, there are a subgroup of cells that are colour sensitive. These cells are distributed throughout V1 in clumps, called 'blobs' and receive input from the P cells (Livingstone and Hubel 1984). From V1, the colour signal from the blobs passes to other visual areas, imaginatively named V2, V3, and V4. Of these, V4 is the most crucial. It is clearly identified as being colour specific in PET and functional MRI[1] images of the human brain, and damage to this area reduces or eliminates the ability to see colour. However, V4 also seems to be involved in other kinds of visual processing. The colour signal does not pass to some areas of the brain, most notably V5, which deals with the detection of motion.

2.3   LIGHT ABSORPTION AND TRICHROMACY

Colour perception begins with light absorption by the photoreceptors in the retina. There are two main kinds of photoreceptors, called rods and cones. Rods, which are long and thin, function only in darkness, and do not contribute to colour perception. Cones, which have a tapered shape, do not respond very well in dim lighting; their purpose is to handle vision during daylight. Colour perception is based on the activity of the cones. There are three different types of cone, called L, M, and S, which stand for Long, Medium, and Short wavelength. Each cone type absorbs light over a broad range of wavelengths, but has its peak absorption at a different wavelength. The L cones peak at ∼570 nm, the M cones at ∼545 nm, and the S cones at ∼440 nm (Fig. 2.2). Cone absorption is minimal outside the range 400–700 nm, which therefore defines the visible spectrum. The cones are occasionally called R, G, and B, instead of L, M, and S, but this nomenclature is misleading. A red sensation is not due to activity in the R cone, but to a comparison of R and G (L and M) cones. The cone absorption curves give the proportion of light absorbed at each wavelength. The total power absorbed by these cones, when illuminated by light with power distribution $P(\lambda)$, is

$$
\begin{aligned}
a_L &= \int L(\lambda)\,P(\lambda)\,d\lambda, \\
a_M &= \int M(\lambda)\,P(\lambda)\,d\lambda, \\
a_S &= \int S(\lambda)\,P(\lambda)\,d\lambda,
\end{aligned}
\qquad (2.2)
$$

[1] PET (Positron Emission Tomography) locates brain activity by following a weak radioactive tracer, usually an oxygen isotope, injected into the cranial blood supply. Decay events are detected and used to build up an image of oxygen concentration (and therefore brain metabolism) in the head. Functional MRI (Magnetic Resonance Imaging) produces three-dimensional images of the blood oxygenation level (and hence brain metabolism), by measuring differences in the decay of spin-coherence of hydrogen nucleii between oxygenated and deoxygenated blood.

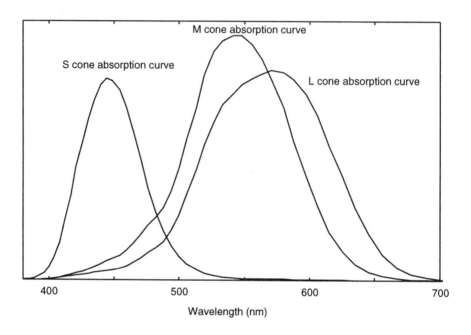

Fig. 2.2 Absorption curves for the L, M, and S cones, from Vos, Estevez and Walraven (1990). The curves have been normalized to produce equal cone absorption rates when shown 'equal energy white' (CIE coordinates $X = Y = Z$).

where $L(\lambda)$, $M(\lambda)$ and $S(\lambda)$ are the absorption curves shown in Fig. 2.2, and a_L, a_M, and a_S are the total absorbed power in the respective cone types. Once light is absorbed by the cones, no further information about the light is available to the visual system. In particular, the cones do not retain information about the wavelengths of the light they absorb. Although it is possible to distinguish different wavelengths by comparing cone absorptions, this is not the purpose of colour vision. The power distributions of light reflected from real objects contain power across the entire visible spectrum, and colour vision is designed to detect and characterize these broad spectra, not individual wavelengths.

2.3.1 Trichromacy

Suppose two power distributions $P_1(\lambda)$ and $P_2(\lambda)$ yield exactly the same cone absorption rates a_L, a_M, and a_S. From equation (2.2), this implies:

$$
\begin{aligned}
\int L(\lambda)\,(P_1(\lambda) - P_2(\lambda))d\lambda &= 0 \\
\int M(\lambda)\,(P_1(\lambda) - P_2(\lambda))d\lambda &= 0 \\
\int S(\lambda)\,(P_1(\lambda) - P_2(\lambda))d\lambda &= 0
\end{aligned}
\tag{2.3}
$$

These constraints do not mean that $P_1(\lambda) = P_2(\lambda)$, so it is possible for different power distributions to yield exactly the same cone absorptions. Such pairs of power distributions would therefore have exactly the same colour, and could not be distinguished under any circumstances. Different but indistinguishable power distributions are called **metamers**. Metamers are the basis for colour matching in film, video, and print: these technologies attempt to produce metamers of natural colours by the combination of a fixed set of mixing colours, called primaries. The way in which this works is outlined below.

Consider a set of three lights, called **primaries**. These lights can be mixed together by, for example, projecting them onto the same point on a screen, or by pointillistic mixing, as in colour TV. Let the normalized power distributions of the primaries be $A(\lambda)$, $B(\lambda)$, $C(\lambda)$, and their respective total powers be a, b, c, so that the power distribution of the mixed light is:

$$P(\lambda) = aA(\lambda) + bB(\lambda) + cC(\lambda)$$

The absorption rates of the L, M, and S cones from the mixed light are then:

$$
\begin{aligned}
a_L &= \int L(\lambda)P(\lambda)\,d\lambda \\
&= a\int L(\lambda)A(\lambda)\,d\lambda \;+\; b\int L(\lambda)B(\lambda)\,d\lambda \;+\; c\int L(\lambda)C(\lambda)\,d\lambda \\
a_M &= \int M(\lambda)P(\lambda)\,d\lambda \\
&= a\int M(\lambda)A(\lambda)\,d\lambda \;+\; b\int M(\lambda)B(\lambda)\,d\lambda \;+\; c\int M(\lambda)C(\lambda)\,d\lambda \\
a_S &= \int S(\lambda)P(\lambda)\,d\lambda \\
&= a\int S(\lambda)A(\lambda)\,d\lambda \;+\; b\int S(\lambda)B(\lambda)\,d\lambda \;+\; c\int S(\lambda)C(\lambda)\,d\lambda
\end{aligned}
$$

Once the primaries are chosen, all the integrals on the right hand side are constants, and only the powers a, b and c can vary. These equations can be used forwards or backwards. Forwards, we specify powers a, b and c and compute the cone absorptions that result. Backwards, we specify a set of cone absorptions and work out the primary powers a, b and c needed to produce them. The latter case is the most interesting. Every light yields only three cone absorptions, so every light can in theory be matched by an appropriate mixture of the three primaries. The match is perfect: the original light and the matching primary mixture have exactly the same colour and they can never be distinguished. The fact that any light can be matched by a combination of three other lights is called **trichromacy**.

There are, however, practical limits to trichromacy. Suppose that one or more of the powers *a*, *b* or *c* needed to reproduce a particular colour is negative. Since negative power is impossible, that colour cannot be reproduced. Only colours which have positive primary powers *a*, *b* and *c* can be reproduced, and the set of all reproducible colours is called the **gamut** of the primaries. The trick is to choose primaries which have the widest possible gamut, but unfortunately no set of three real primaries has a gamut which covers all possible colours[2]. In some applications, it is important to cover all possible colours, which can only be achieved using primaries that are physically impossible. This is the approach taken by the CIE system which is described in section 3.2. Of course, with physically impossible primaries one can specify physically impossible colours, which is a drawback the CIE system lives with.

2.3.2   Response compression

When the cones absorb light they generate a neural signal. The neural signal in photoreceptors (and indeed most other nerve cells) has a limited dynamic range: less than two orders of magnitude separate the smallest from the largest signal. The absorption rate, on the other hand, can vary over six orders of magnitude without any problem. Thus, the range of the absorption rate must be compressed when it is converted to a neural signal. Psychological studies on humans, and physiological studies on monkey retina (which is very similar to human retina) suggest that the neural response *r* from an absorption rate *a*, follows a Michelis–Menten equation:

$$r = r_{max}\frac{(a/A)^n}{(a/A)^n + \sigma} \tag{2.4}$$

[2]Colour matching experiments use another trick to overcome this limitation. In a colour matching experiment, we have two fields, one containing the colour to be matched, and the other containing the primary mixture. If one primary, say primary *A*, needs a negative power $-a$, we simply add $+a$ of primary *A* to both fields. The net effect is that primary *A* disappears from the mixture field and is added to the matching field. If we allow this trick, then *all* colours can be matched by a combination of three primaries.

where σ is the semi-saturation constant ($\sigma \cong 1$) and A is a scaling term called the **adaptation level**; n is just a free variable used to fit the equation; usually, $n \cong 0.7$ (Naka and Rushton, 1966). The adaptation level A is normally modelled as the moving average of the absorption rate, with an integration time in the order of seconds. Since the eye is in constant motion, the temporal moving average of the absorption rate a is much the same as the spatial average of a. Over a small range of absorptions (from $a = 0.1A$ to $a = 3A$) this curve is similar to a logarithmic curve.

The purpose of the adaptation level A is to scale the sensitivity of the photoreceptor to the amount of available light. Adaptation can be seen in action when we move from bright daylight to a dark interior. At first, the cone absorption rates in the dark interior are so much less than the adaptation value that the cone responses immediately fall to very small values, and it is difficult to see anything ($a \ll A$). Slowly, however, the adaptation value approaches the average of the available light, and the responses increase. As a result, the scene appears to lighten somewhat, and we see again. When we move back into bright light, everything appears whited out ($a \gg A$) until the adaptation value A catches up. The pupil of the eye adjusts to help in both circumstances, but the variation in its size (from about 1 mm to say 6 mm in diameter) is insufficient to cope with the full range of possible absorptions on all but the most overcast of days.

2.3.3 Colour blindness

A small number of people have no L-type cones in their retina (they are replaced by more M cones). They cannot therefore distinguish colours which differ only in the amount of L cone absorption, so they are blind to certain colour differences that normal individuals can see. This kind of visual defect is called **protanopia**. Another group of people have no M cones; they are said to have **deuteranopia**. Since distinctions between red and green are based on a comparison of L and M cone activity (section 2.4.1), both protanopes and deuteranopes are colloquially termed red–green colourblind. A rarer form of colourblindness, **tritanopia**, results when the S cones are missing. Calling any of these people colour-blind is a little misleading. Colour blind people can see colours; it is just that they tend to mix up colours that normal people see as different, and the colours they do see are not easy to relate to the colours that normal individuals see. On the other hand, if someone is unlucky enough to be both tritanopic and, say, deuteranopic, they truly are colour blind.

2.4 COLOUR APPEARANCE AND OPPONENT PROCESSES

Trichromacy says that when two lights yield the same cone responses, they have the same colour. It does not tell us what happens when two colours have different cone responses. Obviously, they look different, but how different? And, more importantly, what colour are they? Colours are qualities like red, green, blue, and so on, not vectors (a_L, a_M, a_S) of cone absorptions. Answering these questions needs a different account of colour from that provided by trichromacy. The outlines of the answer were supplied by Hering (1964) in his **Opponent process theory** of colour.

Opponent process theory is based on the observation that some colours are more similar than others. Start with a wide selection of colour samples scattered on a table, and rearrange them so that similar colours sit next to each other. Inevitably, the best way to arrange the colours is in a wheel. (Sometimes they can be arranged in a line, but this is the colour wheel cut at one point and straightened. The opposite ends of the line invariably have similar colours.) Hering noticed that the colours on the wheel can be moved around the rim so that all the colours on one side of the wheel contain some red, while all colours on the other side contain some green. The colours can also be rearranged so that all the colours on half of the wheel have some blue content, while the colours that remain in the opposite half appear yellowish. This second rearrangement can occur independent of the first.

This arrangement of colours yields two crucial points. The first is that all colours can be described as containing red, or containing green, but never as containing both sensations simultaneously: red and green are exclusive colour sensations. Similarly, blue and yellow are exclusive colour sensations; one never sees a colour that appears to be slightly blue and slightly yellow at the same time. (Do not get confused here with mixing paints. A mixture of blue and yellow paint is green; while it might end up bluish-green or yellowish-green, it never ends up bluish-yellowish-green. Green is not a mixture of a blue sensation and a yellow sensation.) The second crucial point is that one cannot say this of any other pair of colours. Thus, classifying the colour as either red or green, and then independently as blue or yellow, is a complete description of the colour.

Opponent process theory says that our sensation of colour is organized along two axes, or processes. The first axis encodes the redness or greenness of a colour. Imagine the axis having positive numbers for red and negative numbers for green. The position along the axis encodes the intensity (or saturation) of the red/green sensation. The second independent axis, which we can draw perpendicularly, encodes the blueness or yellowness of a colour.

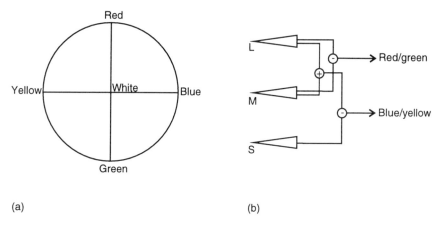

Fig. 2.3 (a) The opponent colour wheel. The upper semicircle contains all colours that appear reddish, the lower semicircle all colours that appear greenish; the left semicircle contains all yellowish colours, and the right semicircle all bluish colours. (b) Diagram of the cone contributions to opponent processes, according to zone theories. L and M cones are differenced to yield a red/green signal. S cones are compared with a sum of L and M cones to yield a blue/yellow signal.

Again, imagine the axis having positive numbers for blue and negative numbers for yellow. The position on this axis encodes the blue or yellow saturation. The complete colour sensation is the red/green and blue/yellow co-ordinates of that colour (Fig. 2.3). Sometimes a third axis is added, which describes the achromatic luminance (whiteness to blackness) of the colour.

Opponent process theory is supported by many observations. If we stare intently at an object, then look away at a blank wall or screen, we see an afterimage of that object. If the object is bright red, the afterimage is green, and *vice versa*. If the object is bright yellow, the afterimage is bluish, and *vice versa*. This strict pairing of colour and afterimage lends some support to the idea that red and green are associated (by being opposites, in this case), as are blue and yellow. Simultaneous contrast also demonstrates a red/green and blue/yellow association. A thin, grey, colourless line running over a red background appears slightly green, while running over a blue background it appears slightly yellow. Opponent process theory also implies that there should be 'pure' colours: a red or a green that has no blue or yellow in it, a blue or a yellow that has no red or green in it. Pure colours can be produced quite easily in the laboratory, and only red, green, blue, and yellow can be made pure. There is no such thing as, for example, a pure orange, since orange always looks like a mix of red and yellow.

2.4.1   Zone theories

Trichromacy and opponent process theory describe the analysis of colour at different stages in the visual pathway. Trichromacy is a model of the retina. Opponent process theory is a model of more central mechanisms in the brain. The opponent processes are a transformation of the cone responses, and Zone Theories (Judd, 1949; Hurvich and Jameson, 1955), describe how this transformation takes place. The simplest zone theory assumes that the opponent processes are just linear combinations of cone responses. The red/green opponent process is a difference of L and M cone responses:

$$\text{red/green signal} = \alpha(r_L - r_M) \tag{2.5}$$

where r_L and r_M are the L and M cone responses, and α is an arbitrary scaling constant. This difference automatically leads to opponency. When $r_L > r_M$, the red/green signal is positive, and this is interpreted to mean that the colour appears reddish. The magnitude of the signal is a measure of the colour saturation (the intensity of the colour experience). When $r_L < r_M$, the red/green signal is negative, and this is interpreted to mean that the colour is greenish. When $r_L = r_M$, the red/green signal is zero, and the colour appears neither red nor green. A colour is never simultaneously red and green; likewise the difference given by equation (2.5) is never simultaneously positive and negative.

The blue/yellow opponent process is a difference between the S cone response and a sum of the L and M cone responses:

$$\text{blue/yellow signal} = \beta(r_S - (\delta r_L + (1 - \delta)r_M)), \tag{2.6}$$

with $\delta \cong 0.7$ and where β is another arbitrary scaling constant. When this is positive, the colour is blue; when negative, the colour is yellow. There remains continuing argument about the correct mix of cone responses in the opponent processes. The phenomenon of short-wavelength redness suggests that there should be some S cone input to the red/green process, and the blue/yellow process is almost certainly not linear; instead, it is better modeled as $r_S - (\delta r_L + (1 - \delta)r_M) + \varepsilon|r_L - r_M|$ for some small constant ε (Burns, Elsner, Pokorny and Smith, 1984). Nonetheless, the simple linear equations

given above provide a convenient way of remembering the opponent processes. Finally, the third axis of a colour description — the light/dark axis —
is usually modeled as a combination of the L and M cone responses:

$$\text{light/dark signal} = \gamma r_L + (1-\gamma)r_M \qquad (2.7)$$

with γ around 0.66.

The zone theory equations (2.5), (2.6) and (2.7) have been written as
combining the cone responses, which is what must actually happen in the
brain. Normally, however, zone theory equations are more commonly written
as combining cone absorption rates, which means that the coefficients will
depend on the choice of normalization of the cone absorption curves, and the
adaptive state of the cones (that is, the values of A in equation (2.4) for all
three cone types). The curves in Fig. 2.2 have been normalized to have equal
area. An alternative normalization is to give the L cone curve twice the area
of the M cone curve (Smith and Pokorny, 1975); this will have the effect of
changing the L and M cone coefficients in the zone theory equations and, in
particular, will change equation (2.7) to simply $a_L + a_M$.

Finally, zone theories do not have an explicit representation for white.
White is the absence of colour, so white must yield a zero signal in both
the red/green and the blue/yellow opponent processes. This can only occur
when $r_L = r_M = r_S$, so any light which yields equal cone responses will be
colourless. In this case, only the light/dark process will respond. In the zone
theories, therefore, white is a derived sensation, caused by the absence of
colour; there is no channel which explicitly signals the amount of whiteness
in an object, although anything which activates the light/dark process only,
must be white.

2.4.2   Theoretical background revisited

Compare equations (2.5) and (2.6) to the theoretical colour descriptor defined
in equation (2.1). If r, defined in equation (2.4) is approximately logarithmic,
then substituting $r = \log(a/A)$ into the red/green equation (2.5) yields:

$$(1/\alpha)\text{red/green} = \log(a_L) - \log(a_M) - [\log(A_L) - \log(A_M)],$$

and into the blue/yellow equation (2.6) yields:

$$(1/\beta)\text{blue/yellow} = \log(a_S) - \delta\log(a_L) - (1-\delta)\log(a_M)$$
$$-[\log(A_S) - \delta\log(A_L) - (1-\delta)\log(A_M)]$$

The first part of each equation has the form of equation (2.1). The second part of each equation, in square brackets, is a kind of colour-balance correction. These measure the average colour of the entire image, assuming that the adaptation terms are space-averages of the cone absorptions. When the image is, on average, white (as is the case when there is a wide sample of colours) the colour-balance terms vanish, since $A_L = A_M = A_S$. When there is a colour cast to the scene, for example at dawn or dusk, or under fluorescent lighting, the colour balance terms can partially compensate for this. They are in fact a kind of colour constancy (described in section 2.5.1).

2.5   OTHER PHENOMENA

2.5.1   Colour constancy

Colours seem to be invariant not just to changes of illuminant intensity, but also to changes of illuminant colour. For example, in direct sunlight a surface is illuminated by light from the sun and by light scattered from the sky. Moving into the shade, direct sunlight is cut off, and the scattered light from the sky and ground is the main illuminant. The scattered light has less long-wavelength light than the direct sunlight, so there is a change in the illuminant colour as well as the intensity. Nevertheless, no colour change is seen when an object moves from sun to shadow. This invariance of colour to general changes in the illuminant spectrum is called **colour constancy**.

Consider the physics behind the problem of colour constancy. The light distribution $P(\lambda)$ that reaches the eye is a multiplication of the illuminant distribution $I(\lambda)$, and the reflectance function of a surface $R(\lambda)$, as follows:

$$P(\lambda) = I(\lambda)R(\lambda) \tag{2.8}$$

The reflectance function, $R(\lambda)$, gives the fraction of light reflected from the object around wavelength λ. If a colour descriptor is invariant to arbitrary changes in $I(\lambda)$, this is in effect saying that it describes the reflectance function $R(\lambda)$. Thus colour constancy implies that the colour vision system is trying to describe reflectance functions independent of illuminant colour. This is obviously useful, since the reflectance function is a property of the material that an object is made of, and can be used to detect and discriminate objects of different material. Unfortunately, the task of extracting $R(\lambda)$ from equation (2.8) is so ill-posed as to be nearly meaningless. However, under certain fairly natural restrictions on $R(\lambda)$ and $I(\lambda)$, a reasonable degree of colour constancy is theoretically attainable (Gershon and Jepson, 1989).

That is theory; in practice, the most important fact about colour constancy is that, in spite of our intuitive feeling that our perceptions are colour constant, they are not. A perceptible change in illuminant colour generally leads to perceptible changes in the colour of surfaces under that illuminant. However, the change in perceived colour is less than would be expected from the change in cone absorptions. There is some process which compensates, albeit imperfectly, for changes in illuminant colour. It is widely believed that most of the colour constancy effects we have are due to the adaptation in cones (so-called **von Kries constancy**) (Walraven, Benzschawel and Rogowitz, 1989). This was noted in section 2.4.2 as a colour-balance term. The remaining constancy may be due to a complex analysis of the spatial layout of the visual image, to prior knowledge of the colours of some surfaces (e.g. one's own skin), and to the fact that colour constancy is usually perceived by comparing a current colour with one in memory, yet the memory for colours is rather poor.

2.5.2 Focal colours, colour memory

Our sense of colour is so sensitive that we can distinguish many millions of different colours. However, we cannot remember these differences. That is, if two colours are not shown simultaneously, our ability to distinguish them diminishes dramatically. It is probable that we normally remember only a few colours, for example the eleven focal colours (in English: red, green, blue, yellow, violet, orange, pink, brown, grey, white and black) (Berlin and Kay, 1991), and other colour memories are stored as a coarse blend between the focal colours. However, under certain conditions, up to thirty can be remembered (Derefeldt and Swartling, 1995) and with training it is possible to remember and distinguish up to fifty different colours. Remembered colours also tend to be more saturated (that is, more intense) than they actually were; this may be the reason why the lurid colour reproduction of Fujichrome™ and Kodachrome Gold™ photographic film is so popular. These films increase the saturation of the colours, bringing them more in line with how we remembered the scene.

2.5.3 Spatial and temporal factors

The red/green signal depends on the difference of the L and M cone responses. Unfortunately, the absorption curves of the L and M cones are so similar that their responses normally differ by only a few percent, so the red/green signal at any point is intrinsically small. It needs to be amplified, and this is done by averaging the signal over a wide area of space, and a long duration of time. The averaging occurs in the visual cortex, not the retina, because the colour sensitive cells of the retina are the P-cells, with the smallest averaging area. The averaging over space means that the spatial resolution of the colour channels is rather low, by comparison with the luminance channel (which does not average as much). The averaging over time means that the colour channels are less sensitive to motion, and indeed cannot 'see' moving objects very well, because they tend to get blurred.

The spatial resolution of the colour system can be measured by means of its contrast sensitivity function (CSF), which is a close relative of the modulation transfer function (MTF). The MTF of an optical system is computed by measuring the output amplitude for a sinusoidal input signal of fixed amplitude. The MTF could equally well be computed by varying the input amplitude to yield a criterion output amplitude. This is effectively how the CSF is measured[3]. The CSF is plotted as a function of the number of sinusoidal cycles per degree of visual angle. The human CSF can be measured for both luminance sinusoids (which vary between black and white) and coloured sinusoids (which vary between, say, red and green). When displaying a colour sinusoid, it is vital that there be absolutely no luminance variation, so the sinusoid must be 'equiluminant'. Techniques exist to guarantee this (Anstis and Cavanagh, 1983). When all confounding factors are accounted for, the human CSF for colour and luminance appears as in Fig. 2.4, replotted from Mullen (1985). The two CSFs are very different. The luminance CSF peaks at about two cycles per degree (roughly two black and two white stripes painted on your thumb and viewed at arm's length) and extends out to over 40 cycles per degree (indeed, all the way to the limit imposed by the optics of the eye). The colour CSF on the other hand dies out at about 5 cycles per degree, and the colour system seems most sensitive at near to zero cycles per degree; that is, broad flat colour fields. The blue/yellow signal is the difference between an S cone response and a combination of L and M cone responses. Since the S cone absorption curve is different from that of the L or M cone, there is less need for spatiotemporal summation. However, the S cones are distributed sparsely in the retina; only about 10% of cones are S cones, and they are en-

---

[3]When human observers are shown an extremely faint sinusoidal pattern, they have trouble detecting its presence. When their probability of detection is at a fixed level, say 80%, we suppose that the internal signal is at a constant level. Thus, we merely vary the contrast of the sinusoidal pattern until 80% correct detection is obtained. The reciprocal of this detectable contrast is the sensitivity, and we plot this as a function of the frequency of the sinusoid in cycles per degree of visual angle.

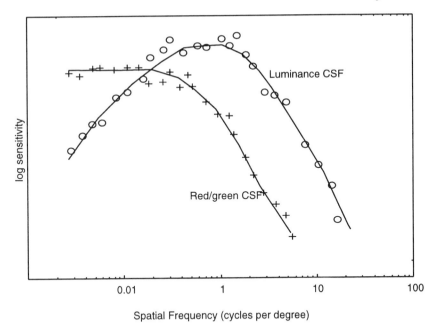

Fig. 2.4   Human CSF (contrast sensitivity function) for luminance (yellow/black) and isoluminant red/green sinusoidal patterns, replotted from Mullen (1985).

tirely absent from the central field of view (the fovea). Thus the blue/yellow signal is only sparsely sampled by comparison with the luminance signal, and indeed aliasing in the blue/yellow channel can be experimentally demonstrated.

Finally, it should be noted that not all parts of the retina have equal sensitivity to colour. The red/green signal is strongest in the fovea, and diminishes towards the periphery of the retina. The blue/yellow signal follows a similar pattern, except that it is completely absent in the fovea, because there are no S cones there.

2.6   THE USES OF COLOUR

Having said what colour is, it might be worthwhile ending by considering what colour is used for. It is clear that colour is not critical to our perception of the world. Colour blind people can survive without it, and people with normal colour vision have no difficulty interpreting black and white photographs. But the existence of colour vision suggests that at some time in our evolution, it was important. It has been suggested that the purpose of colour is to estimate the reflectance functions of objects (in section 2.5.1). Both the intensity invariance of colour vision and partial colour constancy point to this. There are a number of reasons why estimating reflectance might be important, but evolutionarily, perhaps one is central. A common feature of all terrestrial animals with colour vision (insects, birds, and most primates) is that they are involved in the reproductive cycle of plants, either by landing on flowers, or by eating fruit. Equally, the flowers and fruit of plants are the most brightly coloured natural objects. This suggests that colour vision co-evolved with coloured fruit (Mollon, 1989). If the fruit began with a slightly different colour from the rest of the plant, and some animals had a rudimentary colour vision that could, at least to some extent, detect that difference, then firstly those animals with better colour vision could harvest more fruit, and secondly those plants with more contrasting fruit could have more of it harvested. There is then evolutionary pressure for animals with good colour vision, and for brightly coloured plants.

Of course, once colour vision is established, it can be used for other purposes. Many monkeys use colour signals to indicate sexual receptiveness, and birds use brightly coloured plumage as a sexually attractive feature. Colour can also be exploited in more mundane applications, such as scene segmentation (Mullen and Kingdom, 1991). If two parts of a scene have different reflectance functions, then they probably belong to different objects, since they are made of different material. If two nearby parts of the image are the same colour, they are probably part of the same object. Thus we can see objects whose image is partially occluded by associating the same colours. This grouping-by-colour phenomenon is exploited in colour vision tests (e.g. the Ishihara test) and also by animal camouflage.

FURTHER READING

There are some very good texts on colour vision from trichromacy to opponent processes, for example Boynton (1989), Hurvich (1981) and the volume edited by Gouras (1991). A short review by Lennie and D'Zmura (1988) is particularly comprehensive. There is a fascinating literature on colour constancy, for example Maloney and Wandell (1986), Rubner and Schulten (1989). Some interesting views on the purpose of the opponent colour code are given by Buchsbaum and Gottschalk (1983) and colour in general by Rubin and Richards (1982). Mention should also be made of the Web site `http://www-cvrl.ucsd.edu`, maintained by Andrew Stockman, which contains a library of useful colour data bases.

3

Colour science

M. Ronnier Luo

3.1 INTRODUCTION

In 1666, Isaac Newton performed his famous experiment showing that white light can be separated by a prism to form a strip of light, named the **visible spectrum**, which includes all visible colours ranging from red, orange, yellow, green and blue to violet. Subsequently, white light was formed by directing the visible spectrum to pass through a second prism. This experiment shows that spectral colours in the visible spectrum are the basic components of white light. This important finding opened the door for the discipline of Colour Science.

Light is a form of **electromagnetic radiation**. Its physical property is described in terms of **wavelength** (symbol λ) in units of **nanometers** (nm), which is 10^{-9} m. Figure 3.1 shows the range of electromagnetic radiation from the shortest cosmic rays (10^{-6} nm) up to radio waves with wavelengths of many thousands of metres. It can be seen that the visible spectrum from 380 to 780 nm accounts for a very small part of the electromagnetic spectrum. Colour science concerns all aspects of quantifying colour stimuli and it can be divided into many branches such as radiometry, colorimetry, photometry, psychophysics, colour vision. It covers a wide range of sciences associated with how human beings perceive colour. These include physics, chemistry, physiology and psychology. It also covers many engineering subjects associated with the generation of colour and the application of colorants to different materials; such as colour displays, illuminating, printing, dyeing, paint, plastics, colorant manufacturing. As with all science and engineering subjects, measurement plays an important part. This chapter is focused on colorimetry, which numerically specifies a colour stimulus in relation to human colour perception.

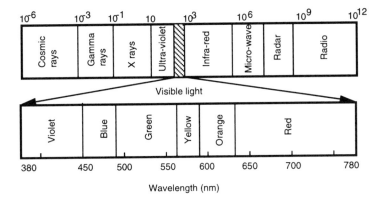

Fig. 3.1 The electromagnetic spectrum. The visible spectrum is expanded to show the spectral colours associated with different wavelengths.

Colorimetry has been widely used in various colour industries. Recently, it has also been increasingly used in the field of image processing due to the widespread development of cheaper and more advanced computer-controlled colour displays, printing and scanning devices. This chapter describes the concepts and methods of colorimetry in a practical way. Readers can easily implement these methods in their own applications.

Colorimetry can be divided into three areas: colour specification, colour difference and colour appearance. The basis of colorimetry is that developed by the **Commission Internationale de l'Eclairage** (CIE), known as the CIE **system**. This is covered in section 3.2. Section 3.3 introduces different types of equipment for measuring colours. An overview of uniform colour spaces and colour difference formulae is given in section 3.4. Finally section 3.5 covers the latest developments in the field of colour appearance modelling. This topic is most useful for readers performing colour image processing and digital colour reproduction.

3.2 THE CIE SYSTEM

Although there are some other colour specification systems, by far the most important of all colour specification systems is that developed by the CIE. The CIE system provides a standard method for specifying a colour stimulus under controlled viewing conditions. It originated in 1931 and was further supplemented in 1964. These are known as the CIE 1931 and CIE 1964 supplementary systems (CIE, 1986).

The CIE system standardizes three key elements of colour perception: light, object and eye. The details are given in the following sections.

3.2.1 Light

Light is an essential element of all colour perceptions. Without light there is no colour. A light source can be quantified by measuring its **spectral power distribution** (SPD, symbol $S(\lambda)$) using a spectroradiometer (section 3.3.2). The SPD described here is the relative spectral power distribution. The absolute SPD is not essential for colorimetric calculation.

There are many different kinds of **light sources**. The most important and common one is daylight, which is a mixture of direct and sky-scattered sunlight. Its SPD varies from quite blue (zenith skylight) to quite red (sunset). This is affected by latitute, position, time of year and day, and weather conditions. Artificial light sources are also widely used such as fluorescent, incandescent, gas discharge and xenon arc lamps. Again, there is a large variation in their SPDs. It is not possible and is not necessary to make colour measurements using all possible light sources. Hence, the CIE recommended a set of standard **illuminants** for industrial applications. (A light source is a physical body capable of emitting light produced by a transformation of energy. The illuminant is a table of SPD values for calculating colorimetric values and may not have a corresponding light source.)

Colour temperature

A light source can also be quantified in terms of **colour temperature** (in kelvins, K). There are three types of colour temperatures. **Distribution temperature** is the temperature of a **Planckian** (or **black-body**) radiator whose radiation is the same as that of the radiation considered. When the Planckian radiator, an ideal furnace, is maintained at a particular temperature, it emits radiation of constant spectral composition. **Colour temperature** is the temperature of a Planckian radiator whose radiation has the same chromaticity (see section 3.2.4) as that of a given colour stimulus. This is mainly used to define a light source whose SPD is very similar to, but not exactly the same as that of a Planckian radiator. **Correlated colour temperature** is the temperature of a Planckian radiator whose colour appears to be the nearest match to a given colour stimulus. This is commonly used for light sources whose SPDs are quite different from that of Planckian radiators such as fluorescent lamps.

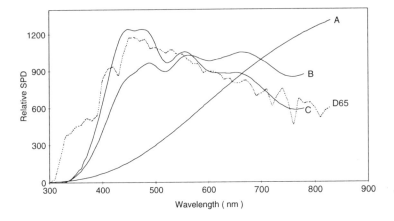

Fig. 3.2 Relative SPDs of CIE standard illuminants: A, B, C and D65.

CIE Standard Illuminants

In 1931, the CIE recommended the use of three standard illuminants, known as A, B and C, which are the representations of incandescent light, direct sunlight, and average daylight respectively. Their colour temperatures are 2856, 4874 and 6774 K respectively and their SPDs are given in Fig. 3.2. The CIE also recommended standard sources to correspond to these illuminants. For source A, a gas-filled tungsten filament lamp is applied to give a colour temperature of 2856 K. The standard sources B and C can be produced by using the standard source A in conjunction with liquid colour filters of specified chemical compositions and thickness. However, at a later stage, it was found that the illuminants B and C were unsatisfactory for some applications. This is because of too little power in the ultraviolet region. This is important for measuring materials with fluorescent colorants. Hence, the CIE also recommended a series of the D illuminants in 1964. The SPD of the D65 illuminant is shown in Fig. 3.2. It clearly illustrates that there are large differences between the illuminants D65 and B or C in the blue end (less than 400 nm). Currently, there is still some use of illuminant C, but illuminant B is obsolete. The most widely used D illuminants are D65 and D50 for the surface colour industries (including textiles, paint and plastics) and the graphic arts industry respectively. In addition, the CIE also provided the SPD values of representative fluorescent lamps. These are divided into 12 types: F1 to F12, in which the F2 (Cool White Fluorescent, CWF), F7 (a D65 simulator), F8 (a D50 simulator) and F11 (TL84) are the most frequently used. Some of these SPDs measured using the actual sources are given in Figs. 3.3 to 3.6.

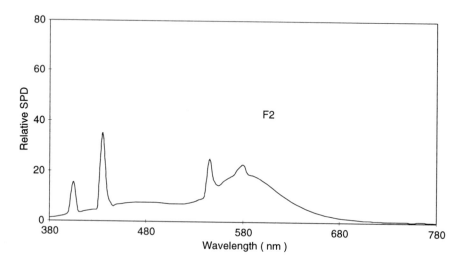

Fig. 3.3   Relative SPD of fluorescent lamp F2 (CWF).

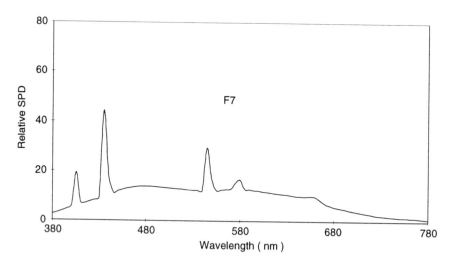

Fig. 3.4   Relative SPD of fluorescent lamp F7 (a D65 simulator).

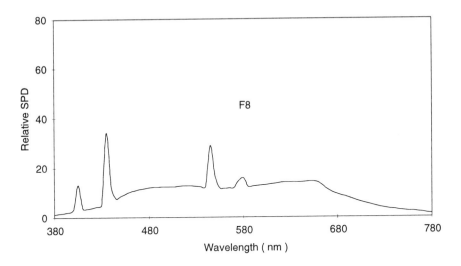

Fig. 3.5 Relative SPD of fluorescent lamp F8 (a D50 simulator).

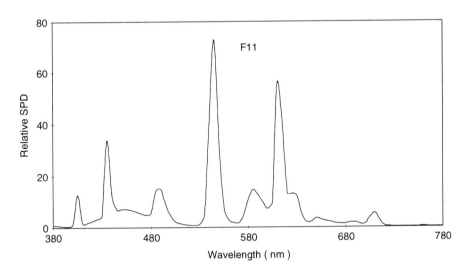

Fig. 3.6 Relative SPD of fluorescent lamp F11 (TL84).

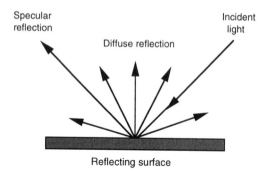

Fig. 3.7 Specular and diffuse reflection.

3.2.2 Object

The colour of an object is defined by the **reflectance** (symbol $R(\lambda)$), a function of wavelength. Reflectance is the ratio of the light reflected from a sample to that reflected from a reference white standard. Reflectance can be divided into two types: **specular** and **diffuse**.   Figure 3.7 shows that if a sample is illuminated at 45°, part of the light will be directly reflected from the surface in the reverse angle, which is called specular reflection. The other part of the light will initially penetrate into the object, and then be partly absorbed and partly scattered back to the surface. The latter is called diffuse reflection. Specular reflection depends upon the characteristics of the surface considered; it will be small for a matt surface, large for a glossy surface. The instrument used for measuring reflectance is called a **spectrophotometer** (section 3.3.3). The CIE specified standard methods for helping to achieve instrumental precision and accuracy. Firstly, a **perfect reflecting diffuser** is recommended as the reference white standard. It is an ideal isotropic diffuser with a reflectance equal to unity. Secondly, the CIE has recommended four different types of illuminating and viewing geometries: 45°/Normal (symbol 45/0), Normal/45° (0/45), Diffuse/Normal (d/0) and Normal/Diffuse (0/d). These are illustrated in Fig. 3.8. An integrating sphere is used for the 0/d and d/0 geometries. The gloss trap in the integrating sphere is used to include or exclude the specular component. The four illuminating and viewing geometries can be divided into two groups: with and without an integrating sphere. The two geometries in the same group should give the same measurement results, but not for different groups. Hence, it is important to report results by referring to the viewing geometry and illumination together with the inclu-

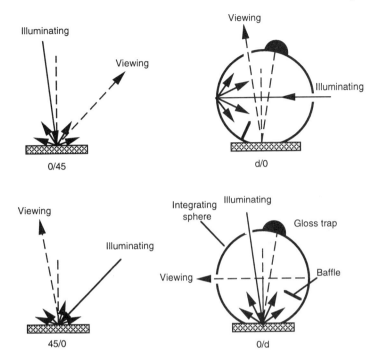

Fig. 3.8 Schematic diagram showing the four CIE standard illuminating and viewing geometries for reflectance measurement.

sion or exclusion of specular reflectance. The CIE also defines the tolerance angles for the axis and spreads for each geometry. The axis is defined as the angle between the central ray of a beam and the normal to the surface to be measured. The spreads define the maximum deviation of the rays of the beam from the central ray.

3.2.3 Eye

The most important component in the CIE system is the **colour matching functions**, which define how human eyes match colour stimuli using a set of red, green and blue reference primaries. They represent the colour matching properties of the CIE Standard Colorimetric Observers.

Additive colour mixing

The derivation of the CIE colour matching functions is based on the laws of **additive colour mixing**. A colour match can be expressed algebraically by:

$$C[C] \equiv R[R] + G[G] + B[B]$$

where the sign $\equiv$ represents a colour match. R, G and B are the amounts of red $[R]$, green $[G]$ and blue $[B]$ reference stimuli (primaries) required to match the C units of $[C]$ target stimulus. Stimuli of the same colour produce identical appearance in mixtures regardless of their spectral compositions.

If a colour is a mixture of two stimuli 1 and 2, it can be expressed by:

Stimulus 1 $\qquad\qquad C_1[C] \equiv \qquad R_1[R] + \qquad G_1[G] + \qquad B_1[B]$
Stimulus 2 $\qquad\qquad C_2[C] \equiv \qquad R_2[R] + \qquad G_2[G] + \qquad B_2[B]$
Resultant colour $(C_1 + C_2)[C] \equiv (R_1 + R_2)[R] + (G_1 + G_2)[G] + (B_1 + B_2)[B]$

A common way to represent a particular colour is to use its **chromaticity coordinates**, r, g and b which are calculated using:

$$r = R/(R+G+B)$$
$$g = G/(R+G+B)$$
$$b = B/(R+G+B)$$

where r, g, b are the proportions of the three primaries $[R]$, $[G]$ and $[B]$; and $r + g + b = 1$. Figure 3.9 illustrates an experimental arrangement for additive colour mixing. The divided inner circle and the shaded annular surround represent the viewing field seen by the observer. The right half of the inner circle provides the target stimulus produced by passing incandescent light through a filter. The left half of the inner circle provides a mixture of red, green and blue stimuli originating from three spotlights. The shaded area is the surround field for adaptation purposes. A panel of observers are asked to match a target stimulus (right half) by adjusting the matching stimulus (left half) using the $[R]$, $[G]$ and $[B]$ primaries.

In 1931, the CIE recommended a set of standard colour matching functions, known as the **CIE 1931 standard colorimetric observer** (or **2° Observer**) for use with visual colour matching of field sizes between 1° and 4°. The functions were derived from two different but equivalent sets of visual results from experimental work carried out by Wright (1928-29) and Guild (1931) based upon 10 and 7 observers respectively. The two sets of results were normalized and averaged to form a combined set. The functions are

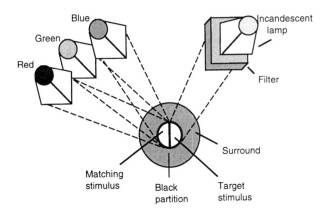

Fig. 3.9 A typical experimental arrangement for additive colour mixing.

designated $\bar{r}(\lambda)$, $\bar{g}(\lambda)$ and $\bar{b}(\lambda)$ which are expressed in terms of colour stimuli of wavelengths 700 nm, 546.1 nm and 435.8 nm for the red, green and blue primaries respectively. Their units were adjusted to match an **equal energy white**, which has a constant SPD over the spectrum. The $\bar{r}(\lambda)$, $\bar{g}(\lambda)$ and $\bar{b}(\lambda)$ functions of the CIE 1931 standard colour matching observer are plotted in Fig. 3.10. Negative amounts indicate that the light was added to the test colour instead of to the red, green and blue mixture. The $\bar{r}(\lambda)$, $\bar{g}(\lambda)$ and $\bar{b}(\lambda)$ functions were later linearly transformed to a new set of functions, $\bar{x}(\lambda)$, $\bar{y}(\lambda)$ and $\bar{z}(\lambda)$ so as to avoid negative coefficients in the former set of functions. The latter functions were recommended for reasons of more convenient application in practical colorimetry. The $\bar{x}(\lambda)$, $\bar{y}(\lambda)$ and $\bar{z}(\lambda)$ functions of the CIE 1931 standard colour matching observer are plotted in Fig. 3.11. A further set of colour matching functions were recommended in 1964 by the CIE for use whenever a more accurate correlation with visual colour matching of fields of large angular subtense (more than 4° at the eye of the observer) is desired. They were derived from the experimental data supplied by Stiles and Burch (1959) and by Speranskaya (1959). This set of colour matching functions is called the CIE **1964 supplementary standard colorimetric observer** (or **10° Observer**), and is denoted as $\bar{x}_{10}(\lambda)$, $\bar{y}_{10}(\lambda)$ and $\bar{z}_{10}(\lambda)$. These functions are also plotted in Fig. 3.11.

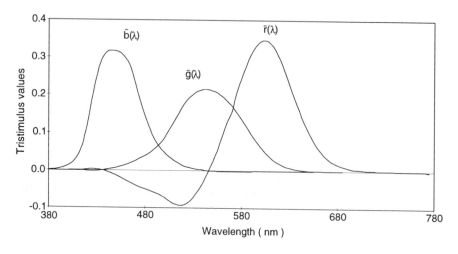

Fig. 3.10   The colour matching functions for the CIE 1931 standard colorimetric observer expressed using R, G and B primaries at 700, 546.1 and 435.8 nm, respectively.

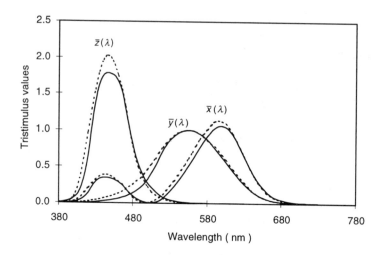

Fig. 3.11   The CIE colour matching functions for the 1931 standard colorimetric observer (solid lines) and for the 1964 supplementary standard colorimetric observer (dashed lines).

3.2.4 Tristimulus values

In the preceding sections, the three key elements of colour perception have been quantified in terms of functions of the visible spectrum. The standard colorimetric observer is defined by the functions $\bar{x}(\lambda)$, $\bar{y}(\lambda)$ and $\bar{z}(\lambda)$; or $\bar{x}_{10}(\lambda)$, $\bar{y}_{10}(\lambda)$ and $\bar{z}_{10}(\lambda)$, to represent the human population having normal colour vision. A number of illuminants are standardized in terms of the SPD, $S(\lambda)$. Additionally, the CIE standardized the illuminating and viewing conditions for measuring a reflecting surface. Each surface is defined by the reflectance, $R(\lambda)$. Thus any colour can be specified by a triple of numbers (X, Y, Z) called **tristimulus values** as given in equation (3.1), which quantify a colour by defining the amounts of the red, green and blue CIE primaries required to match a colour by the standard observer under a particular CIE illuminant. These are the integration of the products of the functions in three components over the visible spectrum.

$$
\begin{aligned}
X &= k \int S(\lambda)R(\lambda)\,\bar{x}(\lambda)d\lambda \\
Y &= k \int S(\lambda)R(\lambda)\,\bar{y}(\lambda)d\lambda \\
Z &= k \int S(\lambda)R(\lambda)\,\bar{z}(\lambda)d\lambda
\end{aligned}
\tag{3.1}
$$

The k constant was chosen so that $Y = 100$ for the perfect reflecting diffuser. If the CIE 1964 supplementary standard colorimetric observer is used in equation (3.1), all terms except $S(\lambda)$ and $R(\lambda)$ should have a subscript of 10.

For measuring self-luminous colours such as light-emitting colour displays, and light sources, equation (3.2) should be used instead of equation (3.1). This is due to the fact that the object and the illuminant are not defined. The P function represents the spectral radiance or spectral irradiance of the target stimulus. The areas of colours considered in display applications usually have quite small angular subtense and the CIE 1931 standard colorimetric observer is the appropriate one to use.

$$
\begin{aligned}
X &= k \int P(\lambda)\,\bar{x}(\lambda)d\lambda \\
Y &= k \int P(\lambda)\,\bar{y}(\lambda)d\lambda \\
Z &= k \int P(\lambda)\,\bar{z}(\lambda)d\lambda
\end{aligned}
\tag{3.2}
$$

The k constant is chosen so that $Y = 100$ for the appropriate reference white.

Colour as described in the CIE system is commonly plotted on the **chromaticity diagram** shown in Fig. 3.12. Chromaticity coordinates are calculated using equation (3.3). The region of all perceptible colours is bounded by the horseshoe-shaped locus of pure monochromatic spectral colours (the

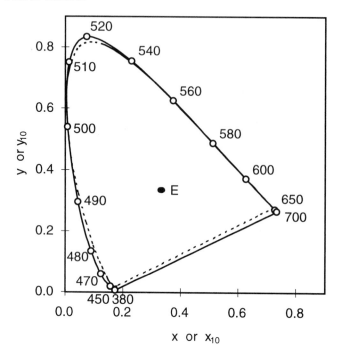

Fig. 3.12 The CIE chromaticity diagrams for the 1931 standard colorimetric observer (solid lines) and for the 1964 supplementary standard colorimetric observer (dashed lines).

spectrum locus), with a straight line connecting the chromaticity coordinates of extreme red and blue (the **purple line**). Whites lie near the centre of the figure (E, equal energy illuminant), and colours become more saturated towards the periphery. It is conventional to use x, y and Y to represent a colour.

$$x = X/(X+Y+Z)$$
$$y = Y/(X+Y+Z)$$
$$z = Z/(X+Y+Z) \quad \text{and} \quad x+y+z=1$$

(3.3)

A feature of the chromaticity diagram is that a point representing an additive mixture of two coloured lights falls on the straight line joining the points representing the chromaticity coordinates of the two lights. As illustrated in Fig. 3.13, a mixture produced by the stimuli N and D will always be located

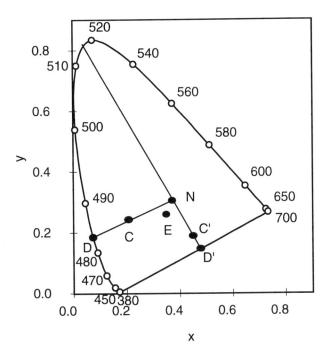

Fig. 3.13 Points plotted on the CIE 1931 chromaticity diagram to illustrate the calculation of the dominant wavelength, complementary wavelength and excitation purity.

on the line ND. By knowing the amounts of N and D, the position of the mixture can be calculated. This feature is important for measuring colour in colour displays, TV and the lighting industry.

Tristimulus values and chromaticity coordinates are measures for colour specification. However, they do not describe colour appearance attributes such as lightness, saturation and hue. In other words, it is difficult to visualize a colour accurately using these measures. The Y tristimulus value is an approximation of the lightness, i.e. a lighter colour has a higher Y value than a darker colour. The CIE also recommended **dominant wavelength** and **excitation purity** to correlate with the hue and saturation appearance attributes. Again, the points in Fig. 3.13 are taken as an example. Consider a surface colour (C) viewed under a particular illuminant (N). A line is drawn from N through C to join the spectrum locus at D for defining its dominant wave-length. In this case, it is about 485 nm corresponding to a greenish blue colour. The excitation purity (Pe) is calculated using NC divided by ND. A value for Pe close to one corresponds to a very saturated spectral colour. In this case, the Pe of colour C is about 0.46 representing a medium saturation colour.

For some colours close to the purple line like the colour C′, the line be-tween N and C′ does not join the spectral locus. In these cases, the line should be extended in the opposite direction until it joins the spectral locus. This is called the **complementary wavelength**. Its excitation purity is calculated using NC′ divided by ND′.

3.3   COLOUR MEASUREMENT INSTRUMENTS

Colour measuring instruments are designed to measure colours in terms of reflectance, radiance, and the CIE colorimetric values such as tristimulus val-ues. A variety of instruments are available. Each instrument is designed for measuring a different form of colour, i.e. self-luminance, surface colours, or both. Also, each includes a set of optical elements such as: a light source for illuminating the sample; a monochromator for dispersing the light into a spectrum; a detector for receiving the dispersed monochromatic light. The arrangement of these elements, such as the viewing and illuminating geome-tries described in section 3.2.2, may be different. These instruments can be categorized into three types:

- Tristimulus Colorimeter

- Spectroradiometer

- Spectrophotometer

Figure 3.14 illustrates the key elements for each type of colour measuring instrument.

3.3.1   Tristimulus colorimeter

Tristimulus colorimeters can only measure colour in terms of tristimulus val-ues under a fixed set of illuminant and observer conditions, e.g. D65/2°. They are very useful for quantifying the colour difference between pairs of

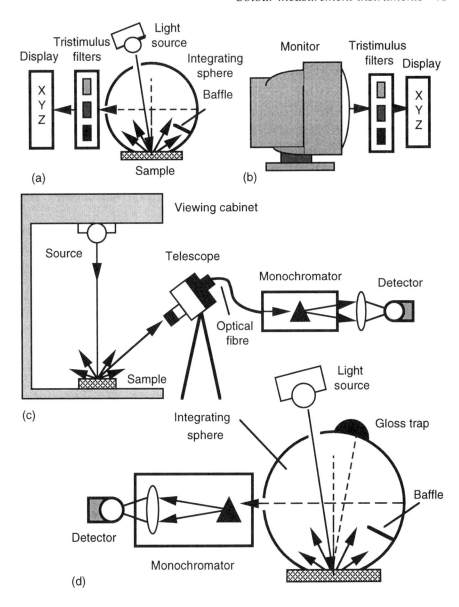

Fig. 3.14 The key elements for each type of colour measuring instrument. (a) a 0/d tristimulus colorimeter measuring surface colour, (b) a tristimulus colorimeter measuring a monitor colour, (c) a tele-spectroradiometer measuring a surface colour in a viewing cabinet, and (d) a 0/d spectrophotometer.

samples for colour quality control purposes and they cost much less than spectroradiometers and spectrophotometers. The uses of the instrument may be divided into two groups: measuring surface colours and measuring self-luminous colours. Fig. 3.14(a) illustrates a 0/d tristimulus colorimeter measuring a surface sample (see Fig. 3.8 for the other viewing and illuminating geometries). The key optical elements include a light source, an integrating sphere and a detector. The latter includes three or four filters intended to have a close match to the CIE $\bar{x}$, $\bar{y}$ and $\bar{z}$ colour matching functions. The instrument responses correspond closely to equation (3.1). Figure 3.14(b) illustrates a tristimulus colorimeter measuring a stimulus from a monitor. A light source is not included for the tristimulus colorimeter measuring self-luminous colours.

In general, the tristimulus colorimeter is easy to use and inexpensive. However, the instrumental agreement and repeatability of the tristimulus colorimeter is poor due to ageing of the detectors (filters) and poor reproducibility of the filters to agree with the CIE colour matching functions.

3.3.2   Spectroradiometer

Spectroradiometers are designed to measure radiometric quantities: **irradiance** (in units of W/m^2) or **radiance** (in units of W/m^2Sr). The radiometric energy is measured over the visible spectrum with a fixed interval such as 5 nm, 10 nm or 20 nm. Their colorimetric values are expressed by **luminance** (in units of cd/m^2) and **illuminance** (in units of lux) for radiance and irradiance units respectively.

The tele-spectroradiometer (TSR) is the most frequently used instrument in this category. Figure 3.14(c) shows a tele-spectroradiometer measuring a surface colour presented in a viewing cabinet. The key components are a telescope, a monochromator and a detector. The TSR is used to measure the colour of a distant object in its usual observing position and common viewing conditions. Any tele-spectroradiometer is capable of measuring both self-luminous and surface colours. For measuring surface colours, an external light source is required as shown in Fig. 3.14(c). There is no need to have a source for measuring self-luminous stimuli.

The advantage of this instrument is that the measurement results can correspond to the actual conditions of viewing. In addition, it can measure all forms of colour (surface and self-luminous), which are particularly important for cross-media colour reproduction, say to match an image displayed on a monitor with the output from a printer.

3.3.3 Spectrophotometer

Spectrophotometers are usually used for measuring surface colours and are designed to measure the ratio between the incident light and light reflected from the measured surface across the visible spectrum with a fixed interval such as 5 nm, 10 nm or 20 nm. The results are expressed by reflectance of a surface colour. Figure 3.14(d) shows a 0/d spectrophotometer. Its optical elements are a light source, a monochromator and a detector. As mentioned in section 3.2.2, its geometry of illumination (light source) and viewing (detector) should conform to one of the four CIE recommendations: 0/45, 45/0, 0/d and d/0. In practice, 45/0 and 0/45 geometries are widely used in the graphic arts industry, and 0/d and d/0 geometries for surface colour (textiles, paint, plastics) industries.

The spectrophotometer is usually the preferred instrument for surface colours. It can be used for quality control and recipe formulation. It may also be used to evaluate the **metamerism** of a pair of samples, i.e. the change of colour differences under different illuminants. (Metamerism is discussed in section 2.3.1 on page 13.) Also, spectrophotometers are quite stable over time and accurate against an absolute standard. However, they are expensive in comparison with tristimulus colorimeters.

3.3.4 Instrumental calibration

Each colour measuring instrument has its own calibration standard such as a white ceramic tile for measuring surface colours, and a standard lamp for measuring self-luminous colours. The white tile has been calibrated against the perfect reflecting diffuser specified by the CIE. They are usually obtained from the instrument manufacturer. This process is essential to calibrate the instrumental results to agree with the international standard in order to ensure a high degree of accuracy and repeatability.

It is also recommended that reference materials be used for checking the instrumental working condition, especially repeatability. The 12 CERAM tiles (Malkin and Verrill, 1983) are commonly used for checking surface colour measuring instruments and standardized filters can be used for checking self-luminous colour measuring instruments.

3.4   UNIFORM COLOUR SPACES AND COLOUR DIFFERENCE FOR-MULAS

The CIE system offers a means of measuring a colour under a standard set of viewing conditions, thus making it an excellent choice as a reference for industrial colour communication. Another important use of colorimetry is to evaluate colour differences for industrial colour quality control. Many sets of experimental results for colour discrimination have been published and show that the CIE colour space has poor uniformity, i.e. some colour differences perceptually equal in size could give a ratio of 20 to 1 in the x, y diagram. In 1976, the CIE recommended a new chromaticity diagram, named the **CIE 1976 uniform colour scale diagram**, or the **CIE 1976 UCS diagram**, which gives a perceptually more uniform space than that of the x, y diagram. It is a projective transformation of the tristimulus values or chromaticity coordinates using the following equation:

$$u' = 4X/(X + 15Y + 3Z) = 4x/(-2x + 12y + 3)$$
$$v' = 9Y/(X + 15Y + 3Z) = 9y/(-2x + 12y + 3)$$

$$(3.4)$$

Figure 3.15 shows the CIE 1976 UCS diagram, which preserves the feature of additive mixing discussed on page 38, in which colours lying on a straight line are the mixtures of two colours represented by two end points.

3.4.1   CIELAB and CIELUV colour spaces

Two uniform colour spaces were also recommended by the CIE in 1976: CIE $L^*a^*b^*$ (or CIELAB) and CIE $L^*u^*v^*$ (or CIELUV). CIELAB is a non-linear transformation of the tristimulus space to simplify the original ANLAB (Adams, 1942) formula which was already widely used in the colorant indus-try. CIELUV is used for industries considering additive mixing such as colour displays, TV and lighting.

Equations defining the CIELAB and CIELUV colour spaces are given be-low. Both colourspaces have the same lightness scale, L^*, and opponent colour axes, approximately red-green versus yellow-blue, i.e. a^* *versus* b^*, and u^* *versus* v^* for CIELAB and CIELUV respectively.

CIELAB colour space

$$L^* = 116f\left(\frac{Y}{Y_n}\right) - 16$$

$$(3.5)$$

$$a^* = 500\left[f\left(\frac{X}{X_n}\right) - f(\frac{Y}{Y_n})\right]$$

$$b^* = 200\left[f\left(\frac{Y}{Y_n}\right) - f(\frac{Z}{Z_n})\right]$$

(3.6)

where

$$f(x) = \begin{cases} x^{\frac{1}{3}}, & x > 0.008856 \\ 7.787x + \frac{16}{116} & \text{otherwise,} \end{cases}$$

and X, Y, Z and X_n, Y_n, Z_n are the tristimulus values of the sample and a specific reference white considered. It is common to use the tristimulus values of the perfect diffuser illuminated by a CIE illuminant or a light source as the

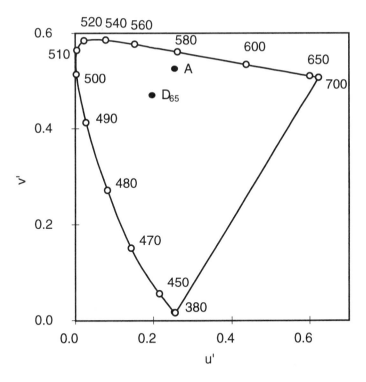

Fig. 3.15   The CIE u'v' chromaticity diagram.

X_n, Y_n, Z_n values. Table 3.1 provides the tristimulus values for some commonly used CIE illuminants and light sources. Correlates of hue and chroma are also defined by converting the rectangular a^*, b^* axes into polar coordinates as given by equation (3.7). These scales together with lightness (L^*) correspond to perceptual colour appearance.

$$h_{ab}^* = \tan^{-1}(b^*/a^*)$$
$$C_{ab}^* = \sqrt{a^{*2} + b^{*2}}$$
(3.7)

Colour difference can be calculated using:

$$\Delta E_{ab} = \sqrt{\Delta L^{*2} + \Delta a^{*2} + \Delta b^{*2}}$$

or

$$\Delta E_{ab} = \sqrt{\Delta L^{*2} + \Delta C_{ab}^{*2} + \Delta H_{ab}^{*2}}$$

where

$$\Delta H_{ab}^* = p\sqrt{2(C_{ab,1}^* C_{ab,2}^* - a_1^* a_2^* - b_1^* b_2^*)}$$

and

$$p = \begin{cases} -1 & \text{if } a_1^* b_2^* > a_2^* b_1^* \\ 1 & \text{otherwise} \end{cases}$$

and subscripts 1 and 2 represent the standard and sample of the pair considered respectively.

Table 3.1 Tristimulus values for some commonly used CIE illuminants and light sources (ASTM, 1996).

	CT	X	Y	Z	X_{10}	Y_{10}	Z_{10}
Illuminant							
A	2856	109.85	100.0	35.56	111.14	100.0	35.20
C	6774	98.07	100.0	118.23	97.29	100.0	116.15
D50	5000	96.42	100.0	82.49	96.72	100.0	81.41
D65	6500	95.05	100.0	108.88	94.81	100.0	107.30
Source							
F2 (Cool white)	4230	99.19	100.0	67.39	103.28	100.0	69.03
F11 (TL84)	4000	100.96	100.0	64.35	103.86	100.0	65.61

Note: *CT* represents Colour Temperature (see section 3.2.1).

CIELUV colour space

L^* in CIELUV colour space is defined identically to L^* in CIELAB colour space as given in equation (3.5). The opponent colour space coordinates are given by:

$$u^* = 13L^*(u' - u'_n)$$
$$v^* = 13L^*(v' - v'_n)$$
$$h_{uv}{}^* = \tan^{-1}(v^*/u^*)$$
$$C_{uv}{}^* = \sqrt{u^{*2} + v^{*2}}$$

where u', v', Y and u'_n, v'_n, Y_n are the u', v' coordinates of samples and a suitable chosen white (if the latter is unknown, the values for one illuminant or source in Table 3.1 can be used as an approximation).

Colour difference can be calculated using:

$$\Delta E = \sqrt{\Delta L^{*2} + \Delta u^{*2} + \Delta v^{*2}}$$

or

$$\Delta E_{uv} = \sqrt{\Delta L^{*2} + \Delta C_{uv}{}^{*2} + \Delta H_{uv}{}^{*2}}$$

where

$$\Delta H_{uv}{}^* = p\sqrt{2(C_{uv,1}{}^* C_{uv,2}{}^* - u_1{}^* u_2{}^* - v_1{}^* v_2{}^*)}$$

and

$$p = \begin{cases} -1 & \text{if} \quad u_1{}^* v_2{}^* > u_2{}^* v_1{}^* \\ 1 & \text{otherwise} \end{cases}$$

and subscripts 1 and 2 represent the standard and sample of the pair considered respectively.

An additional attribute, saturation, is also provided for the CIELUV colour space:

$$s_{uv} = 13\sqrt{(u' - u'_n)^2 + (v'_n - v'_n)^2}$$

A three dimensional representation of the CIELAB colour space is shown in Fig. 3.16. The neutral scale is located in the centre of the colour space. The L^* values of 0 and 100 represent a black and a reference white respectively. The a^* and b^* values represent the redness-greenness, and yellowness-blueness

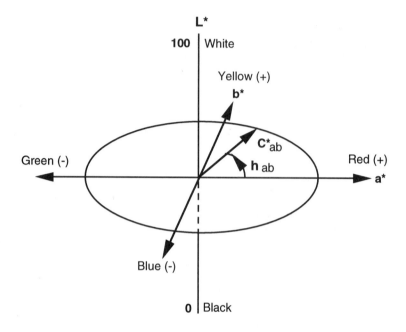

Fig. 3.16   A three-dimensional representation of the CIELAB colour space.

attributes respectively. The C_{ab}^* scale is an open scale with a zero origin (including all colours in the neutral scale, which do not exhibit hue). The hue angle, h_{ab}, lies between 0° and 360°. The colours are arranged following the colour sequence of the rainbow. The four psychological hues (section 3.5.3), which are pure red, yellow, green and blue colours, do not lie exactly at the hue angles of 0°, 90°, 180°, and 270° respectively. The CIELUV space is similar to the CIELAB space, except that it has a saturation scale.

3.4.2   Advanced colour difference formulae

Many sets of experimental results on colour discrimination have been published since 1976. Most of them were conducted using large surface samples viewed under typical industrial viewing conditions. These results show that the two CIE recommended formulae do not accurately quantify small to medium-sized colour differences. Three main sets of experimental data were produced by McDonald (1980a), Luo and Rigg (1986), and by RIT-Dupont

(Berns, Alman, Reniff, Snyder and Bolonon-Rosen, 1991) in 1991. These were used to develop more advanced formulae. Three colour difference formulae are described here: CMC(l:c), BFD(l:c), and CIE94.

CMC(l:c) colour difference formulae
McDonald (1980a) produced over 600 coloured polyester thread pairs surrounding fifty-five colour standards. The visual pass/fail colour matching assessments were carried out by a panel of eight professional colourists. In addition, he accumulated a large set of data involving over 8000 pairs around 600 colour centres assessed by a single observer. These visual results were used to derive the JPC79 formula (McDonald, 1980b; McDonald, 1980c). At a later stage, the JPC79 formula was further modified by members of the Colour Measurement Committee (CMC) of the Society of Dyers and Colourists (SDC) due to the fact that some anomalies were found for colours close to neutral and black. This modified formula is named CMC(l:c) (Clarke, McDonald and Rigg, 1984) and currently is the international standard for the textile industry. The formula is:

$$\Delta E = \sqrt{\left(\frac{\Delta L^*}{lS_L}\right)^2 + \left(\frac{\Delta C_{ab}^*}{cS_C}\right)^2 + \left(\frac{\Delta H_{ab}^*}{S_H}\right)^2}$$

where

$$S_L = \begin{cases} 0.040975L_1^*/(1+0.01765L_1^*) & L_1^* \geq 16 \\ 0.511 & L_1^* < 16 \end{cases}$$

and

$$S_C = 0.0638 C_{ab,1}^*/(1+0.0131 C_{ab,1}^*) + 0.638$$

$$S_H = S_C(Tf + 1 - f)$$

where

$$f = \sqrt{(C_{ab,1}^*)^4/((C_{ab,1}^*)^4 + 1900)}$$

and

$$T = 0.36 + |0.4\cos(h_{ab,1} + 35)|$$

unless $\quad 164° \leq h_{ab,1} \leq 345°$, when

$$T = 0.56 + |0.2\cos(h_{ab,1} + 168)|$$

L_1^*, $C_{ab,1}^*$, and $h_{ab,1}$ refer to the standard of a pair of samples. The l (lightness weight) and c (chroma weight) should equal 2 and 1 respectively for predicting the acceptability of colour differences. For predicting the perceptibility of colour differences, both should equal 1.

BFD(l:c) colour difference formula

Luo and Rigg (1986) accumulated the available experimental data relating to small to medium colour differences between pairs of surface colours to form a combined data set, which contains about 4000 pairs of samples. Supplementary experiments were conducted including over 500 wool textile sample pairs. The new results were used to allow all the previous results to be brought onto a common scale. A new colour-difference formula, BFD(l:c), (Luo and Rigg, 1987a; Luo and Rigg, 1987b) was developed from this set of data. The new formula, given below, is similar in structure to the CMC(l:c) formula in most respects. However, it was found that a new term was required to take into account the fact that when chromaticity ellipses calculated from experimental data are plotted in the CIELAB a^*b^* diagram, they do not all point towards the neutral point as assumed in the CMC(l:c) formula. This effect is most obvious in the blue region.

$$\Delta E = \sqrt{\left(\frac{\Delta L(BFD)}{l}\right)^2 + \left(\frac{\Delta C_{ab}^*}{cD_c}\right)^2 + \left(\frac{\Delta H_{ab}^*}{D_H}\right)^2 + R_T\left(\frac{\Delta C_{ab}^* \Delta H_{ab}^*}{D_C D_H}\right)}$$

where:

$$
\begin{aligned}
L(BFD) &= 54.6\log(Y+1.5) - 9.6 \\
D_C &= 0.035\bar{C}_{ab}^*/(1+0.0365\bar{C}_{ab}^*) + 0.521 \\
D_H &= D_C(GT'+1-G) \\
G &= \sqrt{(\bar{C}_{ab}^*)^4/[(\bar{C}_{ab}^*)^4 + 14000]}
\end{aligned}
$$

$$
\begin{aligned}
T' = \ & 0.627 + \\
& 0.055\cos(\bar{h}_{ab}-254°) \ -0.040\cos(2\bar{h}_{ab}-136°) + \\
& 0.070\cos(3\bar{h}_{ab}-32°) \ +0.049\cos(4\bar{h}_{ab}+114°) - \\
& 0.015\cos(5\bar{h}_{ab}-103°) \\
R_T = \ & R_C R_H \\
R_H = \ & -0.260 +
\end{aligned}
$$

$$0.055\cos(\bar{h}_{ab}-308°)\ -0.379\cos(2\bar{h}_{ab}-160°)-$$
$$0.636\cos(3\bar{h}_{ab}+254°)+0.226\cos(4\bar{h}_{ab}+140°)-$$
$$0.194\cos(5\bar{h}_{ab}+280°)$$

$$R_C = \sqrt{(\bar{C}_{ab}{}^*)^6/[(\bar{C}_{ab}{}^*)^6+7\times10^7]}$$

The terms $\bar{C}_{ab}{}^*$ and $\bar{h}_{ab}$ refer to the mean for the standard and sample. The l (lightness weight) and c (chroma weight) should equal 1.5 and 1 respectively for predicting the acceptability of colour differences. For predicting the perceptibility of colour differences, both should equal one.

CIE94 colour difference formula

Berns, Alman, Reniff, Snyder and Bolonon-Rosen (1991) also conducted visual assessments using glossy acrylic paint pairs. A data set including 156 visual colour tolerances (19 colour centres) perceptually equivalent to a near-grey anchor pair of 1 CIELAB unit was generated. A formula was derived again by modifying the CIELAB formula and was recommended by CIE for field trials in 1994 (CIE, 1995). This is named the CIE94 colour difference formula. It has a similar structure to that of **CMC(l:c)** with simpler weighting functions:

$$\Delta E = \sqrt{(\Delta L^*/K_L S_L)^2 + (\Delta C_{ab}{}^*/K_C S_C)^2 + (\Delta H_{ab}{}^*/K_H S_H)^2}$$

where

$$S_L = 1$$
$$S_C = 1+0.045C_{ab,1}{}^*$$
$$S_H = 1+0.015C_{ab,1}{}^*$$

$C_{ab,1}{}^*$ refers to the $C_{ab}{}^*$ of the standard of a pair of samples. K_L, K_C and K_H are parametric factors to correct for variation in experimental conditions. For all applications except the textile industry, a value of 1 is recommended for all parametric factors. For the textile industry, the K_L factor should be 2, and the K_C and K_H factors should be 1, i.e. CIE94(2:1:1). These factors may be defined by industry groups to correspond to typical experimental conditions for that industry.

3.5 COLOUR APPEARANCE MODELLING

The colour appearance of an object, or an image, changes according to different viewing conditions such as media, light sources, background colours, and luminance levels. This phenomenon causes severe problems in industrial colour control. For example, in the surface industries colourists need to know the degree of colour change across a wide range of illumination conditions. Lighting engineers need to evaluate the colour rendering property between a test and a reference illuminant. Colour reproduction engineers want to reproduce faithfully the original presented on a different medium. The media involved might include original scenes, transparencies, monitors, photographs or reflection prints. All these tasks require visual judgement by experienced workers. This process is subjective and expensive. Hence, there has long been a great demand by industrialists for the ability to quantify accurately changes in colour appearance so as to minimize observer dependencies.

In colorimetry, measures such as tristimulus values, chromaticity coordinates, dominant wavelength and excitation purity, were used to indicate whether two stimuli match each other. On the other hand, the measures used to quantify colour differences such as CIELAB and CIELUV are also developed so that equal scale intervals represent approximately equally perceived differences in the attributes considered. However, all the above measures are limited to being used under fixed viewing conditions: the two stimuli in question should be presented using identical media and be viewed under the same daylight viewing conditions defined by the CIE. Hence, these two classes of measures do not provide a satisfactory means of solving the above problems. The solution is to devise an international standard colour appearance model, which is capable of predicting the colour appearance of colours under a wide range of viewing conditions. Various models have been developed over the years. In general, four colour appearance models are considered to be most promising. These are Hunt (1991b) and (1994), Nayatani (1997), RLAB (Fairchild, 1996) and LLAB (Luo, Lo and Kuo, 1996).

At the CIE *Expert Symposium '96* on *Color Standards for Image Technology*, held in Vienna in 1996 (Cen, 1996), there was demand from industrialists to ask the CIE to recommend a colour appearance model for industrial application. The agreement was reached that Hunt and Luo should examine the existing colour appearance models, try to combine the best features of these models into a high performance model for general use, and test its performance against available experimental data. It was also agreed that the model should be available in a comprehensive version, and in a relatively simple version for use in limited conditions. At its meeting held in Kyoto in 1997, CIE

Technical Committee TC1-34 (Testing colour appearance models), agreed to adopt the simple version, which is named CIECAM97S. The 'S' stands for 'simple'. The inclusion of the year 97 in the designation is intended to indicate the interim nature of the model, depending upon the availability of better models which are expected to emerge in the future. The full CIECAM97S model including its forward and reverse models together with its test data is given in section 3.5.4. It is recommended that the model should be uniformly applied for colour image processing to achieve successful colour reproduction across different media such as prints, monitors and transparencies.

3.5.1 Techniques for assessing colour appearance

Colour appearance models are based on colour vision theories but also must fit experimental results. Hence, CIE Technical Committee 1-34 was formed to gather as many experimental data sets as possible in order to test the performance of models. The best model can then be recommended as an international standard for industrial applications. This section describes the experimental techniques used to generate these data sets. They can be divided into three categories: **haploscopic matching**, **memory matching**, and **magnitude estimation**.

Haploscopic matching is the most widely used experimental technique. This technique requires specially designed viewing apparatus which presents a different adapting stimulus to each of the observer's two eyes. His or her task is to adjust the stimulus at one eye to match that at the other eye. The task is relatively simple and the results in general have higher precision than the other two techniques. However, its validity is dependent on an assumption: that the adaptation of one eye does not affect the sensitivity of the other eye. The technique imposes unnatural viewing conditions with constrained eye movement. In addition, when two eyes are presented with two stimuli under different adapting fields, observers tend to bias towards one field rather than the other. This is known as **binocular rivalry**. The haploscopic technique can be improved to overcome some of the above shortcomings by using a complex field with free eye movement.

Memory matching is carried out under normal viewing conditions using both eyes and without the interposition of any optical devices. This technique provides a steady-state of adaptation with free eye movement. Observers are first trained to describe colours using the three colour appearance attributes of a colour descriptive system such as Munsell or the Natural Colour Order System (NCS). This means that they are able to describe with reasonable accuracy and precision the colour of any object in these terms under any viewing conditions. However, this is not a widely used technique and has drawbacks such as a substantial training period, complicated procedures for data analysis, lower precision than that of the haploscopic technique, limited capacity for retaining information, and memory distortion.

Magnitude estimation has been used increasingly in recent years. Observers are asked to scale some colour appearance attributes such as lightness, colourfulness and hue under fully adapted viewing conditions. Its only disadvantage is in having lower precision than the haploscopic technique. Many advantages are associated with this technique such as: normal viewing conditions using both eyes; steady-state of adaptation; results described in terms of perceived attributes which can be directly compared with the colour appearance models' predictions; and shorter training period than that of memory matching. The author and his coworkers have conducted a large set of experiments using this technique. This is named the LUTCHI data set (Luo, Clarke, Rhodes, Schappo, Scrivener and Tait, 1991a; Luo, Clarke, Rhodes, Schappo, Scrivener and Tait, 1991b; Luo, Gao, Rhodes, Xin, Clarke and Scrivener, 1993a; Luo, Gao, Rhodes, Xin, Clarke and Scrivener, 1993b). The results were used to develop the Hunt and LLAB colour appearance models.

3.5.2   Structure of a colour appearance model

Figure 3.17 shows the generic structure of a colour appearance model. These usually comprise three parts: a chromatic adaptation transform; a dynamic response function; and a colour space for representing the colour appearance attributes. The differences between the four colour appearance models are also described below.

A chromatic adaptation transform is capable of predicting the corresponding colour in terms of a colorimetric specification (such as tristimulus values) from one set of illumination conditions to another. A pair of corresponding colours would look the same when viewed under the two illuminants in question, say standard illuminants A and D65. The earliest, and still a very useful chromatic adaptation transform was that devised by von Kries which is used in the Hunt, Nayatani and RLAB models, however the more recent transform used in the LLAB and CIECAM97s models shows a better performance than that of the von Kries transform in predicting many experimental data sets (Luo and Hunt, 1998).

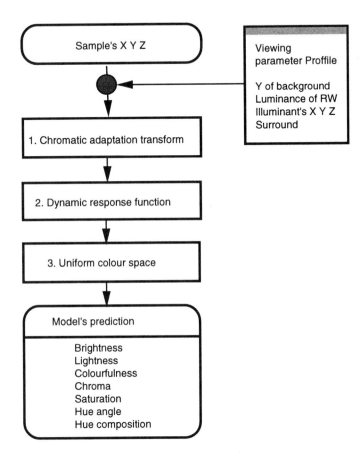

Fig. 3.17 Generic structure of a colour appearance model.

The dynamic response functions are used to predict the extent of changes of responses of stimuli of different luminance factors across a wide range of luminance levels, i.e. from very dark scotopic to very light photopic vision. A cube-root function was used for the RLAB model, hyperbolic in the Hunt, LLAB and CIECAM97S models, and logarithmic in the Nayatani model. The colour spaces used in these models are similar in that they all provide redness-greenness and yellowness-blueness scales to form rectangular coordinates (similar to that in Fig. 3.16 of the CIELAB space), ratios of which are used to derive hue angle from 0° to 360°. The distance away from neutral is used to represent the colourfulness, saturation or chroma, while a non-linear function of an achromatic signal is used to derive a lightness attribute. Some of the models also provide a brightness attribute. The colour space is based upon a reference set of illumination conditions. For the Nayatani model, it refers to the CIE D65 illuminant and 2° conditions. The equal energy illuminant is used for the other colour appearance models.

The differences in the chromatic adaptation transforms, in the dynamic response functions, and in the colour spaces, all contribute to the differences between the predictions made by the different models.

3.5.3   The predicted colour appearance attributes

This section describes the meaning of the colour appearance attributes and what viewing parameters would affect each attribute. Many of the latter phenomena have been taken into account by each colour appearance model. Their definitions can also be found in the CIE International Lighting Vocabulary (CIE, 1989).

Brightness

This is a visual sensation according to which an area exhibits more or less light. This is an open ended scale with a zero origin defining black. The brightness of a sample is affected by the luminance of the light source used. A surface colour illuminated by a higher luminance would appear brighter than the same surface illuminated by a lower luminance.

Lightness

This is the brightness of an area judged relative to the brightness of a similarly illuminated reference white. The lightness scale runs from black, 0, to white, 100 (see Fig. 3.16). The lightness of the background used can cause a change of the lightness of the sample. This is called the **lightness contrast effect**. For example, a colour appears lighter against a dark background than against a light background.

Colourfulness

Colourfulness is the attribute of a visual sensation according to which an area appears to exhibit more or less of its chromatic content. This is an open ended scale with a zero origin defining the neutral colours. Similar to the brightness attribute, the colourfulness of a sample is also affected by luminance. A surface colour illuminated by a higher luminance would appear more colourful than when illuminated by a lower luminance.

Chroma

This is the colourfulness of an area judged as a proportion of the brightness of a similarly illuminated reference white. This is an open ended scale with a zero origin representing neutrals with no hue. Fig. 3.16 shows its relationship with lightness and hue angle in CIELAB colour space.

Saturation

This is the colourfulness of an area judged in proportion to its brightness. This scale again runs from zero representing neutral colours with an open end. It can be considered as the ratio of colourfulness to lightness, like that defined in the CIELUV space (s_{uv}) on page 47. When adding a coloured pigment to a white paint, the resultant colour increases its saturation by decreasing its brightness and increasing colourfulness.

Hue

Hue is the attribute of a visual sensation according to which an area appears to be similar to one, or to proportions of two, of the perceived colours red, yellow, green and blue. Each model predicts hue with two measures: hue angle (see Fig. 3.16) ranged from 0° to 360°, and hue composition ranged from 0, 100, 200, 300 and 400 corresponding to psychological hues of red, yellow, green, blue and back to red. These four hues are the psychological hues, which cannot be described in terms of any combinations of other colour names.

Most of the above attributes are predicted by the colour appearance models considered here. Apart from the above visual phenomena, the chromatic adaptation transformation is considered to be at the heart of each model (section 3.5.2). Also, all models except the Nayatani model predict the effect caused by the luminance of the surround such as a 35 mm projected image viewed against a dark surround, a monitor image against a dim surround, and a printed image against a brighter surround than the above two (designated 'average' surround here). The brightness and colourfulness contrasts are smaller under the dim and dark surrounds in comparison with that under an average surround. There are more specific parametric effects predicted by different models.

3.5.4   Computation procedures for the CIECAM97S colour appearance models

The CIECAM97S colour appearance model is given in this section. The condition of viewing field (called test field hereafter) used in the colour appearance model are shown in Fig. 3.18. A test sample and a reference white are viewed against an achromatic background under a particular luminance of a test illuminant.

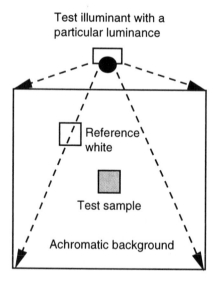

Fig. 3.18 Condition of viewing in test field.

Forward CIECAM97S colour appearance model

Table 3.2 defines the input values required by the model, and Table 3.3 defines the parameters of the model.

Computational procedures

Step 1 Calculate cone responses of test sample (R, G, B), similarly, for the reference whites (R_w, G_w, B_w) and (R_{wr}, G_{wr}, B_{wr}):

$$
\begin{bmatrix} R \\ G \\ B \end{bmatrix} = M_{BFD} \begin{bmatrix} X/Y \\ Y/Y \\ Z/Y \end{bmatrix}
$$

Table 3.2 Variables used in the CIECAM97S colour appearance model.

Symbol	Definition
x, y, Y	Colorimetric values of sample under test field.
x_w, y_w, Y_w	Colorimetric values of reference white under test field.
x_{wr}, y_{wr}, Y_{wr}	Colorimetric values of reference white under reference field (equal energy illuminant, $x_{wr} = y_{wr} = \frac{1}{3}$ and $Y_{wr} = 100$).
Y_b	Y of achromatic background (%).
L_W	Luminance of reference white in cd/m².
L_A	Luminance of achromatic background in cd/m² (calculated by $L_W Y_b / 100$).

Table 3.3 Parameters of the CIECAM97S colour appearance model.

Surround parameters	F	c	F_{LL}	N_c
Average with sample over 4° viewing field	1.0	0.690	0.0	1.0
Average with sample over ≤ 4° viewing field	1.0	0.690	1.0	1.0
Television and VDU displays in dim surrounds	0.9	0.590	1.0	1.1
Projected photographs in dark surrounds	0.9	0.525	1.0	0.8
Large cut-sheet transparency	0.9	0.410	1.0	0.8

$$\text{where} \quad M_{BFD} = \begin{bmatrix} 0.8951 & 0.2664 & -0.1614 \\ -0.7502 & 1.7135 & 0.0367 \\ 0.0389 & -0.0685 & 1.0296 \end{bmatrix}$$

Step 2   Calculate corresponding cone responses of test sample (R_c, G_c, B_c), similarly, for the reference white (R_{wc}, G_{wc}, B_{wc}):

$$R_c = [D(R_{wr}/R_w) + 1 - D]R$$
$$G_c = [D(G_{wr}/G_w) + 1 - D]G$$
$$B_c = \begin{cases} [D(B_{wr}/B_w^p) + 1 - D] \; B^p, & B_c \geq 0 \\ -[D(B_{wr}/B_w^p) + 1 - D] \; |B|^p & \text{otherwise} \end{cases}$$

where

$$P = (B_w/B_{wr})^{0.0834}$$
$$D = F - F/[1 + 2(L_A^{\frac{1}{4}}) + (L_A^2)/300]$$

Step 3   Calculate luminance level adaptation factor (F_L), chromatic background induction factor (N_{cb}) and brightness background induction factors (N_{bb}).

$$F_L = 0.2k^4(5L_A) + 0.1(1 - k^4)^2(5L_A)^{\frac{1}{3}}$$

where

$$k = 1/(5L_A + 1)$$

$$N_{cb} = N_{bb} = 0.725(1/n)^{0.2}$$

where $n = Y_b/Y_w$.

Step 4   Calculate corresponding tristimulus values of test sample (R', G', B'), and similarly, for the reference white (R'_w, G'_w, B'_w).

$$\begin{bmatrix} R' \\ G' \\ B' \end{bmatrix} = M_H M_{BFD}^{-1} \begin{bmatrix} R_c Y \\ G_c Y \\ B_c Y \end{bmatrix}$$

where

$$M_H = \begin{bmatrix} 0.38971 & 0.68898 & -0.07868 \\ -0.22981 & 1.18340 & 0.04641 \\ 0.00000 & 0.00000 & 1.00000 \end{bmatrix}$$

and

$$M_{BFD}^{-1} = \begin{bmatrix} 0.9870 & -0.1471 & 0.1600 \\ 0.4323 & 0.5184 & 0.0493 \\ -0.0085 & 0.0400 & 0.9685 \end{bmatrix}$$

Step 5   Calculate cone responses after adaptation of test sample (R'_a, G'_a, B'_a), similarly, for the reference white $(R'_{aw}, G'_{aw}, B'_{aw})$.

$$R'_a = \frac{40(F_L R'/100)^{0.73}}{(F_L R'/100)^{0.73} + 2} + 1$$

and similarly for G'_a and B'_a. If R'_a is less than zero use:

$$R'_a = \frac{-40(-R'/100)^{0.73}}{(-R'/100)^{0.73} + 2} + 1$$

and similarly for R'_{aw} and for the G'_a and B'_a equations.

Step 6   Calculate the red-green (a), and yellow-blue (b) opponent correlates.

$$a = R'_a - 12G'_a/11 + B'_a/11$$
$$b = \frac{1}{9}(R'_a + G'_a - 2B'_a)$$

Step 7   Calculate hue angle (h):

$$h = \tan^{-1}(b/a)$$

Step 8   Calculate Eccentricity factor (e) and Hue quadrature (*H*):

$$H = H_1 + \frac{100(h - h_1)/e_1}{(h - h_1)/e_1 + (h_2 - h)/e_2}$$

$$e = e_1 + (e_2 - e_1)\frac{h - h_1}{h_2 - h_1}$$

where H_1 is 0, 100, 200, or 300, according to whether red, yellow, green, or blue, respectively, is the hue having the nearest lower value of *h*. The values of *h* and *e* for the four unique hues are:

	Red	Yellow	Green	Blue
H	0	100	200	300
h	20.14	90.00	164.25	237.53
e	0.8	0.7	1.0	1.2

e_1 and h_1 are the values of *e* and *h*, respectively, for the unique hue having the nearest lower value of *h*; and e_2 and h_2 are these values for the unique hue having the nearest higher value of *h*.

H_p is the part of *H* after its hundreds digit, if:

$H = \quad H_p$, the hue composition is H_pYellow, $100 - H_p$Red

$H = 100 + H_p$, the hue composition is H_pGreen, $100 - H_p$Yellow

$H = 200 + H_p$, the hue composition is H_pBlue, $\quad 100 - H_p$Green

$H = 300 + H_p$, the hue composition is H_pRed, $\quad 100 - H_p$Blue

Step 9   Calculate the achromatic response of the sample (*A*) and reference white (A_w):

$$A \quad = \quad [2R'_a \quad +G'_a \quad +\tfrac{1}{20}B'_a \quad -2.05]N_{bb}$$

$$A_w \quad = \quad [2R'_{aw} +G'_{aw} +\tfrac{1}{20}B'_{aw} \quad -2.05]N_{bb}$$

Step 10   Calculate Lightness (*J*):

$$J = 100(A/A_w)^{cz} \qquad \text{where} \quad z = 1 + F_{LL}\sqrt{n}$$

Step 11   Calculate Brightness (*Q*):

$$Q = (1.24/c)(J/100)^{0.67}(A_w + 3)^{0.9}$$

Step 12   Calculate Saturation (s):

$$s = \frac{5000(a^2 + b^2)^{\frac{1}{2}} e(10/13)N_c N_{cb}}{R'_a + G'_a + (21/20)B'_a}$$

Step 13   Calculate Chroma (C):

$$C = 2.44 s^{0.69} (J/100)^{0.67n} (1.64 - 0.29^n)$$

Step 14   Calculate Colourfulness (M):

$$M = CF_L^{0.15}$$

Reverse CIECAM97S colour appearance model

The reverse CIECAM97S colour appearance model is used to calculate the corresponding tristimulus values (X,Y,Z) from the colour appearance attributes computed from the forward CIECAM97S model, i.e. Q or J, M or C, H or h.

Step 1   Calculate Lightness (J) from Brightness (Q):

$$J = 100(cQ/1.24)^{1/0.67} / (A_w + 3)^{0.9/0.67}$$

Step 2   Calculate Achromatic response (A) from Lightness (J):

$$A = A_w (J/100)^{1/cz}$$

Step 3   Use Hue composition to determine h_1, h_2, e_1 and e_2 (see *Step 8* in the previous section).

Step 4   Calculate h:

$$h = \frac{(H - H_1)(h_1/e_1 - h_2/e_2) - 100h_1/e_1}{(H - H_1)(1/e_1 - 1/e_2) - 100/e_1}$$

where H_1 is 0, 100, 200, or 300 according whether red, yellow, green, or blue, respectively, is the hue having the nearest lower value of h.

Step 5   Calculate e as given in *Step 8* of the forward model.

Step 6   Calculate $C = M/F_L^{0.15}$.

Step 7   Calculate s:

$$s = \frac{C^{1/0.69}}{[2.44(J/100)^{0.67n}(1.64 - 0.29^n)^n]^{1/0.69}}$$

where $n = Y_b/Y_w$.

Step 8   Calculate a and b:

$$a = \frac{s(A/N_{bb} + 2.05)}{\sqrt{1 + \tan^2 h}(50000eN_cN_{cb}/13) + s[(11/23) + (108/23)\tan h]}$$

$$b = a \tan h$$

In calculating $\sqrt{1 + \tan^2 h}$ the result is to be taken as positive if $0 \le h < 90$; negative if $90 \le h < 270$; positive if $270 \le h < 360$.

Step 9   Calculate R'_a, G'_a, B'_a:

$$R'_a = \frac{20}{61}(A/N_{bb} + 2.05) + \frac{41}{61}(11/23)a + \frac{288}{61}\frac{1}{23}b$$
$$G'_a = \frac{20}{61}(A/N_{bb} + 2.05) - \frac{81}{61}(11/23)a - \frac{261}{61}\frac{1}{23}b$$
$$B'_a = \frac{20}{61}(A/N_{bb} + 2.05) - \frac{20}{61}(11/23)a - \frac{20}{61}\frac{315}{23}b$$

Step 10   Calculate R', G', B':

$$R' = 100[(2R'_a - 2)/(41 - R'_a)]^{1/0.73}$$

If $R'_a - 1 < 0$, then

$$R' = -100[(2 - 2R'_a)/(39 + R'_a)]^{1/0.73}$$

and similarly for the G' and B' equations.

Step 11   Calculate R_cY, G_cY and B_cY:

$$\begin{bmatrix} R_cY \\ G_cY \\ B_cY \end{bmatrix} = M_{BFD}M_H^{-1} \begin{bmatrix} R'/F_L \\ G'/F_L \\ B'/F_L \end{bmatrix}$$

where

$$M_H^{-1} = \begin{bmatrix} 1.91019 & -1.11214 & 0.20195 \\ 0.37095 & 0.62905 & 0.00000 \\ 0.00000 & 0.00000 & 1.00000 \end{bmatrix}$$

Step 12 Calculate $(Y/Y_c)R$, $(Y/Y_c)G$ and $(Y/Y_c)B$:

$$(Y/Y_c)R = (Y/Y_c)R_c/[D(R_{wr}/R_w)+1-D]$$
$$(Y/Y_c)G = (Y/Y_c)G_c/[D(G_{wr}/G_w)+1-D]$$
$$(Y/Y_c)B = |(Y/Y_c)B_c|^{1/p}/[D(B_{wr}/B_w^p)+1-D]^{1/p}$$

where, if B_c is negative, $(Y/Y_c)^{1/p}B$ is negative; and $Y_c = 0.43231R_cY + 0.51836G_cY + 0.04929B_cY$.

Step 13 Calculate Y':

$$Y' = 0.43231\,RY + 0.51836\,GY + 0.04929\,BY_c(Y/Y_c)^{1/p}$$

Step 14 Calculate X''/Y_c, Y''/Y_c and Z''/Y_c:

$$\begin{bmatrix} X''/Y_c \\ Y''/Y_c \\ Z''/Y_c \end{bmatrix} = M_{BFD}^{-1} \begin{bmatrix} (Y/Y_c)R \\ (Y/Y_c)G \\ (Y/Y_c)^{(1/p)}B/(Y'/Y_c)^{(1/p-1)} \end{bmatrix}$$

where

$$M_{BFD}^{-1} = \begin{bmatrix} 0.9870 & -0.1471 & 0.1600 \\ 0.4323 & 0.5184 & 0.0493 \\ -0.0085 & 0.0400 & 0.9685 \end{bmatrix}$$

Step 15 Multiply each by Y_c to obtain X'', Y'' and Z'', which equal X, Y, Z to a very close approximation.

Working examples for the CIECAM97S colour appearance model

The data in Tables 3.4 to 3.7 is given to assist readers to implement the CIECAM97S model. One sample is viewed under the standard illuminant A and four different levels of adapting luminance, L_A. The viewing conditions are: luminous factor (Y_b) of the background – 18.0; luminances L_A, in cd/m^2 of the achromatic background – 2000, 200, 20 and 2 (4 different levels).

Table 3.4 Colorimetric input data for the test sample, reference whites.

	x	y	Y
Sample under test field	0.3618	0.4483	23.93
Reference white in test field	0.4476	0.4074	90.00
Reference white in reference field	0.3333	0.3333	100.00

Table 3.5 Model parameters for average surround.

F	c	F_{LL}	N_c
1.0	0.69	1.0	1.0

Table 3.6 Results: prediction for the reference white.

Luminances, L_A	2000	200	20	2
Hue angle, h	41.8	57.4	58.8	59.5
Hue quadrature, H	28.3	50.0	52.0	53.0
Lightness, J	100.0	100.0	100.0	100.0
Brightness, Q	70.1	52.7	37.9	26.8
Saturation, s	0.0	0.5	12.6	25.9
Chroma, C	0.1	1.3	12.1	19.8
Colourfulness, M	0.1	1.3	10.8	15.7

Table 3.7 Results: prediction for the test sample.

Luminance, L_A	2000	200	20	2
Hue angle, h	190.2	190.0	183.5	175.7
Hue quadrature, H	239.7	239.4	229.9	218.2
Lightness, J	53.0	48.2	45.2	44.2
Brightness, Q	45.8	32.3	22.3	15.5
Saturation, s	120.0	125.9	114.0	96.5
Chroma, C	52.4	53.5	49.5	44.0
Colourfulness, M	58.8	53.5	44.1	34.9

4

Representations of colour images in different colour spaces

Henryk Palus

Colour spaces (other terms: colour coordinate systems, colour models) are three-dimensional arrangements of colour sensations. Colours are specified by points in these spaces. The colour spaces presented in this chapter are the most popular in the image processing community. Equations describing transformations between different colourspaces and the reasons for using colour spaces other than RGB are presented. Based on examples from the literature, the applicability of individual colour spaces in image processing systems is discussed. Spaces used in image processing are derived from visual system models (e.g. RGB, opponent colour space, IHS *etc.*); adopted from technical domains (e.g. colorimetry: XYZ, television: YUV, *etc.*) or developed especially for image processing (e.g. Ohta space, Kodak PhotoYCC space *etc.*).

4.1 BASIC RGB COLOUR SPACE

The RGB space is the most frequently used colour space for image processing. Since colour cameras, scanners and displays are most often provided with direct RGB signal input or output, this colour space is the basic one, which is, if necessary, transformed into other colour spaces. However, the RGB primaries of these devices are not always consistent. The colour gamut in RGB space forms a cube (Fig. 4.1). Each colour, which is described by its RGB components, is represented by a point and can be found either on the surface or inside the cube. All grey colours are placed on the main diagonal of this cube from black ($R = G = B = 0$) to white ($R = G = B = \text{max}$). When

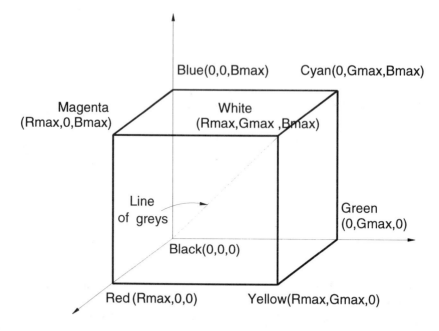

Fig. 4.1 Representation of colour in RGB colour space.

the output signal from a camera is encoded for a TV system, either standard decoders, such as PAL/RGB or NTSC/RGB are used, or the space utilized by a given system is applied, such as YUV space (see section 4.3.1) in the case of the PAL system or YIQ (see section 4.3.2) in the case of the NTSC system. The main disadvantage of RGB colour space in applications involving natural images is a high correlation between its components: about 0.78 for $B - R$, 0.98 for $R - G$ and 0.94 for $G - B$ components. This makes the RGB space unsuitable for compression. Other disadvantages of RGB space are:

- psychological non-intuitivity, i.e. it is hard to visualize a colour based on R, G, B components,

- non-uniformity, i.e. it is impossible to evaluate the perceived differences between colours on the basis of distance in RGB space.

The RGB components for a given image are proportional to the amount of light incident on the scene represented by the image. In order to eliminate the influence of illumination intensity, so-called **chromaticity coordinates** (normalized colours) were introduced in colorimetry (Wyszecki and Stiles, 1982):

$$r = \frac{R}{R+G+B}$$
$$g = \frac{G}{R+G+B}$$
$$b = \frac{B}{R+G+B} = 1 - r - g$$

Since each of the normalized colours is linearly dependent, the rgb space may be represented by two normalized colours. Coordinate b can be calculated for verification. As could be easily shown, the chromaticity coordinates rgb remain unchanged unless the spectral distribution of the scene-illuminating light is varied (Andreadis, Browne and Swift, 1990).

Values of rgb coordinates are much more stable with changes in illumination level than RGB coordinates. This was verified experimentally by Ailisto and Piironen (1987) and Berry (1987). In a computer vision system, translation from RGB space to rgb space more than doubles colour discriminability (Ailisto and Piironen, 1987). When a colour image is segmented, dark pixels can give incorrect rgb values, because they are affected by noise ($R+G+B$ is near 0). These pixels should be omitted during segmentation (Gunzinger, Mathis and Guggenbuehl, 1990). Normalized colours were also tried for analysis of aerial photographs (Ali, Martin and Aggarwal, 1979), edge detection (Nevatia, 1977) and image segmentation (Healey, 1992).

4.2 XYZ COLOUR SPACE

The XYZ colour space was accepted by the CIE in 1931 and is described in detail in section 3.1 and it has been used in colorimetry ever since. It was designed to yield non-negative tristimulus values for each colour. In this system, Y represents the luminance of the colour. The tristimulus values XYZ are related to CIE RGB tristimulus values by the following equations (Wyszecki and Stiles, 1982):

$$X = 0.490R + 0.310G + 0.200B$$
$$Y = 0.177R + 0.812G + 0.011B$$
$$Z = 0.000R + 0.010G + 0.990B$$

The above relationships result from defining three spectral colours with the following chromaticity coordinates as reference stimuli in CIE RGB space:

$x_R = 0.735$ $y_R = 0.265$

$x_G = 0.274$ $y_G = 0.717$

$x_B = 0.167$ $y_B = 0.009$

together with reference white, with coordinates:

$x_W = 0.333$ $y_W = 0.333$

In colorimetry, the following transformation equations for chromaticity coordinates x, y, z were introduced (Wyszecki and Stiles, 1982):

$$x = \frac{X}{X+Y+Z}$$

$$y = \frac{Y}{X+Y+Z}$$

$$z = \frac{Z}{X+Y+X} = 1-x-y$$

In image processing, other versions of RGB space called EBU RGB (European Broadcasting Union) and FCC RGB (Federal Communications Commission, USA) are used, which have been accepted for colour reproduction in colour cameras. These spaces are based on the tristimulus values of cathode ray tube phosphors given in Table 4.1. and reference white with values:

$x_W = 0.313$ $y_W = 0.329$

for illuminant D65 (for EBU RGB) and:

$x_W = 0.310$ $y_W = 0.316$

Table 4.1   EBU and FCC tristimulus values of CRT phosphors.

	EBU		FCC	
	x	y	x	y
R	0.640	0.330	0.670	0.330
G	0.290	0.600	0.210	0.710
B	0.150	0.060	0.140	0.080

for illuminant C (for FCC RGB). Appropriate transformations from RGB to XYZ can be found in TV engineering textbooks (Benson, 1992; Slater, 1991). For the EBU RGB space the transformation is:

$$X = 0.430\,R + 0.342\,G + 0.178\,B$$
$$Y = 0.222\,R + 0.707\,G + 0.071\,B \quad (4.1)$$
$$Z = 0.020\,R + 0.130\,G + 0.939\,B$$

while that for FCC RGB space is:

$$X = 0.607\,R + 0.174\,G + 0.200\,B$$
$$Y = 0.299\,R + 0.587\,G + 0.114\,B \quad (4.2)$$
$$Z = 0.000\,R + 0.066\,G + 1.116\,B$$

The role of XYZ space in image processing systems is rather supplementary. In the work of Slaughter and Harrell (1987), it is required to determine the threshold values of hue and saturation (necessary to recognize an object: an orange). From the spectral reflectance curves for the orange the tristimulus values XYZ were first calculated. Then, applying the inverse transform, RGB values were determined and hence, after further transformations, H and S threshold values. In this colour space, a stereo correspondence algorithm (Koschan, 1993) and colour image segmentation algorithms (Lim and Lee, 1990) were tested. Quite often the XYZ colour space is an intermediate space in the process of determining a perceptually uniform colour space such as CIELAB (sometimes denoted as $L^*a^*b^*$) (Celenk, 1995) or CIELUV (sometimes denoted as $L^*u^*v^*$).

4.3 TELEVISION COLOUR SPACES

The television colour spaces[1] (e.g. YUV, YIQ) are in fact, **opponent** colour spaces (see section 4.4), because they define a luminance component and two chrominance components based on **colour-difference** signals: $R - Y$ and $B - Y$. They were designed to minimize the bandwidth of the composite signals, i.e. to enable transmission of colour-difference signals within the existing bandwidth for black and white television. The human visual system is far less sensitive to spatial details in chrominance than in luminance (Slater, 1991) for reasons discussed in section 2.5.3. Because of this, the bandwidth of the chrominance signals can be, and is, significantly smaller than for the

---

[1] Television colour spaces are also discussed in section 5.3 on page 108.

luminance signal, e.g. in the case of YIQ the required bandwidth is as follows: $Y - 4\,\text{MHz}$, $I - 1.5\,\text{MHz}$ and $Q - 0.5\,\text{MHz}$. The advantages of television colour spaces are their simplicity of transformation and the separability of luminance and chrominance information.

4.3.1   YUV **colour space**

In Europe the YUV colour space (the basis of the PAL TV signal coding system) is used (Slater, 1991):

$$Y = \;\;\;0.299R + 0.587G + 0.114B$$
$$U = -0.147R - 0.289G + 0.437B = 0.493(B - Y)$$
$$V = \;\;\;0.615R - 0.515G - 0.100B = 0.877(R - Y)$$

The Y (luminance) component is identical to the Y component in XYZ space. YUV space can be transformed into an IHS-type space (see section 4.6):

$$H_{UV} \;\; = \;\; \tan^{-1}(V/U)$$
$$S_{UV} \;\; = \;\; \sqrt{U^2 + V^2}$$

The UV plane has been used for colour image segmentation (Ferri and Vidal, 1992). The YUV space is widely used in coding of colour images and in colour video. Because the colour representation of an image sequence requires less detail than the luminance, the data rate can be shared as follows: 80% to the Y component and 10% each to the U and V components (Monro and Nicholls, 1995). Special chips for digital video processing and other image compression hardware often use YUV components with reduced spatial resolution for chrominance (Y:U:V 4:2:2 or Y:U:V 4:1:1).

4.3.2   YIQ **colour space**

The YIQ colour space (the basis of the American NTSC TV signal coding system) is related to FCC RGB coordinates by the equations given below, the Y component being identical to the Y component of YUV space (Benson, 1992):

$$I = 0.596R - 0.274G - 0.322B = 0.74(R - Y) - 0.27(B - Y)$$
$$Q = 0.211R - 0.523G + 0.312B = 0.48(R - Y) + 0.41(B - Y)$$

As with the YUV space, the Y (luminance) component is identical to Y in XYZ space, while the I component (in-phase, an orange-cyan axis) and the Q component (quadrature, a magenta-green axis) express jointly hue and saturation. By introducing cylindrical coordinates, YIQ space is transformed into an IHS-type space (see section 4.6):

$$H_{IQ} = \tan^{-1}(Q/I)$$
$$S_{IQ} = \sqrt{I^2 + Q^2}$$

Transformations such as the above were applied by Slaughter and Harrell (1987) to determine threshold hue and saturation values used for object recognition and also in Solinsky (1985) for edge detection.

In this colour space, colour image segmentation algorithms (Lim and Lee, 1990) were tested. The I component was used for extraction of faces from colour images (Dai and Nakano, 1996). The YIQ space is also useful for colour image coding (Overturf, Comer and Delp, 1995).

4.3.3 YC$_b$C$_r$ colour space

This colour space is independent of TV signal coding systems. It is appropriate for digital coding of standard TV images (525/625 lines) (ITU, 1994b) and is given as follows, the Y component being identical to that for YUV and YIQ:

$$C_b = -0.169R - 0.331G + 0.500B = 0.564(B - Y)$$
$$C_r = 0.500R - 0.418G - 0.081B = 0.713(R - Y)$$

This colour space was used in a process of video sequences encoding on videodisks (Watson, 1994). Other current applications in image compression (e.g. JPEG format) often employ YC$_b$C$_r$ space as a quantization space. Greater compression is achieved using reduced spatial resolution and coarser quantization for C$_b$ and C$_r$ than for the Y component.

4.4 OPPONENT COLOUR SPACE

Opponent colour space has been inspired by the physiology of the human visual system (colour opponent processes) and therefore is sometimes called **physiologically motivated** colour space (Pomierski and Gross, 1996). In the late 19th century the German physiologist Ewald Hering proposed colour

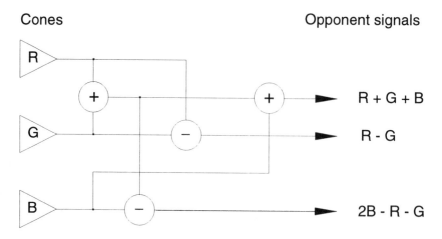

Fig. 4.2 Schematic representation of opponent colour stage of human visual system.

opponency theory[2], which explained some perceptual colour phenomena in-explicable by classical trichromatic theory (Hering, 1964). Illusions called 'negative after-images' and observations during colour naming experiments that reddish-green and yellowish-blue colours are not identified, led Hering to suppose that red must be the 'opposite' colour to green and likewise blue must be 'opposite' to yellow. He presumed the existence of three opponent chan-nels (processes) in the human visual system: red-green **R-G** channel, yellow-blue **Ye-B** channel and achromatic (white-black) **Wh-Bl** channel. Hering's opponent process theory was one of the first approaches to separate lumi-nance from chrominance. At present multi-stage models of colour vision are used with a second opponent stage (De Valois and De Valois, 1993). Fig. 4.2 and equation (4.3) show how RGB 'cone' signals are transformed to three channels — one hypothesized achromatic channel, and two opponent colour channels.

$$
\begin{aligned}
RG &= R - G \\
YeB &= 2B - R - G \\
WhBl &= R + G + B
\end{aligned}
\tag{4.3}
$$

Opponent colour spaces have found numerous applications in different fields of image processing:

[2]Colour opponency theory is also discussed in section 2.4

- interactive computer graphics (Naiman, 1985),

- real-time colour edge detection by using a visual processor (Masaki, 1988),

- colour stereopsis (Brockelbank and Yang, 1989),

- digital colour image sequence coding (Watson and Tiana, 1992),

- detection of colour contrast (Schmid and Truskowski, 1993),

- chromatic adaptation of the image (Pomierski and Gross, 1996).

Much research has been devoted to the use of opponent colour spaces for multi-colour object recognition tasks (Wixson and Ballard, 1989; Swain, 1990; Brock-Gunn and Ellis, 1992). 3-D or 2-D opponent colour histograms were applied as a signature (template) for the representation of object colours. Sometimes modified versions of opponent colour spaces are used. Yamaba and Miyake (1993) presented a system which recognizes characters and their colours using opponent colour space with a modified second equation:

$$YeB = 0.4(R+G) - B$$

Because the responses of the cones in the human visual system are proportional to the logarithm of the stimulus intensity, the following 'logarithmic' version of opponent space was developed (Fleck, Forsyth and Bregler, 1996):

$$
\begin{aligned}
RG &= \log R - \log G \\
YeB &= \log B - (\log R + \log G)/2 \\
WhBl &= \log G
\end{aligned}
$$

There is a long tradition behind this 'logarithmic' approach in image processing (Faugeras, 1979).

4.5   OHTA $I_1I_2I_3$ COLOUR SPACE

The definition of the Ohta colour space is presented in the form of the following equations (Ohta, Kanade and Sakai, 1980):

$$
\begin{aligned}
I_1 &= (R+G+B)/3 \\
I_2 &= (R-B)/2 \\
I_3 &= (2G-R-B)/4
\end{aligned}
$$

The I_1 component represents the intensity information, I_2 and I_3 represent chromatic information. Ohta *et al.* derived these components as a result of a search of completely statistically independent components on a representative sample of images. The Ohta components are good approximations of the results of the Karhunen-Loeve transformation, which is best in respect of decorrelation of RGB components (Pratt, 1991).

Subsequently, Ohta's proposal has been introduced for:

- a colour recognition algorithm for measurement of oxide thickness in microelectronic structures (Barth, Parthasarathy and Wang, 1986),

- colour image segmentation with very low probability of error (Lim and Lee, 1990),

- colour image compression (Deknuydt, Smolders, Van Eycken and Oosterlinck, 1992),

- stereo correspondence based on block matching (Koschan, 1993),

- analysing liver tissue images (Sun, Wu, Lin and Chou, 1993).

4.6   IHS AND RELATED PERCEPTUAL COLOUR SPACES

Although colour receptors in the human eye (cones) absorb light with the greatest sensitivity in the blue, green and red part of the spectrum, the signals from the cones are further processed in the visual system (Levine, 1985). As a result of this processing, an inexperienced and not specially trained person cannot make intuitive estimates of blue, green and red components. In the perception process, however, a human can easily recognise basic attributes of colour: intensity (brightness, lightness) I, hue H and saturation S. The hue H represents the impression related to the dominant wavelength of the colour stimulus. The saturation corresponds to relative colour purity (lack of white in the colour) and in the case of a pure colour it is equal to 100%. For example, for a vivid red S=100% and for a pale red (pink) S=50%. Colours with zero saturation are grey levels. Maximum intensity is sensed as pure white, minimum intensity as pure black as shown in Fig. 4.3. The IHS components are calculated from formulae expressing approximately the psychophysical sense of these notions. The formulae are given in the literature in different forms, in particular for the hue. The reason is a compromise between

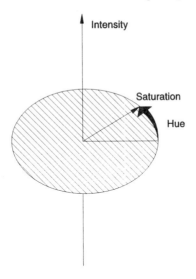

Fig. 4.3 Cylindrical nature of IHS colour space.

transformation accuracy and computational simplicity. The original formula, derived by Tenenbaum, Garvey, Weyl and Wolf (1974) and accepted as basic (Haralick and Shapiro, 1991) is fairly complex and thus often cited with errors:

$$H = \cos^{-1} \frac{2r - g - b}{\sqrt{6[(r - \frac{1}{3})^2 + (g - \frac{1}{3})^2 + (b - \frac{1}{3})^2]}}$$

where, if $b > g$, then $H := 360 - H$, H being expressed in degrees. The version for RGB components is:

$$H = \cos^{-1} \frac{0.5[(R - G) + (R - B)]}{\sqrt{(R - G)(R - G) + (R - B)(G - B)}}$$

and again, if $B > G$, then $H := 360 - H$.

The exact derivation of these equations, resulting from converting the RGB Cartesian coordinate system to the IHS cylindrical one, is given by Ledley, Buas and Golab (1990) and by Gonzalez and Woods (1992). Kender (1976) proposed a fast version of the transformation, containing fewer multiplications and avoiding square root operations:

```
if  R = G = B  then  H := undefined          {achromatic case}
else
    begin                                     {chromatic case}
        if  R > B  and  G ≥ B  then                 {B -- minimum}
```

$$H = \tfrac{\pi}{3} + \tan^{-1}\left[\frac{\sqrt{3}(G-R)}{(G-B)+(R-B)}\right]$$

```
        elsif  G > R  then                          {R -- minimum}
```

$$H = \pi + \tan^{-1}\left[\frac{\sqrt{3}(B-G)}{(B-R)+(G-R)}\right]$$

```
        else                                        {G -- minimum}
```

$$H = \tfrac{5\pi}{3} + \tan^{-1}\left[\frac{\sqrt{3}(R-B)}{(R-G)+(B-G)}\right]$$

```
        end if
    end
end if
```

Some simpler formulae for calculation of hue are cited in the literature, but the results they yield are somewhat different from those obtained with Tenenbaum's and Kender's formulae. Using the version of the transformation from RGB to IHS proposed in 1985 by Bajon, Cattoen and Kim (1985), one obtains numerical results slightly different from the classical ones. In Bajon's transformation, formulae for hue were further simplified and they now contain no trigonometric functions:

```
if  min(R,G,B) = B  then
```

$$H = \frac{G-B}{3(R+G-2B)}$$

```
elsif  min(R,G,B) = R  then
```

$$H = \frac{B-R}{3(G+B-2R)} + \frac{1}{3}$$

```
elsif  min(R,G,B) = G  then
```

$$H = \frac{R-G}{3(R+B-2G)} + \frac{2}{3}$$

```
end
```

There were attempts at applying Bajon's transformation in practice (Bajon, Cattoen and Liang, 1986; Massen, Bottcher and Leisinger, 1988). These papers described applied research (automation in electronics and a biotechno-logical robot respectively) and no attempt was made to evaluate the RGB/IHS transformations. Both were real-time hardware realizations of the transfor-mation. Massen, Bottcher and Leisinger (1988) used a 256 kB × 18-bit look-up table for the transformation, i.e. 6 bits for each RGB component.

As far as the S component is concerned, apart from two basic equivalent versions of the formula:

$$S = 1 - 3\min(r, g, b)$$

$$S = 1 - 3\frac{\min(R, G, B)}{R + G + B}$$

there are a few variants which aim at reducing saturation instability in the neighbourhood of point $S = (0, 0, 0)$ (Gordillo, 1985):

```
if  I ≤ Imax/3  then
```
$$S = I_{max} - 3\frac{\min(I_{max} - R, I_{max} - G, I_{max} - B)}{3 - (R + G + B)}$$
```
else
```
$$S = I_{max} - 3\frac{\min(R, G, B)}{R + G + B}$$
```
end
```

The value I is derived from the formula:

$$I = \frac{R + G + B}{3}$$

or, less frequently, as:

$$I = R + G + B$$

Two other IHS colour spaces are applied in computer graphics, where they are called **colour models**: HSV and HLS (Foley and van Dam, 1982). Figure 4.4 illustrates the geometric interpretation of these models. They differ in the formulae for the values of intensity and saturation given overleaf:

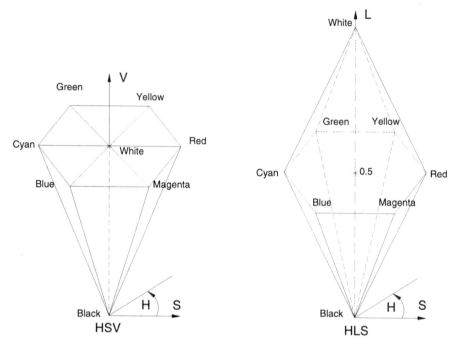

Fig. 4.4 The HSV and HLS models.

HSV

```
min  := minimum(r,g,b)
max  := maximum(r,g,b)
```

$V := \text{max}$

```
if max ≠ 0 then
```

$S := \dfrac{\text{max} - \text{min}}{\text{max}}$

```
else
```

$S := 0$

```
end
```

HLS

```
min  := minimum(r,g,b)
max  := maximum(r,g,b)
```

$L := \dfrac{\text{max} + \text{min}}{2}$

```
if max = min then
```

$S := 0$

```
end if
if L ≤ 0.5 then
```

$S := \dfrac{\text{max} - \text{min}}{\text{max} + \text{min}}$

```
else
```

$S := \dfrac{\text{max} - \text{min}}{2 - \text{max} - \text{min}}$

```
end
```

These models demonstrate that colours become less saturated when the intensity approaches minimal or maximal levels. However not all such perceptual phenomena have been implemented into these models. Fully saturated colours with different hues have the same values $V = 1$ in the case of HSV and the same lightness $L = 0.5$ in the case of HLS space. However, in human perception this is not always true. For example, fully saturated yellow is always lighter than fully saturated blue.

Both transformations, especially HSV, are also used in computer vision: (Taylor and Lewis, 1992; Priese and Rehrmann, 1993; Liu and Yang, 1994). All IHS-based models applied in computer graphics were integrated into one generalised colour model: GLHS (Levkowitz and Herman, 1993).

Important advantages of the IHS space over other colour spaces are: good compatibility with human intuition; separability of chromatic values from achromatic values; and the possibility of using one feature (H) only for segmentation and recognition (Gagliardi, Hatch and Sarkar, 1985). The H and S attributes, particularly useful for recognition, describe a colour object independently of intensity changes. The HS plane has been applied for recognition of technical as well as biological objects. In many cases hue is invariant to shadows, shading and highlights (Perez and Koch, 1994). However, the IHS colour space has some significant drawbacks:

- the irreducible singularities of the RGB to IHS transformation (H for all achromatic colours and S for black),

- sensitivity to small deviations of RGB values near singularities,

- perceptual non-uniformity, in spite of perceptual orientation,

- problems with some operations on angular values of hue e.g. averaging (Crevier, 1993).

The aim of applying IHS space may be the separation of chromatic values from achromatic values using S (Gordillo, 1985; Bajon, Cattoen and Liang, 1986). Sometimes, instead of performing a feature-based segmentation on three channels (RGB), only one feature (H) may be used, allowing the use of much faster algorithms. The IHS space is very convenient for colour image processing. IHS-based algorithms for colour image enhancement make a good example. Median filtering (Zheng, Valavanis and Gauch, 1993) as well as different methods of contrast improvement (Toet, 1992; Kim, Shim and Ha, 1992; Yang and Rodriguez, 1996) use components of IHS space.

Machine vision systems employing the advantages of IHS space have been constructed for the purposes of:

- identification of colour-coded electronic components (Gordillo, 1985; Bajon, Cattoen and Liang, 1986),

- quality control of frozen vegetable mixtures (Gagliardi, Hatch and Sarkar, 1985),

- robotic fruit harvesting (Slaughter and Harrell, 1987),

- bacteria culture manipulations (Massen, Bottcher and Leisinger, 1988),

- fruit sorting and classification (Kay and De Jager, 1992),

- recognition of road landmarks (Kehtarnavaz, Griswold and Kang, 1993),

- discrimination of bottle crates (Milvang and Olafsdottir, 1993).

Due to the vast number of existing variants of the RGB to IHS transformation, we may talk of **IHS-type** colour spaces. This class includes the spaces which result from direct transformation and those created in the effect of a double transformation: e.g. RGB to YUV to IHS (Berry, 1987), RGB to YIQ to IHS (Slaughter and Harrell, 1987) or RGB to $I_1 I_2 I_3$ to IHS (Sandini, Buemi, Massa and Zucchini, 1990). Moreover, new modifications of the IHS space are constantly being created. Yagi *et al.* proposed two new versions of the HSV transformations, which should be more convenient for segmenting of colour images (Yagi, Abe and Nakatani, 1992). For object recognition goals new perceptual attributes, similar to H and S, i.e. **hue ratio** and **colour purity** have been defined (Pritchard, Horne and Sangwine, 1995). The possibility of free choice of colour with $H = 0$ has in some applications certain importance (Palus, 1996). Some state-of-the-art framegrabbers are equipped with real-time RGB to IHS converter chips.

4.7   PERCEPTUALLY UNIFORM COLOUR SPACES

Although perceptual IHS colour spaces provide a more intuitive description of colour than, for example, RGB, they are based on simplifying assumptions and do not constitute perceptually uniform colour spaces (Robertson, 1988). It has been an old idea to construct a colour space in which perceived colour differences recognized as equal by the human eye would correspond to equal Euclidean distances. As perceptually uniform colour spaces of this kind, two spaces recommended by the CIE in 1976 are mainly used. These are the so-called CIELAB space with $L^*a^*b^*$ values (for reflected light) and so-called CIELUV with $L^*u^*v^*$ values (mainly for emitted light) (Wyszecki and Stiles, 1982). A perceptually uniform colour space may also be realised as a set of samples, e.g. the Munsell colour atlas.

4.7.1 CIELAB colour space

The CIELAB space, one of the approximately uniform colour spaces, is defined by the following expressions, where (X_0, Y_0, Z_0) represents reference white:

$$L^* = 116f\left(\frac{Y}{Y_0}\right) - 16 \tag{4.4}$$

$$a^* = 500\left[f\left(\frac{X}{X_0}\right) - f(\frac{Y}{Y_0})\right]$$

$$b^* = 200\left[f\left(\frac{Y}{Y_0}\right) - f(\frac{Z}{Z_0})\right]$$

$$\tag{4.5}$$

where

$$f(x) = \begin{cases} x^{\frac{1}{3}}, & x > 0.008856 \\ 7.787x + \frac{16}{116} & \text{otherwise,} \end{cases}$$

L^* stands for lightness value and is orthogonal to a^* and b^* as shown in Fig. 4.5. L^* takes into consideration the non-linear relation between the lightness and the luminance of a point. The component a^* denotes relative redness-greenness and b^* yellowness-blueness (see section 4.4). The following formula is used to determine colour difference between two sets of $L^*a^*b^*$ coordinates (L^*_1, a^*_1, b^*_1) and (L^*_2, a^*_2, b^*_2):

$$\Delta E_{Lab} = \sqrt{(\Delta L^*)^2 + (\Delta a^*)^2 + (\Delta b^*)^2}$$

where $\Delta L^* = L^*_2 - L^*_1$ etc.. The ΔE_{ab}^* colour difference is widely used for the evaluation of colour reproduction quality in an image processing system e.g. colour image compression processes (Deknuydt, Smolders, Van Eycken and

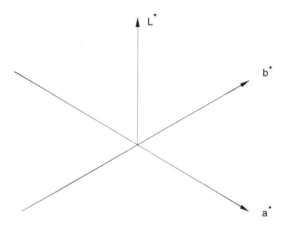

Fig. 4.5 CIELAB colour space coordinates.

Oosterlinck, 1992). It corresponds to human judgements of perceived colour difference, and colour errors represent the human threshold of perceptibility ('just noticeable colour difference' or JND) (Stokes, Fairchild and Berns, 1992).

Sometimes cylindrical coordinates are introduced in CIELAB space:

$$
\begin{aligned}
L^* &= L^* \\
H^\circ &= \tan^{-1}(b^*/a^*) \\
C^* &= \sqrt{a^{*2}+b^{*2}}
\end{aligned}
$$

$$(4.6)$$

Perceptually uniform colour spaces with a Euclidean metric are particularly useful in colour image segmentation of natural scenes using clustering algorithms. This holds true for rectangular coordinates (CIELAB): (Tominaga, 1992; Marcu and Abe, 1995) as well as for cylindrical coordinates (Baker, Hwang and Aggarwal, 1989; Celenk, 1995). Hue H° was applied as the principal feature in a method of detecting regions matching a given colour (Gong and Sakauchi, 1995). Recently a modified version of CIELAB was proposed (Connolly, 1996). It uses the natural logarithmic function instead of the cube root function in the CIELAB chromaticity equations and is anticipated to be useful for the image segmentation of three-dimensional scenes. Tremeau, Lozano and Lager (1995) addressed the problem of non-uniform digitizing of individual components of CIELAB and $L^*H^\circ C^*$ spaces.

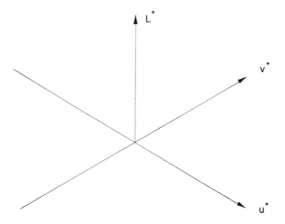

Fig. 4.6 CIELUV colour space coordinates.

4.7.2 CIELUV colour space

The CIELUV space, whose coordinates are depicted in Fig. 4.6, is recommended by the CIE for applications in additive light source conditions (e.g. displays). The definition of L^* is the same as for the CIELAB colour space as given in equation (4.4)). The u^* and v^* coordinates are defined as follows:

$$u^* = 13L^*(u' - u'_0)$$
$$v^* = 13L^*(v' - v'_0)$$

where:

$$u' = \frac{4X}{X + 15Y + 3Z}$$
$$v' = \frac{9Y}{X + 15Y + 3Z}$$

Values with subscript 0 correspond to reference white. It should be noted that the components u^* and v^* are unrelated to U and V in YUV space. Sometimes cylindrical coordinates are introduced in the CIELUV space exactly as for the CIELAB space as given in equation (4.6).

The following formula is used to determine the perceptual colour difference between two luminous sources:

$$\Delta E_{Luv} = \sqrt{(\Delta L^*)^2 + (\Delta u^*)^2 + (\Delta v^*)^2}$$

The CIELUV colour space is very useful in colour image segmentation methods which use a clustering approach (Schettini, 1993; Uchiyama and Arbib, 1994). CIELUV space was recommended for the recognition and localization of two-dimensional objects on a known background (Schettini, 1994). Since it is a device-independent colour space it is also used for generation of colour scales on image displays (Della Ventura and Schettini, 1992). Strachan, Nesvadba and Allen (1990) presented a colour calibration procedure in CIELUV space.

4.8   MUNSELL COLOUR SYSTEM

The Munsell Colour System, known from colorimetry (Wyszecki and Stiles, 1982) is another example of a system based on colour attributes close to human perceptions: **hue** (H) and **value** (V) corresponding to luminance; and **chroma** (C) corresponding to saturation. The vertical axis of the Munsell colour solid is a line of V values, which run from black to white, and it is divided into perceptually equal shades of grey. Hue (H) changes along each of the circles perpendicular to the vertical axis, while chroma (C) starts at zero on the V axis and changes along the radius of each circle as indicated in Fig. 4.7. For a long time no analytical relationship existed between the Munsell attributes HVC and the RGB or XYZ components. Tables expressing Yxy values for Munsell colours under illuminant C may be found in handbooks of colorimetry, for example (Wyszecki and Stiles, 1982). The transformation was done by look-up-tables. In order to avoid implementing such tables, the mapping into the HVC space has been realized experimentally (Tominaga, 1987). In 1988 an algorithmic method of transformation from RGB to HVC was proposed by Miyahara and Yoshida (1988).

However, the Munsell system has a drawback. Instead of a Euclidean metric, Godlove's (1951) colour difference formula usually defines a distance in the Munsell colour system between two coordinates $H_1 V_1 C_1$ and $H_2 V_2 C_2$:

$$\Delta E_g = \sqrt{2C_1 C_2(1 - \cos(\pi \Delta H / 180)) + (\Delta C)^2 + (4\Delta V)^2}$$

where:

$$\Delta H = |H_1 - H_2|$$
$$\Delta V = |V_1 - V_2|$$
$$\Delta C = |C_1 - C_2|$$

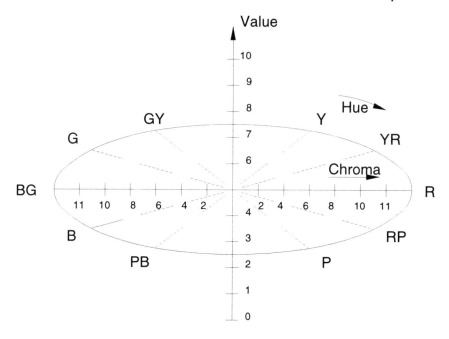

Fig. 4.7 Munsell colour system.

This formula was used for the evaluation of the quality of 2D vector median filters (Bartkowiak and Domański, 1995). Sometimes in practice a simplified formula (e.g. Tominaga (1988)) is used. The Munsell colour system was applied to colour image segmentation by using one-dimensional histograms of Munsell attributes and merging the homogeneous regions (Tominaga, 1987; Tominaga, 1988) and by implementing a k-means clustering algorithm (Hachimura, 1995). In image database applications Munsell attributes can be a convenient tool for matching the object colours to predefined colour zones (Gong, Zhang and Chua, 1994).

4.9 KODAK PhotoYCC COLOUR SPACE

PhotoCDs are becoming a *de facto* standard for low-cost image archiving. Kodak has developed a special colour space known as PhotoYCC for writing and storage of colour images on PhotoCDs (Kodak, 1992).

   The PhotoYCC colour space has its roots in colour television and is a type of luminance-chrominance colour space: Y is a luminance component and the two C components represent chrominance (ITU, 1994b). PhotoYCC allows efficient image compression and the space is based on TV and HDTV standards. PhotoYCC pixel values are the result of a three-step transformation from RGB values from a standardized and calibrated scanner. The RGB colour space uses reference primaries and D65 reference white defined by the ITU (1990). Because the scanned RGB values are not constrained to positive values, not all colour information can be shown on currently used displays.

The three steps to transform RGB values to PhotoYCC values are:

1. gamma correction;

2. linear transformation;

3. quantization of YCC to 8-bit data.

The first step is needed because colour displays have a non-linear characteristic, whereas the RGB scanner is almost linear. The same equations apply to each of the RGB channels: that is x denotes R, G or B.

$$
x' = \begin{cases}
1.099x^{0.45} \quad -0.099 & : \qquad\qquad x \geq \quad 0.018 \\
4.5x & : \quad -0.018 < x < \quad 0.018 \\
-1.099|x|^{0.45} +0.099 & : \qquad\qquad x \leq -0.018
\end{cases}
$$

The gamma-corrected RGB values are then transformed to unquantized YCC values:

$$
\begin{aligned}
Y' &= \quad 0.299R' + 0.587G' + 0.114B' \\
C1' &= -0.299R' - 0.587G' + 0.886B' \\
C2' &= \quad 0.701R' - 0.587G' - 0.114B'
\end{aligned}
$$

Finally, the YCC values are quantized to eight bits:

$$
\begin{aligned}
Y &= \quad (255/1.402)Y' \\
C1 &= \quad 111.40C1' + 156 \\
C2 &= \quad 135.64C2' + 137
\end{aligned}
$$

Values in the PhotoYCC colour space can be transformed to other colour spaces, e.g. RGB display colour space, CIELAB and CMYK for printing. The transformation from PhotoYCC to RGB is not exactly the inverse of the transformation from RGB to PhotoYCC. First, the Y', $C1'$ and $C2'$ components are recovered from their 8-bit representations:

$$Y' = 1.3584Y$$
$$C1' = 2.2179(C1 - 156)$$
$$C2' = 1.8215(C2 - 137)$$

If the chromaticities of the display phosphors are close to the standard chromaticities (ITU, 1990) then the following equations are correct:

$$R_{Display} = Y' + C2'$$
$$G_{Display} = Y' - 0.194C1' - 0.509C2'$$
$$B_{Display} = Y' + C1'$$

As the last step of the transformation the values may be quantized to lie in the range 0 to 255. Sometimes clipping during the quantization process will result in a loss of highlight information.

For details about compression and storage techniques used for the Kodak PhotoYCC colour space see Kodak (1992).

4.10   SUMMARY OF COLOUR SPACE PROPERTIES

The reasons for applying colour space transformations are very varied. The choice of an appropriate colour space can be an important factor determining the results of processing on a colour image (e.g. the quality of image segmentation, compression ratio *etc.*). In practice there is no ideal colour space for all image processing applications. The decision on which colour space to use depends on the processing task. An optimal decision can be very hard to find. A good example is the work on a comparison of colour image segmentation algorithms by Gauch and Hsia (1992). This paper shows that no single colour space exists which would be the best for all three segmentation algorithms tested. However, knowledge of the properties of the various colour spaces makes the choice easier. Some examples of segmentation algorithms applied in different colour spaces are given in Chapter 9 on page 173.

Some of the properties of the various transformations between RGB and other colour spaces described in this chapter are summarized in Table 4.2.

Table 4.2   Properties of transformations between RGB and other colour spaces.

Colour space	Linearity of transformation	Stability of calculations	Perceptual uniformity
rgb	No	No	No
XYZ	Yes	Yes	No
xyz	No	No	No
YUV	Yes	Yes	No
YIQ	Yes	Yes	No
YC_bC_r	Yes	Yes	No
Opponent	Yes	Yes	No
Ohta	Yes	Yes	No
IHS	No	No	No
CIELAB	No	Yes	Yes
CIELUV	No	Yes	Yes
Munsell	No	Yes	Yes
PhotoYCC	No	Yes	No

FURTHER READING

More and more books on image processing include a chapter devoted to colour image processing and in it information about representations of colour images. Among them, Pratt's (1991) book covers the widest range of colour spaces and is recommended.

The book by Wyszecki and Stiles (1982) is a good guide to the CIE colour spaces. Readers interested in the origins of colour television spaces should refer to contemporary television technique handbooks, e.g. Benson (1992) or Slater (1991). The properties of perceptual colour spaces (e.g. IHS) are widely presented in monographs on computer graphics, for example Foley and van Dam (1982).

PART TWO

IMAGE ACQUISITION

5

Colour video systems and signals

Robin E. N. Horne

Colour image processing is usually performed on visual information which has been collected electronically. The principles and practice of this sort of information acquisition derive from techniques which were developed primarily for television applications. Whilst some aspects of television technology are inappropriate, or result in non-ideal images from the processing point of view, nevertheless the fundamental principles of scanning and baseband video signal production are sufficiently important to be included here.

This chapter seeks to show how the scanning process generates timevarying signals representing the parameters of colour and brightness variation over an image. Subsequent sampling of such signals results in digital data which is frequently the raw material of colour image processing operations.

5.1 VIDEO COMMUNICATION

5.1.1 The scanning process

The process of scanning a two-dimensional image usually consists of a left-to-right sweep (line scan) repeated many times during a single top-to-bottom sweep (field scan). By this means, the two-dimensional spatial image information may be converted to a uni-dimensional electronic signal consisting of a time-varying voltage.

For a simple image, it is easy to relate its appearance to that of its representative video signal. Scanning imposes a repetitive structure upon both the waveform of the signal and upon the image which is to be reproduced, and it is instructive to make an intuitive connection between the two.

The greyscale image represented by the waveform shown in Fig. 5.1 consists of a set of vertical bars whose brightness value decreases from peak

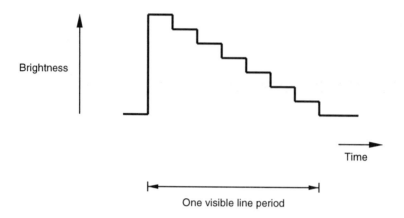

Brightness

Time

One visible line period

Fig. 5.1   Line waveform for greyscale image

white on the left to black on the right. Since the image signal is generated by scanning from left to right, it is clear that the waveform of this signal for the period of one scan line will be a time-varying voltage consisting of a descending staircase of brightness values. The pattern of brightness variation is the same for all the horizontal lines in the picture, so the line waveform shown is what would be displayed on an appropriately triggered oscilloscope. Since each bar has a uniform brightness, the signal for each bar period is a DC level. No high frequencies exist in this case. The technique of scanning to produce a **raster** pattern evolved in conjunction with the development of the vidicon camera and the cathode ray tube (CRT). In each of these devices, an electron beam is focused on to a target and deflected horizontally and vertically in a systematic way such that the image is repeatedly traversed.

   In the case of the camera (source), the target is a light sensitive plate upon which a lens focuses an image of a scene, as shown in Fig. 5.2. A focused electron beam is deflected in the horizontal (X) direction and in the vertical (Y) direction by line and field generators respectively. As the electron beam traverses the image on the target, a variable current is generated in the resistor R, dependent on the instantaneous light value at the point in the image which is being scanned. A time-varying voltage is developed across R, forming a **luminance** signal. The luminance signal representing typical images from a video camera is much more complex than that of the simple pattern of Fig. 5.1, of course, and has a frequency spectrum extending typically to 5.5 MHz.

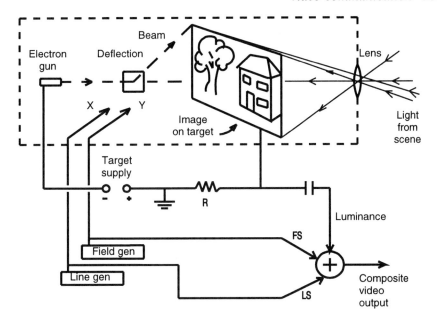

Fig. 5.2 Vidicon camera.

In the case of the CRT (display), the target is a phosphor-coated screen upon which the image from the source is reproduced. A continuous beam of electrons is produced by the electron gun, focused and accelerated towards the screen where it produces a fluorescent dot on the phosphor. Rapid movement of this dot generates visible lines and if sufficient such lines are generated in close enough proximity to each other it is possible to produce an apparently continuous two-dimensional image. The beam is deflected horizontally (the line scan) and vertically (the field scan) by an arrangement of electromagnets situated on the flare of the tube to produce a pattern of horizontal raster lines on the screen. Sequential current waveforms (X and Y drives) generated in the electromagnets effect the repetitive deflections. The intensity of the electron beam varies with the applied luminance signal voltage Z such that the instantaneous light output from the phosphor screen is proportional to the light value of the corresponding point in the scan of the source image. It is necessary of course to ensure that the timing of the X and Y scans at the source corresponds to those at the display, in order for the source image to be reproduced correctly. For this reason it is necessary to convey additional, timing (synchronization) signals as well as the time varying brightness information.

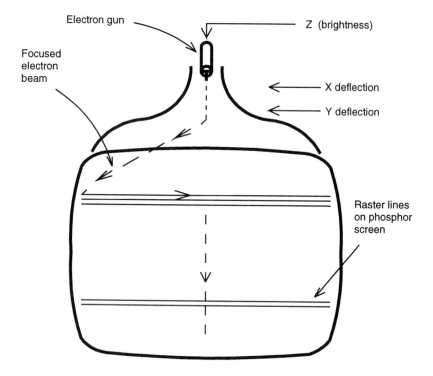

Fig. 5.3 Cathode ray tube (CRT) video display.

5.1.2 Synchronization

This is usually achieved by the incorporation of pulses at points in the wave-form corresponding to the start of the line and field scans. Figures 5.4 and 5.5 show these line (horizontal scan) and field (vertical scan) synchronizing pulses respectively. In this example, the time scales correspond to the European CCIR standard. Those for RS170 (USA) are very similar. Line pulses occur every 64 μs and are responsible for triggering the horizontal retrace of the display. The field synchronizing pulse sequence (FPS) occurs every 20 ms and triggers the vertical retrace. Synchronization pulses ('line sync' and 'field

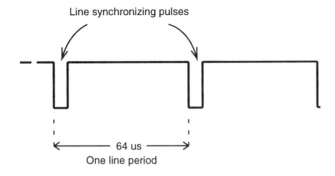

Fig. 5.4 Line synchronization pulses.

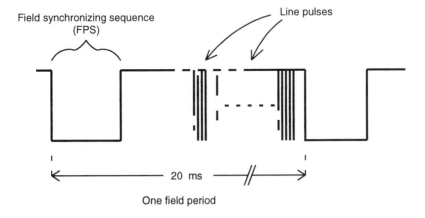

Fig. 5.5 Field synchronization pulses.

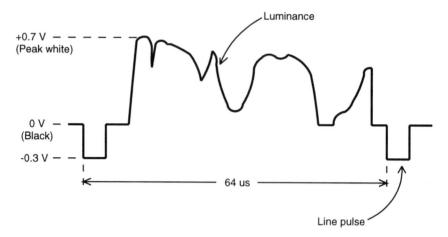

Fig. 5.6 Composite video line waveform.

sync') may be conveyed as separate signals, as in the connection between a PC and its monitor for example, or they may be combined with the luminance and/or chrominance signals in the form of a **composite video** signal.

The waveform shown in Fig. 5.6 represents one line period of a CCIR composite video signal. Notice that the overall amplitude of the signal is 1 V peak-to-peak, with black represented by a nominal 0 V level. Peak white is at +0.7 V and the negative-going synchronizing pulses have an amplitude of -0.3 V. The luminance signal only exists in the **active** period (52 μs in the CCIR standard) between line pulses.

5.1.3   Blanking

The X and Y deflection drives are both ramp waveforms. In each case there is a finite retrace period during which there should be no illumination on the screen. Of course, the composite video signal should not exhibit any level greater than black during these periods (and sync pulses are 'blacker than black'). However, the luminance drive is turned off during retrace to ensure that there there can be no visible interference from spurious noise. Figure 5.7 shows a deflection ramp waveform, and the specific representations of ramps for line and field scan. To clarify the distinction between the two, the field deflection drive waveform is shown as negative-going. These drive

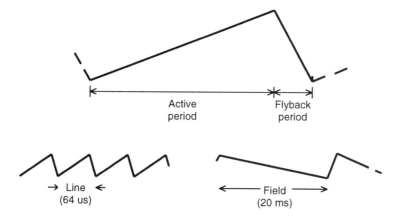

Fig. 5.7 Deflection drive waveforms.

waveforms are generated from line and field rate oscillators at the display and their timing is determined by the received synchronizing pulses. The sync pulses occur in the flyback period of each scan and control the frequency and phase of the oscillators. Geometric distortion occurs if the ramp waveforms depart from linearity.

5.1.4 Interlacing

Some video applications, including broadcast television, employ the technique of interlacing. In order to increase the flicker (picture repetition) frequency, and thus render it less objectionable, the picture is generated in two halves.

Two fields, each with half the total number of lines, are interleaved together to form one complete **frame** (picture). The pattern of interleaving is a continuous process with so-called 'odd' and 'even' fields alternating. The electron beam deflection follows the paths shown in Fig. 5.6, where the arrows indicate a simplified flyback motion. Whilst the field repetition rate (which determines the flicker frequency) in this example is 50 Hz, the whole picture repetition rate (which determines the bandwidth of the signal) is only 25 Hz.

Interlacing is therefore a technique for increasing flicker frequency without increasing the picture frequency which would otherwise extend the signal

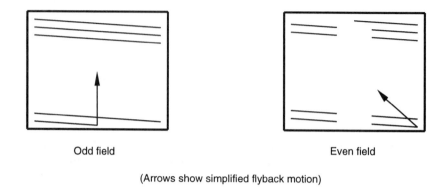

Odd field   Even field

(Arrows show simplified flyback motion)

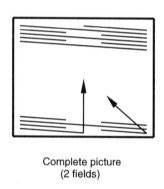

Complete picture
(2 fields)

Fig. 5.8   Raster scan pattern in interlaced fields.

bandwidth. Although this benefit is only of real value when a video signal is to be modulated on to a carrier for broadcast television purposes, it is the case that, for economic reasons, many image processing applications derive images from sources operating to broadcast standards which may incorporate interlacing.

Figure 5.8 shows the pattern of lines for odd and even fields respectively, and the structure of a complete frame with the two sets of lines interlaced together. The advantage of this arrangement (increased flicker frequency with no increase in signal bandwidth) is achieved at the cost of the introduction of some complexity to the signal.

To achieve the interlaced scan pattern, it is necessary to introduce a relative delay of one half-line period between the set of lines in an even field and those in an odd field. As a consequence, one set of lines is displaced vertically to appear midway between the lines in the other set, and the line synchronizing pulses are timed accordingly.

In order to maintain line synchronization throughout the transition from one field to another, it is necessary to 'break up' the field synchronizing pulse so that pulse edges appear as a continuation of line synchronizing pulses. Additionally, groups of pulses occuring at twice line frequency ('equalization pulses') are introduced in the field pulse sequence to facilitate the transition between the two possible nominal positions for the interlaced line pulses. These features of the field pulse sequence are shown in detail in Fig. 5.9.

It should be emphasized that interlacing confers no advantage as far as image processing is concerned, since signal bandwidth is not usually a problem in situations which do not involve radio transmission. However, the availability of relatively inexpensive cameras operating to interlaced standards means that signals of this type may sometimes need to be accommodated.

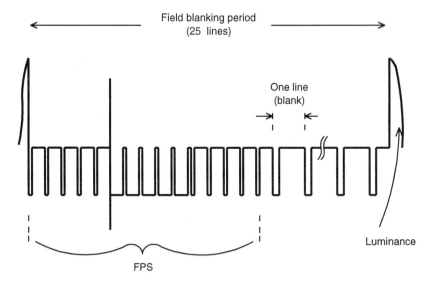

Fig. 5.9 Field synchronizing waveform for interlaced fields.

5.2   COLOUR REPRODUCTION

5.2.1   Three-colour source

A given colour may be synthesized by appropriate admixture of three primary colours as discussed in section 2.3.1 and in section 3.2.4. The simple principle of employing three camera tubes, each of which is made sensitive to a specified primary colour, allows full-colour images to be conveyed or stored. In the camera shown in Fig. 5.10, the three tubes share a common lens system. Light from the scene is split by an arrangement of dichroic mirrors, M1 and M2, into three components: red, green and blue (RGB), and the chromatic response of each tube to its intended spectral colour is refined by individual optical filters. Each light component is directed to one of the camera tubes and three separate signals E_R, E_G and E_B are obtained from variable gain amplifiers. The three tubes scan the scene synchronously, under the control of a common sync generator. **Mixed synchronization** pulses (line and field pulses combined) may be derived from this generator and conveyed as a separate signal. The operation of most colour cameras used for image processing follows this essential principle, though the nature of the video signal output may vary between types.

Solid state cameras using charge-coupled device (CCD) technology have replaced vidicons for most applications. These instruments have arrays of metal oxide semiconductor (MOS) elements upon which the image is focused. Charge is generated in each element proportional to the incident radiation. Subsequent output of the charge packets line-by-line using the shift register principle produces an electrical signal corresponding to the instantaneous light value of the scanned image. RGB versions of this type of camera have triple CCD arrays with colour mosaic filters.

It is possible to produce a colour picture from a single-chip CCD array with a colour stripe filter in front of it, as shown in Fig. 5.11. The filter has equal pitch stripes of green, cyan and white. The output of the CCD array is modified by the effect of the stripe filter on the incoming light and by the effect of low-pass and band-pass filters on the ensuing video signal. The combined effect of the optical and electronic filters is to provide separate output signals corresponding to luminance and two colour difference values $R - Y$ and $B - Y$ which may be encoded as described in section 5.3.3.

The spatially discrete elements of this sort of array can be mapped directly on to computer arrays for image processing purposes. However, the standard outputs from CCD cameras are analogue signals which need subsequent digitization.

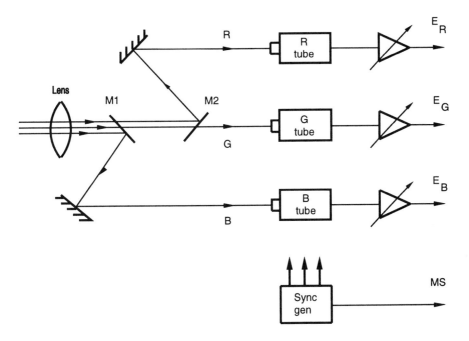

Fig. 5.10 Three-tube colour camera.

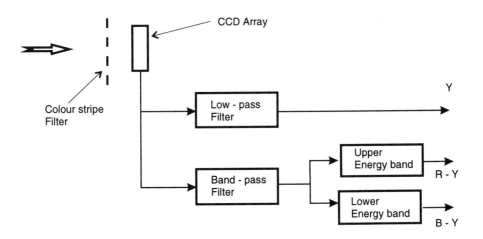

Fig. 5.11 Single-chip CCD camera arrangement.

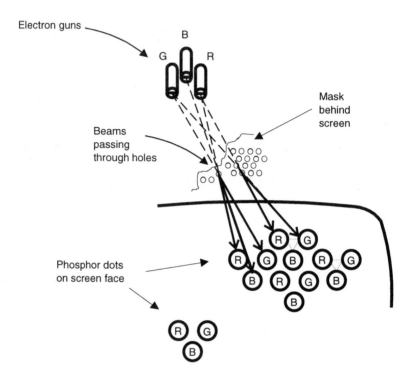

Fig. 5.12 Principle of shadowmask CRT

5.2.2   Three-colour display

Colour displays are based on variants of the shadowmask CRT (RCA) in which three separate rasters (red, green and blue) are produced by three sep-arate scanned electron beams which activate corresponding elements of red, green and blue phosphor patterns deposited in a matrix over the inside of the screen (Fig. 5.12). The instantaneous intensity of each individual beam is determined by the applied R, G or B signal.

The accuracy with which each beam lands on the appropriately coloured phosphor elements is termed **purity** and the accuracy with which all three rasters are superimposed is **convergence**. High resolution displays need to have a large number of very small phosphor elements with a high degree of purity and precise convergence.

5.2.3 RGB video system

The three signal outputs from an RGB source are conveyed as three separate channels. Line and field sychronizing pulses may be output as distinct signals (LS and FS) or combined as mixed sync (MS). The mixed sync combination is sometimes incorporated with the G signal in composite video form for convenience.

If the gains of the respective sources and display amplifiers are adjusted correctly, the system exhibits **chromatic linearity** ; that is to say, any colour produced at the source will be accurately reproduced by the display. Since the three RGB signals are full-bandwidth, each channel must have the appropriate bandwidth capability.

In practical terms, the camera amplifiers are adjusted to give equal amplitude output signals (E_R, E_G, E_B) when a uniform reference white object is being observed, and the display amplifiers are adjusted so that the reference white is *reproduced* when equal amplitude input signals are applied.

5.2.4 Colour mixing (RGB colour bar example)

A simple example of colour mixing is provided by the standard colour bar test pattern. This pattern consists of vertical bars of primary and secondary colours with white and black as shown in Fig. 5.13. The bars have the same arrangement of descending luminance values as was seen in the earlier greyscale and are produced by RGB drives which consist of synchronous square waveforms. Each secondary colour is produced by the combination of two of the primaries, and white results from the combination of all three.

5.2.5 Spatial and colour resolution

An image represented in an analogue video system may be considered to have a spatial resolution of $h \times v$ picture elements (pixels). The vertical resolution, v, is limited by the line structure of the raster while the horizontal resolution h, is determined by the video signal. Since the scanning process corresponds to a sampled-data system, the theoretical maximum number of reproducible

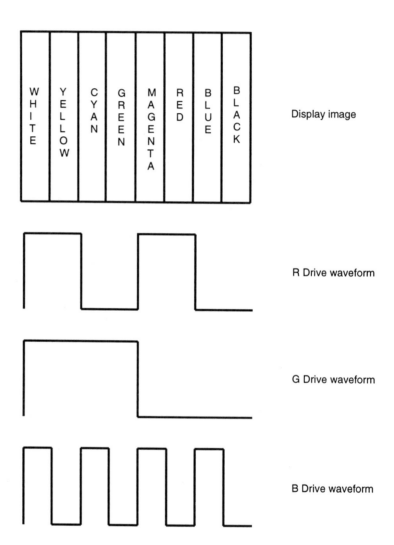

Fig. 5.13 RGB colour bars.

vertical elements is half the number of visible lines. In a 625-line system for example, there are 575 visible lines so the theoretical vertical resolution $v = 575/2$. In practice, a rather larger figure may be achieved for a wide range of typical images. In broadcast television the vertical resolution is calculated from the expression:

$$v = K(N - Ib)$$

where N is the total number of lines per frame (e.g. 625 in the CCIR system), I is the interlace ratio (2), b is the number of lines lost in each field flyback (25) and K is the **Kell factor** (usually specified as 0.7).

The Kell factor is an empirically-derived figure based on subjectively acceptable results for average observers of displays at average viewing distances. Thus 'broadcast quality' images have a vertical resolution of approximately 402 (pixels).

The horizontal resolution is related to the video signal frequency:

$$h = 2fT$$

where h is the maximum number of discernible horizontal elements, f is the maximum video signal frequency (signal bandwidth) and T is the visible line period.

For broadcast quality video the maximum video frequency (for luminance or for R, G, and B components) is 5.5 MHz. In the CCIR system the visible line period is 52 μs. This corresponds to a horizontal resolution of 575 pixels. The parameters for RS170 are similar.

The spatial resolution of **digitally** derived images is not subject to the above limitations since discrete pixel positions are determined by coordinates in image arrays. Maximum vertical resolution in this case is equal to the total number of visible lines (usually slightly less than the total number of raster lines due to the need to impose a border so as not to lose any image detail). Horizontal resolution of a digital image is limited by the physical characteristics of the display, which may be expected to be significantly superior to one intended for analogue use since computer and processing applications require reproduction of text and pictorial images in fine detail. Additionally, in the digitally-derived image, the brightness and/or colour values of each pixel take discrete values, unlike the analogue case where there is a continuous grey level or set of colour component levels. The higher the number of quantization levels the better the image quality. 24-bit colour (16 million possible values for each pixel) represents a reasonably high specification for colour images at the present time.

5.3   ENCODED-COLOUR SYSTEMS

5.3.1   Luminance/chrominance representation

Representation of colour video in RGB form is inconvenient for many applications such as broadcast television since three separate full bandwidth signals are involved. Instead, the source information may be encoded ultimately into the form of just a single, composite, baseband signal representing combined brightness, colour and synchronization information. This combined baseband signal may then be used to modulate a main carrier whose frequency is appropriately chosen for terrestrial or satellite radio transmission, or it may be used in baseband form for convenience.

The three main analogue encoded-colour systems used throughout the world are:

NTSC   (National Television Standards Committee, USA)

PAL   (Phase Alternation by Line, Germany)

SECAM   (Séquential Couleur à Mémoire, France)

Although somewhat long in the tooth, each of these systems has proved to be reliable, robust and popular. The huge distributed investment in consumer systems tends to inhibit radical change and it is expected that these stalwarts of the analogue age of broadcast television will be around for many years, notwithstanding the introduction and parallel operation of digital systems for both terrestrial and satellite television. An incidental advantage of the mass nature of the television market is that cameras and other equipments which may be used for image processing purposes are available cheaply. For this reason, commercial framegrabbers, processors *etc.*, frequently accommodate encoded colour signals although, as will be seen, the format is not ideal.

Each of the above systems converts original RGB component information into a combination of **luminance** and **chrominance**. The advantage of this representation is that luminance and chrominance components may be treated as separate entities; a situation which accords more closely with the process of human visual perception in which two types of sensor exist, **rods** and **cones** as discussed in section 2.3.

5.3.2 YUV **format**

The main requirements of an encoded colour system are that the chrominance signal should be incorporated as a separate 'optional extra' signal, thus preserving compatibility with monochrome displays, and if possible it should be transmitted *within the same frequency band* as the luminance in order to conserve transmission bandwidth. Since the eye/brain combination does not perceive fine detail in colour, as discussed in section 2.5.3, it is advantageous to assign a lower bandwidth to the chrominance than to the luminance. Perhaps the most common form of this type of coding is YUV, in which a full-bandwidth luminance signal, Y, is combined with two subsidiary lower bandwidth chrominance signals U and V which together convey the colour (hue and saturation) information. Equations for the YUV colourspace are given in section 4.3.

5.3.3 The PAL **system**

The YUV format was originated for the PAL broadcast television system. In the PAL system, a full-bandwidth luminance signal, E_Y, is generated by combining the RGB source components in appropriate proportions:

$$E_Y = lE_R + mE_G + nE_B$$

where l, m and n are **luminosity coefficients** whose values are chosen to simulate the relative sensitivity of the standard human eye to the respective chosen primaries, thus ensuring a panchromatic representation. This signal is equivalent to that derived from a monochrome camera and therefore provides a complete greyscale image.

Chrominance is conveyed separately by two **colour difference** signals. These are generated by the subtraction of E_Y from two of the original RGB components:

$$E_R - E_Y \quad \text{and} \quad E_B - E_Y$$

These signals have no particular physical significance but they are used since they represent colour only and play no part in the transmission of luminance information. Additionally, since the colour-difference values fall to zero for black, white or grey and are of low amplitude for conditions of low saturation, they may be combined with the luminance signal with minimum cross-interference.

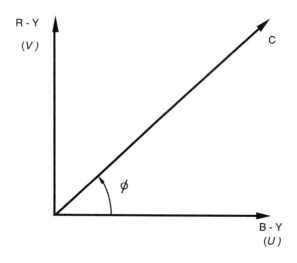

Fig. 5.14   Phasor representation of quadrature-modulated chrominance subcarrier.

In the PAL system, these two colour-difference signals modulate **sinusoidal subcarriers** which are identical in frequency but exist in phase quadrature. Combination of the two modulated carriers results in a single subcarrier which is effectively modulated simultaneously in amplitude and in phase. The instantaneous amplitude of the modulated signal corresponds to the saturation at a given point in the image and the instantaneous phase corresponds to the hue at that point.

The phasor diagram of Fig. 5.14 illustrates the production of the modulated subcarrier by the two colour-difference signals $E_B - E_Y$ and $E_R - E_Y$ which are designated U and V respectively for the modulation process. The magnitude of the resultant subcarrier C (saturation) is proportional to $(V^2 + U^2)$ and the phase (hue) is proportional to $\tan^{-1}(V/U)$. The quadrature modulation is achieved by arranging for the V signal to modulate a sinusoidal subcarrier $E_{sc} \sin \omega_{sc} t$, while U modulates a quadrature subcarrier $E_{sc} \cos \omega_{sc} t$. The expression for the single modulated chrominance signal is therefore:

$$U(E_{sc} \cos \omega_{sc} t) + V(E_{sc} \sin \omega_{sc} t)$$

The acronym 'PAL' (Phase Alternation by Line) is derived from an additional mechanism whereby the phase of the V modulated subcarrier is inverted on alternate scan lines, and subsequently reinverted at the receiving end. This operation enables the cancellation of possible hue errors resulting from phase shifts in the transmission path.

To reduce cross-modulation effects, the subcarrier is **suppressed** before transmission. This means that it is necessary to incorporate a sample of un-modulated subcarrier in the signal at regular intervals to provide a phase reference and to facilitate demodulation. A 'colour burst' is inserted in each line flyback period immediately after the line synchronizing pulse and consists of 10 cycles of unmodulated subcarrier in a reference phase corresponding to a position along the -U axis. Absolute phase (hue) is determined with reference to this angular position.

Incorporation of the modulated subcarrier into the luminance signal frequency band is made possible by the existence of 'gaps' in the frequency spectrum which result from the scanning process. These gaps occur throughout the signal spectrum at harmonics of line frequency (f_L, $2f_L$, *etc.*) with luminance frequency components grouped at field frequency intervals. Careful specification of the subcarrier frequency so as to incorporate fractional offsets of line and field frequency ensures that the similarly-spaced chrominance components are positioned in these gaps across the spectrum. This frequency multiplexing is imperfect and results in some cross-modulation (for example, artificial colour scintillation in images containing regular patterns such as checks and stripes). However, the minimization of the amplitude and bandwidth of the modulated chrominance signal achieves a subjectively acceptable result. The diagram of Fig. 5.15 shows a detail of the frequency spectrum of the luminance signal, and the positioning of the chrominance components in the luminance gaps.

5.3.4 PAL colour bars

It is instructive to consider the PAL line waveform for the colour bar image of Fig. 5.13 produced by RGB coding as discussed on page 105. Again, every line in the field is the same, resulting in a display of 8 vertical bars each of duration approximately 6.5 μs. The waveform of Fig. 5.16 clearly illustrates the distinction within the signal of the luminance and chrominance components. The luminance is the greyscale staircase of descending values from white to black (as observed in Fig. 5.1), and onto this is superimposed the modulated chrominance subcarrier. In PAL, the subcarrier has a frequency of 4.43361875 MHz which is precisely specified to incorporate the fractional offsets of line and field frequency referred to earlier.

The six coloured bars of this test signal exhibit portions of subcarrier of equal amplitude (since they all have the same saturation), but different phase

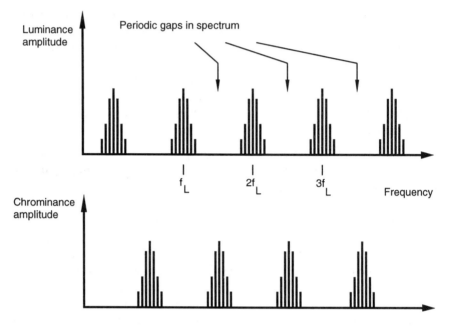

Fig. 5.15 Frequency spectrum of luminance/chrominance.

(they all have a different hue). The colour burst is seen in the line flyback period.

5.3.5   NTSC chrominance signal

The NTSC chrominance signal is similar to that of PAL in that it consists of a quadrature–modulated subcarrier. The quadrature modulating components are termed I and Q and again are derived from colour difference signals but this time in a slightly more complex way:

$$I = 0.74(E_R - E_Y) - 0.27(E_B - E_Y)$$

$$Q = 0.48(E_R - E_Y) - 0.41(E_B - E_Y)$$

The rationale for this arrangement is that, since the eye is more sensitive to colour detail in orange and cyan hues than it is to those in green and magenta,

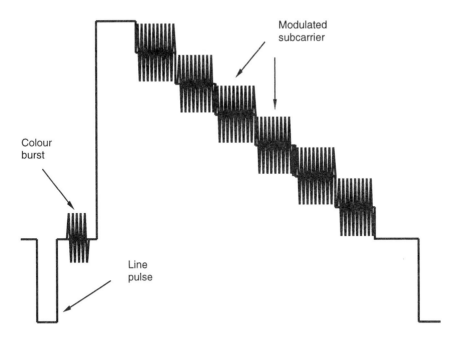

Fig. 5.16   PAL colour bar line waveform.

it is possible to optimize the signal bandwidth allocation for the two components. Thus the *I* signal is transmitted with a significantly higher bandwidth than *Q*.

The phase reference for the subcarrier is again carried as a colour burst in the line flyback period and the appearance of an NTSC signal waveform for standard colour bars is superficially indistinguishable from that for PAL.

5.3.6   Other analogue encoding methods

Other encoded-colour formats which may be encountered in colour imaging systems are SECAM, Separated Video (S-Video) and Multiplexed Analogue Components (MAC).

In the SECAM system, which is in use in some parts of Europe, separate colour difference signals are conveyed *sequentially* on alternate lines using frequency modulation of a chrominance subcarrier. Subsequent recombination is achieved with a one-line period delay-line to allow both signals to be available simultaneously. The advantage claimed for this modulation method is that phase distortion in the transmission path is not a problem so it is not subject to hue error as PAL and NTSC are. However, the luminance/chrominance intermodulation effects are significantly worse than they are in the quadrature modulation systems.

S-Video is employed in the SVHS videotape recording system, which is a possible source of colour images for processing. It is another quadrature modulation system but in this case a subcarrier frequency is chosen which results in the chrominance band being situated outside that of the luminance (i.e. 'separated'). This means that a higher overall bandwidth is required (not an embarrassment if the signal does not have to be broadcast) but there are no intermodulation effects and luminance and chrominance detail is better preserved. The separation of the signals results in a corresponding separation of luminance and chrominance **channels**, so multi-way connectors and cables are needed.

The MAC format was developed specifically for satellite broadcasting and achieves good separation of luminance and chrominance by multiplexing these components in the time domain rather than in the frequency domain. Its use in image processing is rare.

None of the above systems are compatible in terms of the signals used between source and display. With the advent of digital methods (e.g. MPEG) for encoding luminance and chrominance, it is likely that all these analogue systems will become redundant. However, as has already been mentioned, a great deal of analogue equipment will be around for many years to come and much processing will be performed on colour images derived therefrom.

6

Image sources

Christine Connolly

6.1 OVERVIEW OF SOURCES FOR IMAGE PROCESSING

The basic function of an image source is to represent the distribution of light
energy in an 'image' as an electronic signal which may be communicated
to a computer for storage and subsequent analysis or enhancement. Thus a
two-dimensional light intensity function $f(x, y)$ ultimately undergoes conversion to an array of numbers representing light amplitudes at a set of points.
The output of the source is usually a unidimensional, time-varying signal $v(t)$
which results from the process of scanning as described in Chapter 5. Application areas for image processing are very wide (ranging from microscopy to
astronomy) and source types are correspondingly varied.

Sources may be classified as:

- area scan or line scan;

- standard or non-standard scan format;

- visible or non-visible spectral response.

6.1.1 Area scan or line scan

Although the great majority of commercially manufactured sources use two-
dimensional sensors, linear arrays represent a means of obtaining high resolution (typically 8000 pixels) at relatively low cost. A single-pixel wide array
of x elements produces image data by relative movement of the object in the y
direction. The continuous motion of items on a production line, for example,
may provide the necessary transverse 'scan'. Alternatively, the linear array

itself may be progressively moved across a static object as in the case of a document scanner. Colour line scan devices use triplets of sensing elements and optical filters.

6.1.2   Standard or non-standard scan format

Standard scan formats correspond to those derived from broadcast television systems. Sources which operate with standard scan parameters are cheap and readily available but are unsuitable for certain applications. Where an image is to be built up in non-real time (as in slow-scan geographic surveillance, for example) the source will produce trigger or scan synchronizing pulses at non-standard rates and the acquisition system must be able to accommodate them. Other non-standard applications include monoshot operation and multi-frame scanning.

6.1.3   Visible or non-visible spectral response

Visible sources are covered in the following section on cameras. Non-visible sources (responsive to infra-red, ultra-violet, X-radiation etc) may or may not operate to standard scan parameters. CCD devices are inherently sensitive to infra-red and this part of the spectrum may be deliberately included or excluded by the addition of appropriate filters to standard cameras. However, specialist non-visible sources (for example body scanners) may be required to have very high resolution and will nearly always have non-standard scan rates. To allow for human interpretation of images derived from sources operating in non-visible regions of the spectrum, it is common for processing systems to assign specific colours to values of intensity for display. This technique is known as 'false colour'.

6.2   CAMERAS

6.2.1   Types of camera

There is a wide selection of cameras available from many manufacturers, and consequently many decisions to make when considering a purchase, for example:

- analogue or digital output;

- continuous or monoshot operation;

- single chip or three-chip CCD sensor;

- composite video or RGB outputs;

- size of sensor (number of pixels horizontally and vertically);

- type of lens mount.

The choice will be influenced by budget, and it is possible currently to spend anything from a few hundred pounds to £30 000 or so on a camera. Choice is also influenced by technical needs. Each of the options listed above will now be considered, and the technical ramifications set out.

Analogue or digital format

Currently, the choice of analogue or digital output has the greatest impact on camera cost. Digital cameras are relatively new, and still very costly. However, if high spatial resolution is needed, a digital camera is the only way of achieving it, short of using multiple cameras or repeated image capture by a camera on a scanning rig. Standard analogue cameras produce about 700×500 pixels; digital cameras give 4000×3000 or more. Another advantage of digital cameras is their ability to provide more than 8 bits per channel of colour resolution. Most colour framegrabbers digitize to 8 bits per channel, and this effectively limits the colour resolution which is obtainable with an analogue camera (see Table 7.1 on page 137). If 8 bits per channel is sufficient for the application, an analogue camera with a signal-to-noise ratio (SNR) in the incoming video signal of 48 dB will suffice. Cameras of higher SNR are readily obtainable, but are more expensive, and it is difficult to obtain any advantage since the system SNR is effectively limited by the framegrabber. Digital cameras often produce 12 bits per channel, the equivalent of a 72 dB SNR. The pixel data is input directly to the computer's RAM through a digital interface, and it is not degraded by the extra circuitry of a framegrabber. This is achieved, of course, by putting the ADC in the camera, close to the CCD sensor. Even in a digital camera, the output from each CCD element is analogue (a quantity of charge which can be converted into a voltage ready for analogue-to-digital conversion). It is sound practice in digital signal processing to digitize any analogue signal close to its source because this gives minimum opportunity for noise to corrupt the signal.

Continuous or monoshot

This choice is decided by the nature of the scenes to be captured. Continuous cameras are suitable for capturing images of stationary scenes, or of slow-moving objects. Each image is built up by a raster scan taking typically 1/25 second, so if there is significant movement during this time, the image will be blurred. For applications involving rapidly moving objects, such as the on-line inspection of items moving past the camera at 3 ms^{-1}, a camera with a high-speed electronic shutter will be needed. A monoshot camera is useful if the image capture is to be triggered by an external event, such as an item interrupting a light beam on a conveyor belt. Using external triggering, the item to be imaged will always be in the same position within the captured image, and this simplifies subsequent processing. An application at the opposite extreme is the imaging of slides on a microscope. This could require the integration of many frames in order to cope with the low light levels. Specialist cameras are available for this purpose.

Single chip or three-chip sensor

For accurate colour reproduction coupled with good spatial resolution, it is necessary to use a three-chip camera. The reasons for this are explained in section 6.2.5. This adds considerably to the cost of the camera and makes it bulkier and heavier.

Composite video or RGB

This is treated in detail in section 6.2.5. If a three-chip camera has been selected, it is necessary to ensure that the separate RGB signals can be accessed before they are encoded into composite video. With a single-chip camera, composite video will probably give a better signal-to-noise ratio than RGB.

Size of sensor

CCD arrays come in one-third, half and two-third inch sizes. The larger the chip, the better the spatial resolution, but the larger the camera and its lens. This subject is well explained by JVC (1996).

Type of lens mount

C-mount lenses were designed to give interchangeability between cameras produced by different manufacturers. They screw into the camera. They are cheap and in widespread use. However, since they were originally designed for 16 mm motion film cameras, they produce significant lens aberrations if they are used with large sensors. Their chief selling point is their low cost, and it is difficult to find a C-mount lens which is of high quality.

Bayonet lenses attach to the camera by a push-and-turn technique. Each manufacturer has its own fitting, so these lenses may not be interchanged between different manufacturers' cameras. Good bayonet lenses are very robust; but small cheap ones are poorly secured, and can defocus if subject to vibration. In both C-mount and bayonet, a wide variety of lenses are available, ranging from fixed focus to motorized zoom, wide angle and narrow angle. The use of close-up lenses makes it possible to enlarge the view of objects, and this means that small areas of colour can be studied. This gives colour image processing an advantage over colorimeters and spectrophotometers, which have observation ports of 3 mm diameter or more.

6.2.2 Settings

Whichever type and model of camera is selected, care should be taken in its adjustment and positioning relative to the scene to be studied. The camera needs to be properly focused and positioned, and the zoom setting of the lens (if applicable) appropriately adjusted so that the object of interest is wholly within the field of view and important features appear at adequate size. Precise measurements cannot be expected if too few pixels are used. The geometry of camera, light source and surfaces within the scene greatly affect the appearance of the image, in particular the rendering of colour. Specular reflections (where the angle of reflection equals the angle of incidence) are usually better avoided, since they carry no information about the colour of the surface. In specular reflection, the light reflects from the outermost boundary of the surface, and is not affected by the pigments or dyes within as explained in section 3.2.2. The colour of specular reflection ('highlight' colour) is the colour of the light source, not the colour of the reflecting surface. Specular reflections may be avoided by arranging the geometry of lighting and camera so as to avoid mirror-like angles. For example, the light can be incident normal to the reflecting surface, and the camera at 45° to the normal - this imitates the arrangement in colorimeters. If this is not practicable, or if the scene contains surfaces at many different angles, polaroid filters are a possible way of blocking specular reflections: if a polaroid filter is placed between the light source and the reflecting surface (e.g. fitted over the light source itself), and a differently oriented polaroid filter is placed in front of the lens specular reflections may be attenuated.

For some applications, a set of objects at different distances from the camera must be in focus – this requires a large **depth of field** which is achieved by using a small iris setting. This then necessitates either bright lighting or a long exposure time.

The publication by JVC (1996) presents a clear summary of lens defects. The chief problems are spherical and chromatic aberration. These problems are more severe when the optical path is short, necessitating greater curvature in the lenses. In some small cameras, chromatic aberration is corrected by electronic means. Whilst this produces an image which is acceptable to the eye, the chromatic errors are still significant at pixel level. So for colour image processing, it is best to specify a larger lens system.

6.2.3   Applications

The camera is the only image-acquisition device which is able to capture images of three-dimensional scenes or of heavily textured surfaces. Since the camera captures images in the same way as the human eye - complete with ambient lighting effects - it opens up new fields of research. For example, recent studies of the reflecting properties of objects and the response of the human vision system have led to fascinating findings concerning the evolution of human vision, and have led to a more logical approach to image processing. The field is called 'physics-based vision' (Klinker, 1993; Healey, Shafer and Wolff, 1992). Of particular interest is the relationship between our perception of colour, and the spectral properties of the light entering the eye. This relationship was quantified empirically in the 1960's using spectrophotometer measurements, and the resulting 'colour difference equations' provided a much-needed tool for the dyeing and paint industries to assess acceptability of their product batches. A brief account of these developments is given by McDonald and Smith (1995). Using colour video cameras, this work has been extended by Connolly (1996).

Until recently, the only way of measuring colour has involved the use of contact instruments - colorimeters and spectrophotometers. However, the camera captures images without contact with the objects involved. This makes it very useful for some specific application areas:

- In the measurement of a multicoloured object, such as food packaging material or printed fabric.

  To use a colorimeter for this purpose would require it to be brought into contact with each different coloured area in turn, and this takes time and considerable care by the operator; on the other hand, the camera captures the whole scene at once, and can process all the different areas rapidly.

- In measuring the colour of items in a production environment.

  Using a contact instrument, a sample must be taken from the production line and perhaps be carried to a laboratory to make the measurements. The delay between sampling and obtaining the colour measurement result can lead to high product wastage rates. By using a
  camera with a high-speed electronic shutter, images can be captured as
  the products move past on the conveyor. With today's high speed computers, decisions about the colours present in the product are available
  within seconds.

- In the inspection of very small areas.

  Colorimeters have a sensing port of about 3 mm diameter or more. This
  makes it impossible to measure very small areas. However, with a
  close-up lens, a camera can enlarge the view of the object and make
  colour inspection possible.

- In the inspection of non-solid materials.

  Some objects are unsuitable for measurement by contact instruments,
  e.g. dog food, chocolate, blancmange.

6.2.4   Camera configurations

There are four basic approaches:

1. Use of a digital camera, with a SCSI interface to the computer – this
   inputs digital values directly into the computer's RAM. This is simple,
   fast and relatively free from electronic noise. However, digital colour
   cameras are currently very expensive, and the computer requires large
   amounts of RAM. This is the best approach if high spatial resolution is
   required (e.g. 4000×3000 pixels, requiring about 128 MB of RAM for
   the image and the processing software).

2. Use of an analogue video camera, connected to a framegrabber within
   the computer. The framegrabber samples and digitizes the video signal,
   quantizing the signal spatially into pixels and also quantizing the magnitude of the signal into (usually) 256 levels. This approach produces
   images of about 750×500 pixels. Historically, this technique has long
   been available. Colour framegrabbers are now available from a variety
   of manufacturers, and are consequently quite cheap. Three-chip CCD
   cameras cost about £4000 (about $6500) at the time of writing.

3. Use of a still camera, with film developed and printed and the print scanned to produce a digital image. This approach introduces colour distortions arising from the different colour primaries used in the film, the print and the scanner. A more direct use of the still camera is to have the transparency or negative scanned using a slide scanner. This avoids the problems of the photographic print. A commercial service of this type is available from Kodak, marketed as the Photo CD process, giving a digital image on a CD ROM disk – but this is still subject to mismatches between film and scanner primaries. The Kodak Photo CD colourspace is discussed in section 4.9. The still camera provides the cheapest method of acquiring digital colour images and it has the advantage of portability and freedom from electrical power supplies.

4. Use of a digital still camera based on the 35 mm format popular for film-based imaging, but using a CCD sensor and digital storage into RAM or magnetic disk inside the camera, with later transfer of the images to a computer.

6.2.5   Analogue camera quality

Analogue colour cameras have been available for a long time, and their design has evolved in line with the requirements of television broadcasting. They provide a relatively cheap source of colour images, but their design was not originally intended for image processing, and the user should be aware of their limitations and peculiarities. Three specific issues will be addressed here:

- the difference between composite video signals and RGB signals;

- single chip *versus* three-chip cameras;

- signal-to-noise ratio.

Composite video requires some historical explanation. A colour camera produces three times as much information as a monochrome camera, but the information somehow has to be broadcast in the same bandwidth. A description of how encoded colour signals are produced is given in section 5.3. The image is later reconstructed on a monitor, by using each camera channel to drive a separate electron gun. The first complication is that the display phosphors in the monitor produce light whose intensity is proportional to the square of the signal. To compensate for this, the RGB signals are given a square root transfer function, by a special gamma correction amplifier within the camera. For colour image processing, this is a nuisance, since the relationship between different colours in the scene is distorted. Gamma correction is still used in solid state cameras, although it is often possible to turn it off, or switch to a gamma value of unity, although even when unity gamma is selected there is still usually some residual non-linearity which may require further correction, as discussed in section 7.3.3.

A second complication arises from the fact that the camera's signals are intended to be broadcast. When colour television was first introduced, it was necessary to work within the following constraints:

- the colour broadcast signals must fit into the same bandwidth as monochrome signals;

- the colour broadcast signals must drive both monochrome and colour TV receivers.

To fulfil these historical requirements, the gamma-corrected RGB signals are encoded into YUV form, for example, as described in section 5.3.2. In the PAL system (discussed in section 5.3.3), Y, U, and V are derived as follows:

$$
\begin{aligned}
Y &= 0.222R + 0.707G + 0.071B \\
U &= 0.493(B - Y) \\
V &= 0.877(R - Y)
\end{aligned} \tag{6.1}
$$

Y is the luminance signal, and it is allocated the same 5.5 MHz bandwidth as for monochrome signals. Chrominance signals U and V carry the colour information in a 2 MHz sub-carrier within the same bandwidth. Thus, the colour information is 'compressed' during the processing to composite video.

The weightings in equation (6.1) for luminance are the UK PAL system weightings. From 1953 until 1974, NTSC weightings 0.299, 0.587, 0.114, were used in the calculation of luminance for NTSC, PAL and SECAM television systems, but the PAL weightings were then changed. Sproson (1978) explains why. Unfortunately, some television and image processing textbooks published since this change attach the incorrect weightings to the PAL system, and many papers do not make it clear whether they are using PAL or NTSC equipment. Care should be taken to avoid confusion in this matter. If possible, R, G and B signals should be used for image acquisition for colour image processing. Because of the chrominance compression, composite video signals carry much less detailed colour information than the original RGB signals. It is important that the RGB signals are output from the camera before processing - some camera designs process the signals immediately to produce composite video, and later reprocess them to give separate RGB signals again! This is an unsatisfactory arrangement for colour image processing applications, since the signal-to-noise ratio is severely degraded by this double processing.

Nowadays, charge-coupled device (CCD) cameras have replaced tube cameras in virtually all applications. The camera's target consists of an array of discrete charge-coupled devices (approximately 0.25 million of them on a chip half an inch square), so the image data is inherently spatially quantized. However, because CCD cameras needed to be compatible with the tube cameras which they replaced, they produce standard video output by scanning the CCD cells and interpolating the signal in the gaps between the cells. This continuously varying signal is then quantized again by the sampling at the front end of the framegrabber! This obviously gives scope for errors to arise, and it is important to match the sampling rate of the framegrabber to the bandwidth of the camera.

Another factor influencing the quality of colour images obtained from a camera is the number of CCD chips in the camera. The best colour cameras have three separate CCD arrays, one responding to red light, one to green and one to blue, very similar to the three-tube camera of Fig. 5.10. This allows the colour of a precise spot to be described in terms of its R, G and B components. Cheaper cameras use one CCD array, with finely striped colour filters to produce the YUV signals, as shown in Fig. 5.11.

6.2.6   Colorimetric principles

A video camera is designed to produce colour images which appear correct when displayed on a video monitor or television screen. The chief requirement is that the spectral response curves of the colour camera match the chromaticities of the display phosphors. In the early days of colour television, different manufacturers used different phosphors in their television screens, but in 1970 a standard was agreed for use in the PAL system described in section 5.3.3 (Sproson, 1978). The xy chromaticities of the phosphors and the white point were defined, using the CIE 10° standard observer data, as listed in Table 4.1 on page 70. (In the USA FCC standards apply, as also listed in this

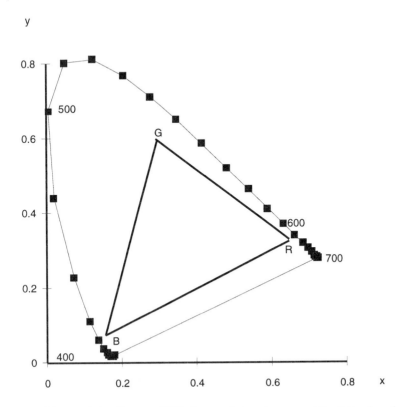

Fig. 6.1 Chromaticity diagram (10° observer) showing gamut for PAL phosphors.

table.) Figure 6.1 shows the gamut of colours which may be reproduced by these standard phosphors in the PAL system. The three phosphors are plotted at points R, G and B, and a triangle is drawn joining these points. All colours whose chromaticities lie within this triangle can be reproduced by the additive mixing of light from the three phosphors. The reproduction of a colour whose chromaticity coordinates lie outside the triangle would require a 'negative' amount of light from one or two of the phosphors. For example, to reproduce the appearance of light from a monochromatic source of wavelength 500 nm, positive light output would be required from the blue and green phosphors, and negative output would be required from the red phosphor. Therefore this colour cannot be reproduced. This is not a serious limitation in practice, since most surface colours have broad spectral reflectance curves with significant contributions from many wavelengths, and under these conditions the major-

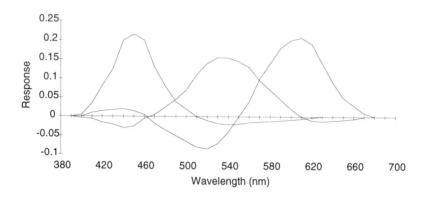

Fig. 6.2   Ideal spectral response curves of PAL camera.

ity of colours can be reproduced accurately. It is only the reproduction of near-spectral colours that gives problems.

Now that the phosphor chromaticities are standardized, the required response curves of the three channels of the camera may be calculated. Cameras can be designed in the knowledge that they will produce good colour rendering on all television and monitor screens using the standard phosphors. The camera response curves are shown in Fig. 6.2, following the diagram given by Sproson (1978). Each of the three curves has both positive and negative response regions. The actual transmission characteristics of the analysis filters in a camera have only positive lobes, and a technique of matrixing is used to mix the signals to imitate the correct response. This ensures that all in-gamut colours are accurately reproduced on the screen.

So the major goal of camera and display designers has been achieved: coloured scenes are acceptably reproduced. In colour image processing, we want to go further, and make use of the colour data present in the image pixels to give a numerical measure of colour in different parts of the image. Our numerical values must correspond to international standards for colour measurement. So we need to convert the red, green and blue signals to CIE values. From the phosphor chromaticities and the D65 whitepoint, it is possible to calculate a matrix relating the camera's theoretical RGB values to CIE tristimulus values, as shown in equations (4.1) and (4.2) on page 71 in Chapter 4.

6.2.7   Signal-to-noise ratio (SNR)

The precision of a camera, or its freedom from random errors, is measured by its signal-to-noise ratio. This is the root mean square (RMS) signal voltage divided by the RMS noise voltage, with the ratio expressed in decibels:

$$\text{SNR} = 20\log_{10}\left(\frac{\sqrt{\sum v_s^2/N}}{\sqrt{\sum v_n^2/N}}\right)$$

where v_s is the signal voltage and v_n is the noise voltage. In the standard computation of signal-to-noise ratio, the RMS voltage is calculated by averaging over time. However, an equivalent approach in an imaging situation is to capture an image of a uniformly coloured surface under uniform illumination, and find the average over a collection of N pixels. The standard deviation (σ) of the pixel values provides a practical measure of noise in an image:

$$\sigma = \sqrt{\frac{\sum(v_s - \bar{v}_s)^2}{N}}$$

In fact, the standard deviation is the RMS value of the deviation from the mean – in other words, the RMS of the noise. By measuring the standard deviation in the image, a rapid check can be made on whether the signal-to-noise ratio claimed by the camera's specification is borne out in the practical imaging set-up being employed. Sometimes, the camera's quoted precision is degraded as the video signals pass through other parts of the system – the cables, the framegrabber, even the computer bus.

For example, a 54 dB camera produces a ratio approximately equivalent to 9-bit precision:

$$\frac{\text{RMS signal}}{\text{RMS noise}} = 10^{\frac{54}{20}} = 501.1$$

In an 8-bit image, where the signal may range from 0 to 255, the RMS value of the signal is equal to 147.2. The standard deviation among pixel values should therefore be $147.2/501.1 = 0.29$. However, if the camera's signals are digitized by an 8-bit per channel framegrabber, they are likely to be degraded to 8-bit precision by the introduction of electronic noise from the framegrabber circuitry, since analogue to digital converters are usually designed to give noise just less than their quantization error. This would give an effective system signal-to-noise ratio of only:

$$20\log_{10}(256) = 48.2\,\text{dB}$$

(a decrease of 6 dB corresponds to a doubling of the noise). The corresponding standard deviation among pixels would be $147.2/256 = 0.57$.

Most cameras have a signal-to-noise ratio of between 48 and 60 dB. Manufacturers use two methods of improving the signal-to-noise ratio:

1. cooling the CCD array to reduce the 'dark current' (that is, the noise);

2. extending the exposure time to increase the signal at low light levels.

Cooled cameras are more expensive than conventional CCD cameras, but they have significantly better signal-to-noise ratio.

7

Practical system considerations

Christine Connolly and Henryk Palus

7.1   IMAGE ACQUISITION TECHNIQUE

7.1.1   Introduction

The importance of careful image acquisition technique is often neglected in image processing. Since it is so easy to manipulate the image, it is sometimes assumed that all acquisition errors may be corrected in later processing. As a general rule in any branch of signal processing, however, it is better to avoid errors at source. This gives the advantage of controlling errors more effectively, and also simplifies and speeds up the processing stage. There are in fact some acquisition errors which cannot be corrected accurately, for example **colour clipping**. This occurs when the aperture or gain settings of the camera are such that some areas of the image are overexposed, and one or more of the colour channels reaches saturation, giving an unbalanced rendering of the colour. It is not possible to correct this error properly, since there is no means of knowing by how much the signal exceeded the saturation value.

To acquire a good colour image, it is necessary to set up the camera and the lighting environment so as to use the full dynamic range of the camera, i.e. black objects in the scene should produce very low pixel values, and white objects should produce pixel values just below the maximum permitted. It is also necessary to balance the colour channels correctly. Experiments should be carried out on the camera adjustments and lighting conditions in order to achieve a good image. It should be noted that framegrabbers often have programmable gain and offset settings which may be adjusted independently on the three colour channels.

In this chapter, the characteristics of cameras, framegrabbers and lighting are examined and advice given on how to use them in such a way as to give good colour images. There is a brief section about image file formats, since it is often required to port image data to other computers, or to commercial application packages. Colorimetric calibration of acquisition hardware is discussed.

7.1.2   Capturing images with good colour quality

Whilst the design of colour video cameras is underpinned by sound colorimetric principles, various precautions are necessary to ensure optimum colour fidelity in the captured image. This section discusses the sources of error and the practical steps which can be taken to optimize colour accuracy in video camera images. Unless special precautions are taken, video cameras may produce different signals on different occasions when pointing at the same object. A detailed study of the errors in colour video cameras has been undertaken (Connolly and Leung, 1995). Some errors may be readily avoided, once their source is understood; others are unavoidable, but their effects may be minimized by appropriate techniques.

Avoidable errors
Modern video cameras are sophisticated devices with many built-in features which, whilst making the camera easier to use, do not necessarily lead to accurate colour reproduction. Readers should be aware of the features listed below.

Gamma   Although CCD cameras produce a linear response to light intensity, they are usually equipped with a gamma correction amplifier to output signals which are proportional to the square root of the light intensity. This compensates for the fact that display phosphors produce light whose intensity is proportional to the square of the voltage signal. For accurate colour reproduction, the gamma correction facility must be turned off, since in addition to the non-linear response it also causes an increase in electronic noise.

Automatic iris   If fitted, this aims to ensure that the camera is always correctly adjusted to the lighting conditions, and is useful when filming out of doors. The iris is usually controlled by a signal derived from the integrated response over the whole field of view. Unfortunately, this often means that small areas of bright colour are prone to colour clipping, that is the saturation of one or more of the three colour signals. A related effect is **blooming**, where CCD cells near such saturated areas are also affected. These effects are noted by Klinker (1993). To avoid clipping and blooming, the camera should be used with manual iris, set by experiment to avoid saturation in all regions of the captured image (section 7.3.3).

Automatic white balance   If fitted, this aims to compensate for changing illuminants. The relative gains of the red and blue channels of the camera are automatically adjusted in line with a signal obtained by integrating the red and blue responses over the whole field of view. Unfortunately, this means that if the same object is viewed first against a blue background and later against a red background the signals from the object itself will change. For accurate colour reproduction, an automatic white balance facility should not be used.

Unavoidable errors

Pixel – pixel noise   Whilst the overall appearance of an image may be acceptable to the human eye, the RGB values of neighbouring pixels may differ markedly from each other, although all represent areas of the same uniformly coloured object. Experiments have also shown that the RGB value of an individual pixel varies randomly with successive exposures. This variation is related to the signal-to-noise ratio of the system. Whenever it is necessary to establish the RGB value of a region of interest in the image, a more precise value will be obtained if a mean is computed over as many pixels as possible, whilst avoiding variations caused by non-uniformity of illumination. Statistical theory shows that the error of the mean decreases in proportion to the square root of the number of pixels averaged.

Camera drift   The signals from a video camera are affected by the temperature of the camera and tend to vary with time. The greyscale values over the whole image drift up and down together, although the camera may be pointed at an unchanging scene, under fixed lighting conditions. The errors are worst during the first 30 minutes after the camera is switched on, but never settle down completely. A way round this problem is to include in the scene an object whose RGB values are known, and to measure their variation every time an image is captured, and to correct the whole image in proportion to the change. This can be done conveniently using a framegrabber's look-up tables, if fitted. (Some framegrabbers have input look-up tables between the ADCs and RAM which can be loaded with correction data. If there are no look-up tables in the framegrabber, the equivalent operation can be performed in software.) This technique is based on the 'double beam' system of

colour analysis which is well-established in colorimetry and spectrophotom-
etry. The drawback is that if the correction is made after image capture, the
camera may be operating outside its optimum range, and the corrections may
not fully overcome all errors. A better technique is to adjust the camera so
as to hold the RGB values on target. This system for stabilizing the camera
measurements has been patented (Connolly and Leung, 1994).

7.1.3   Lighting

Lighting is a very important component in the rendering of colour. At any
particular wavelength, the light entering the eye is the product of the intensity
of the illuminant and the reflectance of the surface. If changes occur, either
in the surface colour or in the lighting, then the colour sensation in the eye is
affected. The human vision system automatically compensates for changes
in lighting when we move, say, from a room with artificial light out into
daylight, and we are not normally aware of the effect of lighting on surface
colour[1]. In contrast, a camera makes no such compensation, and the pixel
RGB values in a colour image are drastically affected by changes in lighting.
It is therefore important to provide stable lighting for the objects or scenes to
be studied. The lighting should be constant both in spectral composition and
in intensity.

The first way to provide constant lighting is to build an enclosure with
light sources inside. It is possible to purchase standard lighting cabinets,
specially designed for the visual assessment of colours, and equipped with
standard illuminant tubes D65, A, *etc.* (GretagMacbeth's 'Judge'™ lighting
cabinets, for example).

However, these are quite expensive, so many researchers prefer to build
their own lighting enclosures. The important thing is to ensure that the scene
is lit by a controlled light source, and stray ambient light is cut out. This
avoids the variations inherent in natural illumination, where the intensity and
colour of the ambient light is affected by the position of the sun and the
cloudiness of the sky. The lighting enclosure also helps avoid shadows of
moving observers, and the changing reflectivity of their clothing. When light
reflects from the object viewed, only a fraction of it reflects towards the sensor. The rest reflects in many different directions, and goes on to reflect again
from the walls of the room, and other objects within it, and some of this
eventually gets back to the object and thence into the camera. A few simple

---

[1]This topic is discussed in section 2.5.1.

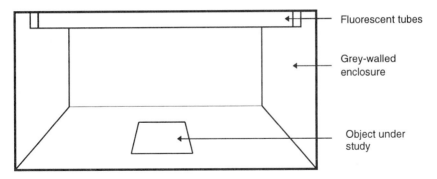

Fig. 7.1 Controlled lighting enclosure: looking in from front.

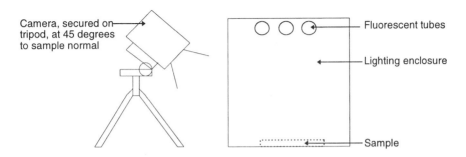

Fig. 7.2 Side view of controlled lighting enclosure.

experiments will convince you that the illumination of the scene viewed by the camera is affected by the proximity of reflecting objects – try standing close to the scene wearing first a white shirt, and then a dark jumper. Whilst it is not possible to see any change in illumination a change will be seen in the colour image pixel values. In standard colour-matching lighting cabinets, the walls are painted with a neutral grey of approximately 50% reflectance, in order to diffuse the light and control the colour effect of multiple reflections. Figure 7.1 shows a lighting cabinet, viewed from the front. The light is incident normal to the sample on the floor of the cabinet. To avoid specular reflections, the camera may be placed outside the cabinet, tilted down at an angle of 45°, as shown in Fig. 7.2.

The operator should be aware of the effects caused by the ageing of lamps. Most light sources change both their intensity and their spectral balance during their lifetime. GretagMacbeth, for example, recommend that lamps be changed after 5000 hours running time. It is also worth noting that stroboscopes or flash sources have a different intensity and spectral balance between one flash and the next.

Uniformity of lighting is very difficult to achieve, and an object may change its appearance when moved relative to the lighting. This effect is difficult to detect by eye, because the human vision system seems to compensate automatically for this type of change; however, it is obvious in the pixel values of any image. To achieve reasonable uniformity, the light source should be large compared with the object under study - for example, a bank of 1.2 m fluorescent tubes would be suitable for lighting an A4 size sheet of paper. Fluorescent tubes are reasonably uniform in the centre, but their intensity falls off significantly towards their ends. Davies (1997) uses the analogy of Helmholtz coils (which produce a fairly uniform magnetic field) to propose lighting arrangements whereby good uniformity may be achieved by the careful positioning of two parallel fluorescent tubes. Specialist companies are now offering fibre optic solutions to uniform lighting problems, but generally they attempt to light only small regions.

It has been noted above that the human vision system compensates automatically for changes in lighting intensity across the field of view. We also tend not to notice changes in intensity across the surface of a uniformly coloured three-dimensional object – we tend to see a red coffee mug as a single, well-defined object, rather than as separate vertical regions of dark red and bright red. In a coloured image, these different areas are easily detectable, and cause problems in image segmentation in most colour space representations. Studies have been carried out on the ability of different colour spaces to segment natural scenes (Connolly, 1996), and some interesting relationships have been noted between the behaviour of the human vision system and the reflective properties of different materials (Healey, 1992; Tominaga, 1994; Connolly, 1996).

7.1.4   Digitization and framegrabbers

An analogue video signal from a camera is quantized in two ways when it enters a framegrabber:

1. sampling produces **spatial quantization**;

2. analogue-to-digital conversion produces **magnitude quantization**, converting an analogue voltage to a particular digital 'greyscale' level.

The sampling frequency determines the number of pixels across each horizontal line of the image. This should be chosen to match the horizontal resolution of the camera (see JVC (1996) for a helpful explanation of camera resolution). The pixels produced in the digitized image will, even so, probably not coincide with the original CCD elements, and this mismatch degrades the pixel data to a certain extent.

ADC resolution

The majority of framegrabber boards use 8-bit analogue-to-digital converters (ADCs) on each input channel. This produces 256 possible greyscale levels per channel, or over 16 million possible colours. However, this large number of colours is not uniformly distributed perceptually. Whilst some areas of colour space – neutral greys and pastel shades – are over-populated with colours too similar to be distinguished by the human eye, other areas are sparsely populated. In particular, there is insufficient resolution in the bright and dark colours. It has been shown by Ikeda, Dai and Higaki (1992a) that the maximum quantization error for 8-bit digitization of the R, G and B signals is 10 CIELAB units. This has two repercussions:

1. a pair of very dark shades, differing from each other by 10 CIELAB units, could in certain circumstances produce identical pixel RGB values;

2. a colour measured on different occasions can appear to vary by 10 CIELAB units, due to very small changes in the image capturing conditions.

Methods of improving colour resolution

There are several ways in which the resolution of dark and bright colours can be improved; including averaging; and enhancing the ADC resolution, combined with non-linear compression.

Even though individual pixels are quantized into 256 levels per channel, it is possible by averaging over many pixels to obtain a more accurate measure of the R, G and B values. The average should be calculated for a block of pixels from the same area of surface colour, uniformly illuminated.

Monochrome framegrabbers are often available with 10-bit ADCs. If the image is captured channel by channel, switching the input from the red channel of the colour camera to the green channel and finally to the blue channel, a 30-bit colour image can be built up. The 10-bit ADC produces 1024 possible levels per channel, a 64-fold increase in the total number of colours.

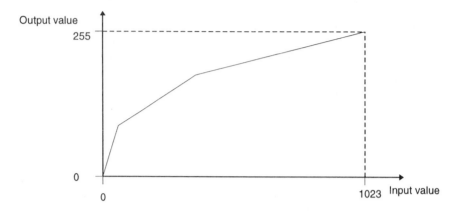

Fig. 7.3 Nonlinear compression of 10-bit per channel image.

However, because the red, green and blue images must be captured sequentially, this technique is only suitable for stationary scenes. A better option, but considerably more expensive, is to use a digital camera, giving say 12 bits per channel. The digital camera option has the additional advantage that each pixel in the final image corresponds spatially with a particular CCD element in the camera's array, so no spatial smoothing and resampling errors are introduced.

Most image-processing software, and all graphics display cards, are designed for a maximum of 8 bits per channel. So if the camera and ADC produce 10 or 12-bit per channel images, the data has to be compressed before it can be displayed or processed. This compression may be performed in a nonlinear way, keeping the high resolution at low signal values and sacrificing resolution at high values, as shown in Fig. 7.3. In this way, the distribution of the 16 million colours is made perceptually more uniform.

Limitations on colour resolution

Spectrophotometers make use of increased measurement resolution at low reflectance values, to imitate the eye's response to light, and it is necessary to do something similar with a camera if serious errors are to be avoided. However, the resolution is restricted by the camera's signal-to-noise ratio, which varies in practice from about 48 dB to 72 dB. Table 7.1 indicates the relationship between signal-to-noise ratio and bit resolution and the corresponding quantization error in CIELAB units. For example, 10 bits of resolution provide 1024 possible levels. If the error is less than 1 level, this corresponds to

Table 7.1 Effect of ADC resolution and signal-to-noise ratio on colorimetric error. (The information in the first two columns is taken from JVC (1996), and that in the final column from Ikeda, Dai and Higaki (1992a).)

Resolution (bits)	Signal-to-noise ratio (dB)	Worst case colorimetric error (CIELAB units)
8	48	10
9	54	5
10	60	2.5
11	66	1.3
12	72	0.6

a signal-to-noise ratio of 1024, or 60 dB. The graphs in Ikeda, Dai and Higaki (1992a) show that the maximum quantization error for 10-bit RGB data is 2.5 CIELAB units. Table 7.1 is a very powerful summary of performance, and is helpful in specifying camera and framegrabber requirements for different applications.

7.2 IMAGE STORAGE

7.2.1 Image files

Images take up a lot of memory. A 768×512 pixel colour image with 8 bits per channel, occupies:

$$768 \times 512 \times 3 = 1\,179\,648 \text{ bytes}$$

that is, most of a 1.44 MB floppy disk. Many applications save image files as raw byte data, but there are many ways of doing it. For example, R, G and B pixel values may be stored consecutively in the order RGB (or sometimes BGR), followed by the next pixel, and so on. Alternatively, the whole red image can be stored, followed by the whole green image, *etc.* Pixels can be stored in raster scan order, odd lines followed by even lines, or ordered by column position.

If just one image processing package is being used, this is acceptable. The file layout is known, and programs may be written to read the file if needed.

However, another user would need the file plus a lot of detailed information – number of pixels, number of channels, arrangement of data on the file.

There is frequently a requirement to pass image information to other researchers, or to access the image with commercial image manipulation and editing packages. To make images portable, standard image file formats have been developed. The existence of standards also allows images to be displayed by different devices (e.g. monitors, plotters and printers) and sourced from different hardware (cameras or scanners). Sharma and Trussell (1997) give an excellent account of the development of different imaging input and output devices, with many useful references and an overview of the development of colour knowledge. Scanners, digital cameras and digitizer pads provide bitmapped or raster image information, and monitors and graphics printers can handle this. Computer aided design (CAD) packages produce vector files, where shapes are made up from line segments, in order to drive plotters. The most common image file formats encountered in image processing are TIFF and GIF.

The TIFF (Tagged Image File Format) was developed by Aldus (TIFF, 1992) and is widely used in the PC market. The tags are extra header information added to the image data, containing the type of detailed information identified above as necessary. This makes the file slightly longer, but much more portable.

GIF (Graphics Interchange Format) was developed by Compuserve in 1987. Full details of the file layout are given by Held and Marshall (1996). For up to date information on new standards currently being developed, see Nier (1996). Nier also reviews the imaging needs of various industries including prepress, photographic, graphic arts, high definition TV, multimedia and video conferencing.

Images often have to be compressed for storage and/or transmission and Chapter 13 discusses this subject in detail. Many of the common file formats, e.g. TIFF, support compression of the pixel data.

7.2.2   Trends in host computers

For a long time, serious image processing was done on computers designed around the VME bus – Sun, Apollo, *etc.* Framegrabbers compatible with the VME bus tended to be large robust devices, and in the 1980's only one company produced a colour framegrabber. The recent development of the PCI bus in the Pentium™ computer has brought a new generation of framegrabber boards which give excellent performance on the PC platform, making cheap image processing available at last. Many manufacturers produce colour PCI framegrabber boards, with the result that the boards have also become much cheaper. For real-time machine vision applications, there are PCI framegrabbers with on-board signal processing hardware, such as the Matrox Genesis™. In response to the need for pipelined architecture and RISC operation, the Pentium processor has brought this into practical, affordable form. There has never been a better time for getting into colour image processing.

7.2.3 General observations

Cameras bring their own particular advantages to the study of colour. With colorimeters and spectrophotometers, the only instruments available for the measurement of colour until recent years, only flat objects can be studied, and the object has to be in contact with the measuring instrument, lit by the instrument's light source. Cameras, on the other hand, see scenes as humans do, without contact, and lit by the ambient illumination. Therefore it is possible to study the role which colour plays in our perception and interaction with the physical world.

Recent work in physics-based vision shows that, under certain conditions, the light reflected by a three-dimensional object with uniform surface colour has constant spectral balance, but varies in intensity. This provides the basis for distinguishing between different objects in a three-dimensional scene. The spectral balance idea has been tested explicitly by Lee, Breneman and Schulte (1990) using a tele-spectroradiometer (section 3.3.2), and examined in regard to the theories of physical optics. Healey (1992) developed a general 'approximate colour-reflectance model', applicable to dielectrics and metals, which allowed geometric effects in a scene to be 'factored out' of the colour pixel values. In the case of dielectrics, Klinker (1993) used the dichromatic reflection model to develop an algorithm to segment scenes whilst ignoring colour changes due to highlights and shading, and Tominaga (1994) applied the dichromatic reflection model to a wider variety of materials. The effectiveness of various colour difference equations in segmenting three-dimensional scenes has been investigated. It emerges that equations based on band ratios are good for the segmentation of three-dimensional scenes.

Physics-based vision is a very valuable development. Image processing traditionally involves a lot of statistically-based techniques – histogram equalizations, smoothing, *etc.* Images are regarded as a collection of random data containing signals and noise, requiring statistics and signal processing techniques to make sense of them. Physics-based vision recognizes that images actually contain a lot of information which varies in a logical, structured way and relates very precisely to the geometry of the surface from which it reflects. So by looking for expected patterns in the image we can get information about the composition of the scene and the shape of individual objects.

Of particular interest is the study of colour in three-dimensional objects. Colour provides a very rapid way of segmenting objects in a scene (Chapter 9). A logarithmic colour space has been proposed, which combines the advantages of rapid colour segmentation of three-dimensional scenes with the convenience of being uniform relative to human colour vision (Connolly, 1996).

7.3   COLORIMETRIC CALIBRATION OF ACQUISITION HARDWARE

7.3.1   Introduction

Fidelity of colour in an image acquisition process is a necessary condition of further image processing. Generally, colour cameras are not built for measuring colours: they are designed to render an acceptable image of a scene for human viewing.

In many applications of colour image processing systems, repeatability of measured RGB values is very important (Strachan, Nesvadba and Allen, 1990; Connolly and Leung, 1995). Sometimes the requirements are stringent. For example, the absolute measurement of object colour in an image needs a colorimetric calibration i.e. a mapping transformation between the colour space of the camera and a standardized colour space (e.g. XYZ, CIELAB, CIELUV). The general idea of colorimetric calibration is presented in Fig. 7.4. Using calibration equipment, e.g. colour targets   (charts), the RGB colour values are measured and mapped to new, calibrated values in the same space or in one of the CIE standardized spaces.

7.3.2   Sources of error in colour image acquisition

Any element of an image processing system (colour camera, framegrabber, illumination system *etc.*) can be a source of errors during colour image acquisition. The non-stability of the parameters of colour cameras (noise, drifts, residual autogain *etc.*) are widely described (Dalton, 1988; Connolly and Leung, 1995). In the case of a framegrabber, a colorimetric error may be caused

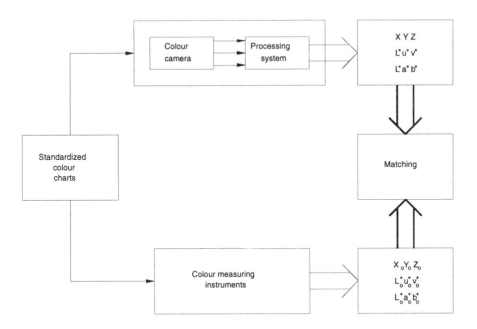

Fig. 7.4 The principle of colorimetric calibration of a colour image processing system.

by quantization – $10\Delta E_{ab}$ for 8-bit resolution of RGB components (Ikeda, Dai and Higaki, 1992a).

Colour image acquisition should be preceded by adjustment of the camera and framegrabber parameters to make the system work within its dynamic range and to avoid adverse effects such as clipping, blooming or colour imbalance. Correction of these effects after the image acquisition stage is very difficult. Clipping is a result of colour signal limitation by reason of exceeding the dynamic range of the system. Since clipping does not appear in all three colour channels simultaneously, this effect can degrade colour reproduction. For example, for a colour with maximal G component, clipping appears first in this channel. Continued increase in the other two components will cause colour errors. When the colour signal exceeds the clipping limit, blooming can occur, although many new CCD cameras incorporate anti-blooming circuits. For an image processing system to work within its dynamic range, the lens aperture should be controlled.

7.3.3   Procedure for calibration of an image acquisition system

The following arrangement allows compensation for the imperfections of a colour image processing system. (Initially the automatic gain control (AGC) of the camera and other correctors should be turned off.)

Black and white balance
The output level which results from analogue-to-digital conversion of a camera's dark current is designated **black level**. Black levels can be measured for each of the RGB channels with a closed or covered camera lens. For some colour camera/framegrabber combinations, black levels may be found equal to 0 but it may be that the signal from the camera is less than the zero voltage level for which the ADC output would be zero ('blacker than black'). This situation requires a reduction of the reference voltage of the ADC or, if possible, an increase in the black level. Sometimes, a final black balance requires some adjustment inside the camera. Some framegrabbers have an option of software control over the reference voltages of the ADCs. This is a great advantage for precise black and white balance. Black balance, unlike white balance (described below), is independent of illumination.

White balance implies equality of the three RGB output signals from a colour camera obtained for an achromatic object in the camera's field of view. White imbalance can stem from non-uniform sensitivity of the CCD elements and different transmission characteristics of the colour mosaic filter in different parts of the spectrum. The white balance of the three channels of a camera may be achieved by manual or automatic gain control. However, a change in illumination conditions makes new white balancing necessary. Various white balance methods are presented in the literature. Some colour cameras have built-in automatic white balancing. In the work of Strachan, Nesvadba and Allen (1990) and in the MAVIS instrument described in Chapter 16 a reference white tile was used as a target achromatic object for white balance. Neutral filters were applied for white balance of a monochrome camera with RGB filters (Andreadis, Browne and Swift, 1990). The black and white levels of an image processing system are used (Dalton, 1988; Strachan, Nesvadba and Allen, 1990) to correct the measured RGB components in the following way (for 8-bit resolution):

$$R = \frac{R_m - R_{bl}}{R_{wh} - R_{bl}} 255$$

where: R_m is the measured value of the R component, R_{bl} is the value of the black level in the R channel, and R_{wh} is the value of the white level in the red channel; and similarly for the green and blue channels.

Linearization of transfer functions (photometric calibration)
Although the photoelectric effect used in CCD image sensors is linear, colour cameras usually have non-linear transfer characteristics. As discussed earlier, cameras sometimes have a gamma correction circuit, which introduces a complementary non-linearity to compensate for monitor non-linearity (Novak, Shafer and Wilson, 1990) and also discussed earlier in this chapter:

$$L = kU^\gamma$$

where L is the luminance of the monitor's screen, U is the monitor input voltage, k is a constant, and γ is a number between 2.0 and 3.0.

The gamma correction circuit can be turned off ($\gamma = 1$) in some types of camera. However the linearity of such a camera is poor and applied linearization is useful. For linearization it is necessary first to determine the transfer characteristics for the three camera channels. Transfer characteristics are obtained as a result of measurement of the RGB components for several viewed samples with different grey levels (grey charts or neutral filters) and known Y values. The following equations allow the calculation of the values of non-linearity coefficients γ_R, γ_G and γ_B (Strachan, Nesvadba and Allen, 1990):

$$Y = R^{\frac{1}{\gamma_R}}$$

and similarly for green and blue. Hence:

$$\gamma_R = \frac{\log R}{\log Y}$$

and similarly for green and blue.

On the basis of system transfer characteristics the correction coefficients can be calculated. These coefficients, when loaded into a framegrabber's input look-up table (LUT) or applied by software after image acquisition, will make linear operation of an image processing system possible as illustrated in Fig. 7.5.

7.3.4   Methods of colorimetric calibration

Lee (1988) was the first to propose a procedure for colorimetric calibration in CIE XYZ colour space. However, this procedure needs a knowledge or

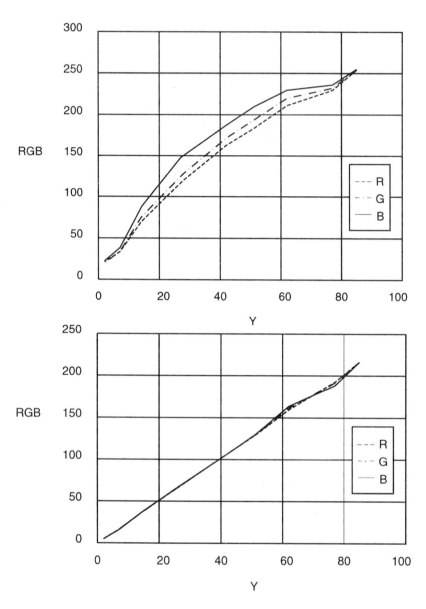

Fig. 7.5 Transfer functions of a colour image processing system using a one-chip camera. Top: before linearization; bottom: after linearization using a look-up table.

measurement of the spectral characteristics of a lighting system and colour chart. If the Macbeth[2] ColorChecker™ (18 coloured and 6 grey samples) is used as the chart, then the spectral characteristics can be found (McCamy, Marcus and Davidson, 1976). The measurement of the spectral distribution of a lighting system requires a spectroradiometer, which is a disadvantage.

With regard to spectral characteristics, a method for colour calibration in CIELUV colour space presented in (Strachan, Nesvadba and Allen, 1990) is much simpler. Here the spectral characteristics were not used. Instead, the $L^*u^*v^*$ values were measured by colorimeter for samples from the Macbeth ColorChecker™ chart. The transformation matrix from RGB space to XYZ space was found as a result of minimization of the average colour difference calculated for colour samples in CIELUV uniform colour space. Similarly, in the work by Kollhoff and Kempe (1995) the correction matrix for RGB components was calculated iteratively in a process of minimization of the average colour difference for colour samples in CIELAB uniform colour space.

The procedures for colorimetric calibration discussed above apply to image processing systems working in fixed illumination conditions. Marszalec and Pietikäinen (1994) have recently developed a real-time procedure for colour camera calibration that is independent of changing illumination. This work incorporated the results of research on colour constancy. The new aspect of this procedure is the use of a colour chart for the reconstruction of the spectral distribution of a given illumination system. However, at the start of the procedure, the spectral characteristics of the image processing system and spectral reflectances of the samples must be known. First experimental results indicate smaller errors than in the case of a calibration procedure which assumes fixed illumination.

Other equipment used for colour image reproduction also requires calibration (colour scanners, printers *etc.*). These procedures can be adopted for colour image processing systems. Colour scanners (Kang and Anderson, 1992; Schettini, Barolo and Boldrin, 1995) are, like cameras, image acquisition devices and sometimes need translation from their non-colorimetric RGB spaces into one of the defined CIE colour spaces (most often CIELAB). The use of neural networks to realize calibrating transformations is playing an increasingly important role in scanner calibration (Kang and Anderson, 1992; Schettini, Barolo and Boldrin, 1995). Calibration of printers is discussed in Chapter 15.

In some approaches, colorimetric calibration is not realized but repeatability of RGB measurements in changing illumination conditions can be as-

[2]http://www.macbethdiv.com/

sumed (colour calibration: (Austermeier, Hartmann and Hilker, 1996)). Alternatively, correcting the variations in RGB values caused by components of a system (RGB calibration: (Chang and Reid, 1996)) is possible.

7.3.5   Colour targets and charts for calibration

Apart from the Macbeth ColorChecker™ already mentioned, other colour targets, charts and samples from colour atlases may be used in colorimetric calibration procedures. For example, in the work of Frey and Palus (1993) using a light box, Kodak neutral gelatine filters for linearization and chips from the Richter colour atlas for colour balance were applied. Asano, Kenwood, Mochizuki and Hata (1986) used 10 grey Munsell papers for software white balance. For their on-line calibration procedure Marszalec and Pietikäinen (1994) used Labsphere samples in addition to the Macbeth chart. The ANSI IT8 colour target was specially developed for input scanner calibration (Schettini, Barolo and Boldrin, 1995; ANSI, 1993a) and could also be used for camera calibration.

Conclusions

A colorimetric calibration procedure can be performed as a preprocessing stage in some image processing systems. Together with the possibility of absolute colour measurement it gives additional positive results. For example, in IHS colour space, hue (H) and saturation (S) are independent of intensity (I) only if the camera works linearly (Frey, 1988). Frey and Palus (1993) showed that colorimetric calibration improves the *rg* and *xy* independence of changes in intensity of lighting.

PART THREE

PROCESSING

8

Noise removal and contrast enhancement

John M. Gauch

The objective of noise removal is to detect and remove unwanted noise from a digital image. The difficulty is in determining which features in an image are genuine and which are caused by noise. In general, it is assumed that variations in intensity and colour will be gradual in an image, so points which are significantly different from their neighbours can often be attributed to noise. Hence, the central idea behind many noise removal algorithms is to replace anomalous pixels with values derived from nearby pixels. Local averaging is commonly used for this purpose, which has the side effect of smoothing the output image. Many noise removal algorithms have parameters which can be adjusted to trade off noise level versus smoothing, so the ideal image for subsequent processing can be interactively selected.

The goal of contrast enhancement is to process images to increase the visibility of the features of interest. Since the human visual system has limited ability to detect small variations in pixel intensity or colour within homogeneous regions in an image, as discussed in section 2.5.3, these variations will be very difficult to perceive. For this reason, most contrast enhancement algorithms operate by amplifying local variations in colour or intensity within an image. One side effect of this sharpening process is that any noise which is present in an image is typically amplified too. Fortunately, most contrast enhancement methods have parameters which can be manually adjusted to specify the amount of enhancement to be performed, so the appropriate contrast level for image display can be obtained.

Processing colour images presents several challenges for noise removal and contrast enhancement methods which were originally devised for monochromatic images. Care must be taken to retain chromatic information, particularly hue, whenever possible. Methods which operate on colour channels separately are in most danger of introducing unwanted shifts in colour. Techniques which manipulate image intensity, or work directly with vector-valued pixels are most effective in this regard.

8.1 NOISE REMOVAL

The process of digitizing colour images often introduces random variations in pixel values due to imperfect electronics or optics. These errors are collectively known as **noise**. When the magnitude of image noise is low, it is typically not a problem for most image analysis applications. When noise is moderately high, images begin to look rough or grainy. When the magnitude of noise is greater than the magnitude of the image itself, it often becomes difficult to recognize anything in the image.

The most obvious way to reduce the effects of noise is to improve the digitizing process. Manufacturers of digital cameras and scanning devices have made significant improvements in this area over the last ten years. Unfortunately, many image analysis applications require the use of low illumination or high magnification where problems associated with image noise persist. For this reason, noise removal continues to be an important image processing task. Before we can describe algorithms for noise removal, we must have a better understanding of what noise looks like in an image.

8.1.1 Noise models

The simplest noise model is **additive noise**. Each pixel in the observed image $g(x,y)$ is modelled as a combination of the true image $f(x,y)$ plus a noise image $n(x,y)$. In most cases, noise can be modelled as a random variable with zero mean and a Gaussian distribution. Other possible noise models use uniform, Poisson, or exponential distributions. Impulse noise consisting of isolated intensity spikes can also occur in an image. For colour images, noise is typically added to each colour component separately. In RGB space, the resulting model would be:

$$g_r(x,y) = f_r(x,y) + n_r(x,y)$$
$$g_g(x,y) = f_g(x,y) + n_g(x,y)$$
$$g_b(x,y) = f_b(x,y) + n_b(x,y)$$

The level of noise is typically expressed by its variance. For example, if a colour image has been digitized with RGB values in the range $0..255$, additive Gaussian noise with $\sigma_n^2 = 1$ would be almost impossible to detect visually. Moderate noise levels with $\sigma_n^2 = 100$ begin to look grainy, while high levels with $\sigma_n^2 = 10\,000$ obscure the original image considerably. Similar effects are observed for other noise distributions as variance increases. This is demonstrated in Plate I.   The range of pixel intensities in the underlying image plays an important role in how additive noise is perceived. For this reason the **signal-to-noise ratio** (SNR) is often used to characterize noise. SNR is defined in decibels (dB) as

$$\text{SNR} = 10\log_{10}\left(\frac{\sigma_f^2}{\sigma_n^2}\right)$$

where the variance of the image $f(x,y)$ is given by σ_f^2. When this value is not known, it can be approximated by squaring the range of intensities in the image. Using this definition, the variance for the images in Plate I would be $65\,025$ and the SNRs of the three images with Gaussian noise would be 48, 28 and 8 dB respectively.

8.1.2   Binomial smoothing

Averaging is one way to remove noise in an image. When is possible to obtain K independent observations of an image $f(x,y)$ which has been corrupted with zero mean Gaussian noise with variance σ_n^2, the average of these observations

$$\bar{g}(x,y) = \frac{1}{K}\sum_{k=1}^{K} g_k(x,y) = f(x,y) + \bar{n}(x,y)$$

where

$$\bar{n}(x,y) = \frac{1}{K}\sum_{k=1}^{K} n_k(x,y)$$

will have zero mean Gaussian noise with variance $\sigma_{\bar{n}}^2 = \frac{1}{K}\sigma_n^2$. The SNR of the averaged image will increase by $10\log_{10}(K)$, which in many cases may make the noise unobservable.

When separate observations of $g(x,y)$ are not available, noise can still be reduced by averaging K values in the neighbourhood of each point (x,y) in the noisy image. This will smooth the image and remove part of the noise. Weighted averages of pixels can also be easily computed. For example, if we use an $M \times M$ array of weights $m(x,y)$, a smoothed image can be obtained using the following:

$$\bar{g}(x,y) = \sum_{i=0}^{M-1}\sum_{j=0}^{M-1} m(i,j)g(x-i,y-j) = m(x,y) * g(x,y)$$

This double summation is known as **convolution**, and $*$ is called the convolution operator. The matrix $m(x,y)$ is typically known as the convolution **mask** in this context. The size and contents of the mask will determine how much smoothing takes place. Normally, a small mask will remove part of the noise and cause little smoothing, whereas a large mask will remove most of the noise and produce a very smooth image. Hence, it is possible to trade off noise and smoothness to suit the needs of specific applications.

Variable amounts of smoothing may be achieved by repeatedly convolving an image with a convolution mask whose weights correspond to binomial coefficients:

$$m(x,y) = \begin{array}{|c|c|c|} \hline 1 & 2 & 1 \\ \hline 2 & 4 & 2 \\ \hline 1 & 2 & 1 \\ \hline \end{array}$$

Convolution is associative, that is:

$$m(x,y) * [m(x,y) * g(x,y)] = [m(x,y) * m(x,y)] * g(x,y)$$

so smoothing an image twice with $m(x,y)$ would be the same as convolution with the following 5×5 mask:

$$\begin{array}{|c|c|c|} \hline 1 & 2 & 1 \\ \hline 2 & 4 & 2 \\ \hline 1 & 2 & 1 \\ \hline \end{array} * \begin{array}{|c|c|c|} \hline 1 & 2 & 1 \\ \hline 2 & 4 & 2 \\ \hline 1 & 2 & 1 \\ \hline \end{array} = \begin{array}{|c|c|c|c|c|} \hline 1 & 4 & 6 & 4 & 1 \\ \hline 4 & 16 & 24 & 16 & 4 \\ \hline 6 & 24 & 36 & 24 & 6 \\ \hline 4 & 16 & 24 & 16 & 4 \\ \hline 1 & 4 & 6 & 4 & 1 \\ \hline \end{array}$$

Notice that the central values of these masks are higher than their neighbours, and they gradually fall off towards the perimeter. Binomial convolution masks are also approximately circularly symmetric which is desirable because it ensures that smoothing will be invariant to image rotation.

To ensure that the average intensity in an image remains the same after smoothing, the weights in a convolution mask need to be normalized by dividing by their sum. For example, the 3×3 and 5×5 masks should be normalized by $\frac{1}{16}$ and $\frac{1}{256}$ respectively. Since the weights in the 3×3 mask are powers of two, bitwise shift operations can be used instead of multiplications and division when applying the weights to pixels in the image. Thus, software and hardware implementations of binomial smoothing are typically very fast. The application of binomial smoothing to Gaussian and impulse noise removal is shown in Plate II.

8.1.3 Gaussian smoothing

Operations which can be defined using convolution alone are called **linear**. Among the many important properties of linear operators is the fact that convolution in the spatial domain can be implemented by using filtering in the frequency domain. If we let $M(u,v) = \mathcal{F}\{m(x,y)\}$ represent the Fourier transform of the mask $m(x,y)$ and let $G(u,v) = \mathcal{F}\{g(x,y)\}$ represent the Fourier transform[1] of the noisy image $g(x,y)$, the smoothed image can be computed as:

$$\bar{g}(x,y) = \mathcal{F}^{-1}\{M(u,v)G(u,v)\}$$

It is important to note that the $N \times N$ images $M(u,v)$ and $G(u,v)$ are multiplied together on a point-by-point basis, not using matrix multiplication. Hence, the bulk of the computational effort associated with this approach is spent performing the Fourier transforms. The fast Fourier transform (FFT) algorithm is a divide-and-conquer method which performs $O(N^2 \log_2 N)$ steps to process an $N \times N$ image. Convolution with an $M \times M$ mask requires $O(N^2 M^2)$ operations. Hence, for large M, it is often faster to perform filtering in the frequency domain than it is to use convolution. The precise break-even point depends on implementation details, but M is commonly in the range $[10..15]$.

One interesting property of repeated convolution with any smoothing mask is that the effective convolution mask eventually converges to a Gaussian function. Hence, instead of applying 25 iterations of binomial smoothing, we can convolve the image once with a Gaussian convolution mask. Gaussians have another property which makes them a popular choice for frequency domain smoothing. The Fourier transform of a Gaussian with standard deviation σ is another Gaussian with standard deviation $\alpha = 1/2\pi\sigma$ and

---

[1] Fourier transforms are discussed in Chapter 12.

amplitude $A = \sqrt{2\pi\sigma^2}$. Gaussian smoothing can be performed using the following frequency domain filter:

$$M(u,v) = A\exp\left(\frac{u^2 + v^2}{-2\alpha^2}\right)$$

Different amounts of smoothing can be obtained by varying the standard deviation. Unlike binomial convolution, the time needed to perform Gaussian filtering remains fixed.

To remove noise from colour images using either convolution or filtering, the natural choice is to apply a linear smoothing operation to each colour component separately. This is very effective when the additive noise can be modelled using a uniform or Gaussian distribution. This approach is less successful for impulse noise. When an image contains impulse noise, most of the image is noise free, and isolated points are significantly different from the underlying image. Linear smoothing in this case averages noise pixels with adjacent noise free pixels. This reduces the magnitude of the noise pixel, but it corrupts the colour and intensity of adjacent points. The application of Gaussian smoothing to Gaussian and impulse noise removal is illustrated in Plate III.

8.1.4   Median filtering

One problem with linear smoothing is that it tends to spread out errors caused by impulse noise rather than removing them from an image. In addition, the edges of objects are often over-smoothed while noise is removed. Both of these issues can be addressed by applying **median filtering** (Tukey, 1974). As the name suggests, median filtering operates by calculating the median value within a specified neighbourhood in an image and using this value in the output image. This is accomplished by collecting K pixel values in the neighbourhood of pixel (x,y) and sorting these values. The value in location $K/2$ in the sorted list is the median.

For images with uniform or Gaussian noise, the median value is often very close to the weighted average obtained using linear smoothing, so median filtering reduces noise and increases SNR as desired. When impulse noise is present in an image $g(x,y)$, noise pixels are almost always at the beginning or end of the sorted list, and seldom appear at location $K/2$. Consequently, the median value is a better estimate of the original image $f(x,y)$ than any local average which contains the noise pixel. For this reason, median filtering is the preferred method for impulse noise removal.

Median filtering also does a better job of retaining the edges of objects in an image than linear smoothing. If we consider the K pixels in the neighbourhood of an edge point (x, y), some fraction A will lie inside the object and the remainder $(1 - A)$ will be from the object's background. When A is greater than 0.5, the median of the K pixels will be a value which is representative of the object. Correspondingly, when A is less than 0.5, the median will be chosen from among the pixels which lie outside the object. Thus, as we process points in the neighbourhood of an edge, the transition between object pixel values and background pixel values is retained in the median filtered image.

Median filtering can be extended to colour images in a number of ways. The easiest approach is to perform median filtering on each colour component separately. Unfortunately, the resulting RGB combination may not correspond to any colour pixel in the filtering neighbourhood. Chromatic shifts are often visible, particularly near edges.

A better approach is to treat RGB values at each point as vectors and calculate the **vector median** (Astola, Haavisto and Neuvo, 1990). Since there is no natural way to sort vectors in RGB space, the vector median is computed using the property that the sum of distances between all vectors and the median is less than the sum of the distances to any other vector in the set:

$$S_m = \sum_{i=1}^{K} \|v_m - v_i\| < \sum_{i=1}^{K} \|v_j - v_i\|, \quad \text{for} \quad \forall j \neq m$$

When vectors are in RGB space, the Euclidean distance between values is given by:

$$\|v_i - v_j\| = \sqrt{(v_{i,r} - v_{j,r})^2 + (v_{i,g} - v_{j,g})^2 + (v_{i,b} - v_{j,b})^2}$$

where $v_{i,r}$ is the red component of v_i and so on. Since N^2 distance calculations are needed to determine the minimum value for S associated with v_m, computing the vector median can be relatively time consuming. One way to save time is to approximate the vector median using the pixel which is closest to the mean value $\bar{v}$ in the specified neighbourhood (Valavanis, Zheng and Gauch, 1991):

$$S_m = \|v_m - \bar{v}\| < \|v_j - \bar{v}\|, \quad \text{for} \, \forall j \neq m$$

As with convolution, the shape and size of the sorting neighbourhood determines how smoothing is performed. For example, 'plus sign' shaped neighbourhoods are often more effective for retaining object corners than circular neighbourhoods. Also, larger neighbourhoods result in more smoothing than small neighbourhoods. For Gaussian noise, median filtering is comparable to binomial smoothing or Gaussian smoothing. On the other hand, median filtering often outperforms binomial smoothing and Gaussian smoothing for removing impulse noise, particularly for colour images. This is demonstrated in Plate IV. Vector filtering is discussed in more detail in Chapter 10 and also in Chapter 11.

8.1.5   Other smoothing methods

Median filtering is one example of a nonlinear image processing technique. Since there is sorting involved, there is no way to express median filtering in terms of convolution. Because median filtering is very successful in excluding noise points when obtaining smoothed pixels, there has been considerable effort to develop other nonlinear smoothing methods with similar objectives. Although a full history of nonlinear algorithms is beyond the scope of this book, two approaches are particularly noteworthy: **anisotropic diffusion** and **mathematical morphology**.

The central idea behind anisotropic diffusion is to vary the size and shape of smoothing neighbourhoods in different parts of the image based on image content. Since edges are important for visual and automatic image analysis, it is desirable to prevent smoothing which blurs edges. This can be accomplished by applying smoothing parallel to edges and not perpendicular to edges. Although it is possible to define anisotropic Gaussian masks at each point to smooth an image, this approach is computationally infeasible. Alternative formulations of anisotropic smoothing have been proposed which operate by iteratively solving nonlinear partial differential equations (Perona and Malik, 1990; Shah, 1991). This nonlinear smoothing approach has also been extended to vector-valued images, with very good results (Pien and Gauch, 1995; Whitaker and Gerig, 1994).

Mathematical morphology encompasses a broad range of nonlinear image processing operations which are based on set theory (Serra, 1982; Dougherty, 1993). Operations which are related to image smoothing are **erosion** and **dilation**, which operate by treating the image as a surface and passing a **structuring element** over the image to fill in the dents or remove peaks. Combinations of erosion followed by dilation, or dilation followed by erosion can be very effective for noise removal if the size and shape of the structuring element are carefully chosen. Mathematical morphology was originally developed to process binary or greyscale images. Extending this approach to vector valued colour images requires special care to avoid hue shift artifacts.

An approach to morphological processing of colour images is presented in Chapter 11.

8.2 CONTRAST ENHANCEMENT

Image contrast is a measure of sharpness. An image with high contrast will typically have large intensity or colour variations separating different objects in an image, so it is easy to visually locate object boundaries and distinctive features within objects. An image with poor contrast has variations which are gradual and difficult to detect visually. Contrast enhancement methods amplify the local intensity or colour variations within an image, thereby increasing feature visibility. Since there are a number of causes for poor image contrast, a variety of enhancement algorithms have been devised. In this section, we describe representative algorithms from three very different approaches: point operators; linear operators; and nonlinear operators.

8.2.1 Windowing

A point operator is an image processing transformation that determines the value of a pixel in the output image $g(x,y)$ by the value of a single pixel in the input image $f(x,y)$. Contrast enhancement can be performed by defining a mapping function $g(x,y) = m(f(x,y))$ such that local intensity or colour variations in the output image are greater than in the input image.

When the objects of interest in an image all lie within a known range of intensities $[I_{min}..I_{max}]$, there is little need to display image pixels which lie outside this range. Pixels whose intensities are greater than I_{max} can be mapped to the maximum display value D_{max}. Similarly, pixels below the minimum intensity I_{min} are mapped to the minimum display value D_{min}. Intensities within the specified range can be **windowed** to utilize the entire display range using the following formula:

$$m(i) = D_{min} + (i - I_{min}) \left(\frac{D_{max} - D_{min}}{I_{max} - I_{min}} \right)$$

This has the effect of amplifying the differences between adjacent pixels which lie within $[I_{min}..I_{max}]$, so contrast is enhanced for these regions. Pixels above and below this range are mapped to a single intensity so contrast is clearly lost in these regions. Low contrast details in different intensity ranges in an image can be explored by interactively adjusting I_{min} and I_{max}.

Windowing can be applied to colour images by selecting a desired volume of colour space $[R_{min} .. R_{max}, G_{min} .. G_{max}, B_{min} .. B_{max}]$ and performing windowing on each of the colour channels separately. One potential problem with this approach is that the colours in the output image may be very different from the colours in the input image. For example, different shades of green may be mapped to a range of colours from yellow to red, depending on which window of RGB values is chosen. Images of natural scenes such as trees would look very unnatural as a result.

When hue or saturation in an image have been incorrectly captured or accidentally corrupted, it is common to transform the image into IHS space and perform interactive windowing or scaling of hue or saturation to restore the image. This is one application where changing chromatic information is beneficial.

8.2.2 Histogram equalization

The mapping function for windowing is linear. This causes the contrast to be uniformly enhanced only within the specified range. In certain images, there may be several intensity ranges which need to be enhanced by different amounts. To process these images, it is possible to define the necessary mapping functions interactively using piecewise linear functions or low order polynomials. A less labour-intensive approach is to define the mapping function automatically based on properties of the intensity histogram.

For a monochromatic image, the intensity histogram $h(i)$ is defined to be the number of pixels in the image with intensity equal to i.

$$h(i) = \sum_{(x,y)} \begin{cases} 1 & \text{if} \quad f(x,y) = i \\ 0 & \text{otherwise} \end{cases}$$

If we consider an image containing several large objects with nearly uniform intensities, the corresponding histogram would have several peaks near the mean intensities of these objects. To enhance contrast within these objects, we must apply a mapping function which increases the separation of intensities near peaks in the histogram. This will widen and lower the peaks in the histogram and make small intensity variations within these objects more visible. If this contrast stretching process is applied to all peaks in the initial histogram, the histogram of the output image will be flattened. Hence, this contrast enhancement approach is called **histogram** equalization (Hummel, 1975).

To implement histogram equalization, we must define a mapping function $m(i)$ which yields a flat histogram. The first step is to calculate the cumulative histogram $H(i)$, which is defined to be the number of pixels in the image with intensity less than or equal to i. $H(i)$ can be easily obtained by numerically integrating $h(i)$:

$$H(i) = \sum_{j=0}^{i} h(j)$$

When a particular $h(i)$ is flat the corresponding $H(i)$ will be linear. Thus, to perform histogram equalization, we need to make the cumulative histogram linear. This can be accomplished using the following mapping function:

$$m(i) = D_{min} + D_{max} \frac{H(i)}{H(I_{max})}$$

When output pixels are computed using $g(x,y) = m(f(x,y))$, the range of display intensities is used more effectively, making small image variations easier to see.

To amplify contrast even further, local properties of an image must be considered. One way to do this is to divide the image into small regions and calculate the $h(i)$, $H(i)$, and $m(i)$ for each region separately. Enhancement can then be done in each region separately or by interpolating region histograms to obtain a mapping function $m(i)$ for each point in the image. The amount of enhancement performed at each point is determined by the size of the regions used to calculate $h(i)$ and $H(i)$. Smaller regions yield greater enhancement. This contrast enhancement approach is called **adaptive histogram equalization** (AHE) (Pizer, 1983; Gauch, 1992).

When histogram equalization and AHE are applied to each of the colour channels in an image separately, chromatic information in an image is often distorted. Therefore, it is preferable to convert the image to IHS space and apply enhancement only to image intensity I. The image can then be transformed back to RGB space for display. Colour image enhancement using histogram equalization and AHE are shown in Plate V.

8.2.3   Unsharp masking

Earlier we saw that linear operators can be used to remove noise and small details in an image by smoothing. As the amount of blurring increases, more and more detail is removed from the image until eventually the output image is almost uniform.

Since the goal of contrast enhancement is to increase the visibility of small details in an image, the information removed when smoothing an image is precisely what we want to see in an enhanced image. Thus, one way to perform contrast enhancement is to subtract a smoothed version of an input image from the original data. This approach is known as **unsharp masking**. When a single application of binomial smoothing is used for this purpose, the resulting 3×3 convolution mask is given by:

$$m(x,y) = \frac{1}{16} \begin{array}{|c|c|c|} \hline -1 & -2 & -1 \\ \hline -2 & 12 & -2 \\ \hline -1 & -2 & -1 \\ \hline \end{array}$$

Notice that the sum of the convolution mask weights in this case is equal to zero. This causes the average intensity of an image after unsharp masking to have zero mean. For colour images with RGB values in the range $[0..255]$, output values are normally in the range $[-128..127]$. Windowing must be applied to bring these values back into the desired range. To avoid potential hue shifts associated with windowing, it is important to apply the same windowing function to all three colour channels, rather than defining a mapping function for each channel separately.

Multiple applications of binomial smoothing also yield convolution masks with a positive central value surrounded by negative weights. Since more image detail is removed as blurring increases, subtracting a very blurred image from the original will yield an image very similar to the input image. Conversely, if an image which is only slightly smoothed is subtracted, the resulting image will contain mostly noise and very fine image details. Therefore, the amount of contrast enhancement associated with unsharp masking is inversely proportional to the amount of blurring performed prior to subtraction. The use of unsharp masking for contrast enhancement is illustrated in Plate VI.

8.2.4   Constant variance enhancement

The amount of contrast enhancement provided by unsharp masking can be further amplified by considering the local statistical properties of pixels in an image. If we let our input image be given by $f(x,y)$, an image smoothed with a convolution mask containing all 1's can be viewed as an estimate of the mean intensity at each point. Hence, unsharp masking produces an output image $g(x,y)$ using:

$$g(x,y) = f(x,y) - \bar{f}(x,y)$$

The visual significance of local changes in intensity may vary considerably in different parts of an image. For example, if one object in an image is highly textured while another object is relatively smooth, a value $g(x,y) = 10$ might represent a modest change or a significant difference in intensity depending on the pixel location. If we compare the value of $g(x,y)$ with a local estimate of the standard deviation of intensity values, we can determine how significant a particular value of $g(x,y)$ is. This is the principle behind **constant variance enhancement** (CVE).

If we calculate local variance in the neighbourhood of each point using

$$\sigma(x,y) = \sum_{(i,j)} \left(f(x-i, y-j) - \bar{f}(x,y) \right)^2$$

an image with high contrast can be obtained by calculating:

$$g(x,y) = \frac{f(x,y) - \bar{f}(x,y)}{\sigma(x,y)}$$

Since this formula tends to over-enhance an image, the amount of enhancement can be reduced by adding back a fraction k of the average image. Thus, the general formula for CVE becomes:

$$g(x,y) = \frac{f(x,y) - \bar{f}(x,y)}{\sigma(x,y)} + k\bar{f}(x,y), \qquad k \in [0..1]$$

Other statistical methods are also possible which take the global mean or standard deviation of the input image into consideration (Wallace, 1976). When any of these CVE approaches are applied to colour images, it is desirable to apply the same contrast boost to each colour component in the image. Averaging the local standard deviations in each colour channel is one way to obtain $\sigma(x,y)$. Separate estimates for the means should be used for each colour component. The use of CVE to enhance colour images is shown in Plate VII.

8.2.5 Other enhancement methods

In addition to windowing and histogram equalization, a number of other point operations are useful for enhancing contrast. For example, the mapping function $m(i)$ can be a function which emphasizes low values, such as

$m(i) = \log(i)$, or high intensities, such as $m(i) = i^p$ with $p > 1$. These formulas are best applied in situations where the distribution of pixel values is known *a priori*.

By looking at linear image processing methods in the frequency domain, any filter which reduces the contribution of low frequencies and boosts high frequencies will emphasize small details in an image. A number of high-pass and band-pass filtering techniques have been developed which achieve this objective. Detailed descriptions of these filtering approaches can be found in any digital image processing reference, e.g. Gonzalez and Woods (1992). Overall, the behaviour of these linear methods is comparable to unsharp masking provided care is taken to select filters which do not cause ringing or other artifacts in the spatial domain.

Certain combinations of linear methods and point operations have also proved to be successful for enhancing image contrast. Of particular note is **homomorphic filtering**, which applies a point operation to the Fourier transformation of an image prior to filtering to reduce the effects of varying illumination in an image (Stockham, 1972). Although this approach is very successful for monochromatic images, it produces unsatisfactory results when applied to each colour component in an image separately. This is again a situation where the image must be converted to IHS space to enhance only image intensity.

9

Segmentation and edge detection

John M. Gauch

The goal of image segmentation is to partition an image into disjoint regions which correspond to objects of interest in the image. For example, if we have a digital image of an apple and several leaves lying on the ground, a natural partitioning would be to have all red pixels labeled as 'apple', all green pixels labeled as 'leaf' and the remaining brown pixels labeled as 'ground'. If the objects in the image have very distinct colours and are well separated, this task is relatively straightforward. On the other hand, if the scene has many complex objects with less distinct colours, the problem becomes very challenging. For example, when leaves change colours in autumn this approach might incorrectly classify red leaves as 'apple' and brown leaves as 'ground'. Objects may also contain gradual variations of colour, which must be considered when performing segmentation.

In many ways, edge detection can be considered the dual of image segmentation. Instead of finding the regions associated with various objects, the goal of edge detection is to find the boundaries of objects of interest. Once the edges have been found, the interior can be filled in to obtain the region associated with an object. The primary hypothesis used in edge detection is that there is a change in pixel colour or intensity at pixels on the boundary of the object. When this change is large relative to pixel variations within objects, edge detection is relatively easy. For example, the colour change from pixels on an apple to pixels on a leaf or the ground would be very large. When adjacent objects have similar colours and intensities the problem of locating boundary points becomes much more difficult. For example, finding the edges of two overlapping leaves of approximately the same colour can be quite challenging.

Once image segmentation or edge detection has been performed, it is often possible to perform computer-based analysis of the position, size or shapes of objects in an image. This information can be directly useful in applications where object measurement is the primary task, for example, grading the sizes of fruit while they are moving down a conveyor belt. In order to make use of this type of quantitative information, it is important to evaluate 'quality' of the image segmentation or edge detection algorithm you use with the particular class of images you are processing. As we shall see, many algorithms have one or more parameters which control their sensitivity to regions or edges. These parameters can be tuned to achieve the highest quality image segmentation or edge detection results for the images being processed.

9.1   PIXEL-BASED SEGMENTATION

As discussed above, colour is one of the easiest ways to identify which pixels belong to which object. When we associate a range of colours with each of the expected objects in the image, we can perform a colour lookup operation to determine what object label should be recorded for each pixel in the image. There are a number of ways to calculate and represent the colour lookup function and each has its relative advantages and disadvantages.

9.1.1   Object colour specification

Before we can perform pixel-based segmentation, we need to examine the distribution of colours within an image. The colour histogram is the most common approach used for this purpose. It is a discrete approximation to the probability density function for colours in an image. A colour histogram $h(c)$ is calculated by counting the number of pixels in the image $I(x,y)$ with each distinct colour c:

$$h(c) = \sum_{(x,y)} \begin{cases} 1 & \text{if } I(x,y) \equiv c \\ 0 & \text{otherwise} \end{cases}$$

To normalize for image area it is common to divide each value of $h(c)$ by the number of pixels in the image. The size of the colour histogram depends on the number of colours in the input image and also the needs of the application using the colour histogram. In the case where we are given an 8-bit pseudo-colour image as input, the number of colours in the image has been drastically reduced *via* colour quantization. We can represent the colour histogram of such an image using an array of 256 counters.

When we are given a 24-bit RGB image with 8 bits for each of red, green and blue, a complete colour histogram would require a $256 \times 256 \times 256$ array of counters. This would require far more memory than is typically available, so it is common to discard the least significant bits of each colour byte to reduce the size of the colour histogram. For example, by using the most significant four bits for each red, green and blue value, a $16 \times 16 \times 16$ array can be used (4096 counters). Another popular choice is to use three bits for red and green and two bits for blue, resulting in an $8 \times 8 \times 4$ colour histogram (256 counters). This reduction in space does come at a cost. By calculating the histogram using quantized colour values we lose information about the subtle variations in colour which occur in most natural images.

In applications where user input can be easily obtained, it is possible to have the user specify the associations between colours and different objects in the image. There are two basic approaches used to obtain this information: the user can select colours of interest from the colour histogram, or the user can specify typical colours by outlining objects in the input image. Both approaches yield excellent colour information, but require user input.

If the colour histogram is displayed visually as a bar chart or a scatter plot, the user can manually select histogram entries which correspond to objects of interest. For example, in our apple and leaf image we would expect significant peaks in the colour histogram near 'apple red' and 'leaf green'. Images with multiple objects with distinct colours will result in images with clearly separated histogram peaks, but images with very small objects or non-uniformly coloured objects will produce colour histograms whose peaks will be less distinct, and more difficult to identify visually.

Another way to specify the colour associated with each object is to have the user manually outline regions in the image which contain representative pixel colours for each object of interest. It is not essential to outline the whole object since the goal is to obtain a sample of typical object colours. This approach avoids problems associated with colour histogram display and also enables users to specify colours of small objects whose histogram peaks are hard to identify.

9.1.2   Distance-based pixel classification

Once the range of colours in the colour histogram or in the image have been specified for each object in the image, a distance-based pixel classification function can be defined. The first step of this process is to calculate the mean colour $\bar{c}_o$ for each object in the image. If we are given an image region R_o

containing a sample of pixels for each object o, the mean can be calculated using:

$$\bar{c}_o = \sum_{(x,y) \in R_o} I(x,y)$$

If a collection of histogram entries H_o has been selected for object o, the mean colour can be calculated using:

$$\bar{c}_o = \sum_{c \in H_o} c \cdot h(c)$$

Once mean colours $\bar{c}_o$ have been calculated for all objects, the process of classifying pixels in the image reduces to the problem of finding which mean colour is closest in colour space to the given pixel. If pixel $I(x,y)$ has colour c in RGB space, this distance is given by:

$$d_o = \sqrt{(c_r - \bar{c}_{o,r})^2 + (c_g - \bar{c}_{o,g})^2 + (c_b - \bar{c}_{o,b})^2}$$

The point (x,y) is classified as being part of object o if its colour is closer to $\bar{c}_o$ than to any other object colour. As the number of object colours to be considered increases, the classification time increases linearly. Notice that the classification of nearby pixels has no effect on the classification of pixel (x,y). Hence, if we observe a single red pixel in the middle of a large green leaf region, it will be incorrectly identified as part of an apple.

This classification method will yield slightly different results if different colour spaces are used to represent pixel colours. For example, if IHS colour space is chosen, the expected range of values for hues are $H \in [0 .. 2\pi]$ whereas intensity has a range of $I \in [0 .. 1]$. As a result, differences in hue have more weight in the distance between colours. For this reason, a perceptually uniform colour space such as CIELAB may be more appropriate.

Distance-based classification essentially divides colour space into O disjoint regions where all points in region o are closest in colour space to $\bar{c}_o$. In areas where the mean colours are relatively close together, these regions can be quite small. On the other hand, certain distributions of mean colours may result in very large portions of colour space being assigned to a single object class. This may produce very large differences in classification error for different objects based on their colours and the colours of other objects.

One solution to this problem is to restrict the size of the colour space region about each colour mean using either a distance cutoff or a bounding box. For example, we could say that only colours within a distance D of colour $\bar{c}_o$ belong to object o. Any points whose minimal distance $d_o(c) > D$ could then be classified as 'unknown'.

Alternatively, we could say that all colours within the range

$$c \in [(\bar{c}_{o,r} - D, \bar{c}_{o,g} - D, \bar{c}_{o,b} - D) .. (\bar{c}_{o,r} + D, \bar{c}_{o,g} + D, \bar{c}_{o,b} + D)]$$

are associated with object o. Points which are not within one of the bounding boxes in colour space would again be classified as 'unknown'. The bounding box approach can be implemented using only six comparisons, so it is typically faster than using a distance cutoff, even when the square root calculation is avoided. Another advantage is that the width of the bounding box in each colour dimension can be adjusted to reflect the dynamic range of each colour channel. For example, if colours are represented in IHS coordinates, the width of the bounding box in the hue dimension can be made 2π wider than in the intensity dimension.

9.1.3   Maximum likelihood pixel classification

In order to automate pixel-based segmentation, a method is needed to select the colour ranges associated with each object in an image. In cases where the colour histogram has distinct peaks, a natural choice would be to select the colours at these peaks as our estimates of $\bar{c}_o$. Unfortunately, the direct approach of scanning the colour histogram for the N largest local maxima does not always work. Consider our apple and leaf example. If the three largest peaks in the colour histogram are $h(c_1) = 40$, $h(c_2) = 35$ and $h(c_3) = 20$, this algorithm would choose c_1 and c_2 as the colours of the two major objects. If c_1 and c_2 are very similar colours of green, and c_3 is red, then this automatic colour selection method will not separate the image into leaf and apple regions.

To avoid this problem, we need to explicitly consider the position and shape of peaks in the colour histogram when selecting our estimates of $\bar{c}_o$. This can be accomplished using **maximum likelihood** (ML) pixel classification (Duda and Hart, 1973). Assume that an image contains a mixture of K objects with relative areas given by w_k. If the colour range of each object is modeled as a Gaussian distribution with mean μ_k and colour covariance matrix $\mathbf{V}_k$, then the expected distribution of colours in the image can be represented as a sum of weighted Gaussians as:

$$g(c) = \sum_{k=1}^{K} w_k p_k(c)$$

where

$$p_k(c) = a_k \exp\left[-\frac{1}{2}(c - u_k)^T \mathbf{V}_k(c - u_k)\right]$$

and

$$a_k = (2\pi)^{-\frac{3}{2}}|\mathbf{V}_k|^{-\frac{1}{2}}.$$

The colour covariance matrix is defined as:

$$\mathbf{V}_k = \begin{bmatrix} V_{r,r,k} & V_{r,g,k} & V_{r,b,k} \\ V_{g,r,k} & V_{g,g,k} & V_{g,b,k} \\ V_{b,r,k} & V_{b,g,k} & V_{b,b,k} \end{bmatrix}$$

where the diagonal terms specify the variance of each RGB colour component in object k, and the off-diagonal terms give the cross correlation of colour components.

When the number of objects K and the distribution parameters w_k, u_k, and $\mathbf{V}_k$ have been chosen correctly, the observed colour histogram $h(c)$ should be very similar to the estimated colour distribution $g(c)$. Now we must quantify this similarity. If we assume that all pixel values in the image are independent, the **likelihood** for the image given the distribution parameters is given by:

$$G\left(I| \vec{w}, \vec{u}, \vec{\mathbf{V}}\right) = \prod_{(x,y)} g(I(x,y)).$$

When $g(c)$ matches $h(c)$ the value for G will be maximized. Thus, the search for distribution parameters can be written as an iterative optimization problem. Given an initial estimate of w_k, u_k, and $\mathbf{V}_k$, it is possible to refine parameters for each class k and colour component $i, j \in (r, g, b)$ using the following:

$$w_k = \sum_{(x,y)} p_k(I(x,y))$$

$$u_{i,k} = \sum_{(x,y)} p_k(I(x,y)) I(x,y)i / w_k$$

$$V_{i,j,k} = \sum_{(x,y)} p_k(I(x,y))[I(x,y)i - u_k i][I(x,y)j - u_k j]/w_k$$

The stopping criteria for this iterative process can be based on the rate of change of the likelihood function G or some fixed number of iterations can be used.

Once the distribution parameters w_k, u_k, and V_k have been calculated, pixels in the image can be classified by determining which object k has the highest probability $p_k(I(x,y))$. This is very similar to selecting the object k whose mean value u_k is closest to the intensity $I(x,y)$ with one very important difference. The covariance matrix V_k specifies the shape and width of the region in colour space which contains colours most likely to belong to object k. This feature is particularly useful when segmentation is performed in a colour space with different ranges of values in different colour dimensions (e.g. IHS colour space). For this reason, ML pixel classification is a big improvement on the bounded box classification approach, although at greater computational expense. The use of ML segmentation is demonstrated in Fig. 9.1 using an image of a red ball on a table. Both the ball and the background have variations in illumination which make colour distance-based pixel segmentation very difficult. On the other hand, ML segmentation can be quite effective, particularly when the correct colour space is chosen. In Fig. 9.1(b) the image is segmented into three ML classes based on intensity alone. Notice that the bright red portions of the ball and bright blue pixels in the background are assigned to the same class. Figure 9.1(c) demonstrates the use of a colour ML algorithm in RGB space. The ball and the background are no longer assigned to the same class, although the ball is divided into bright and dark regions. The use of CIELAB colour space is shown in Fig. 9.1(d). Here the ball and its reflection on the table are assigned to a single segmentation class.

9.2   REGION-BASED SEGMENTATION

Pixel-based segmentation suffers from the fundamental problem that spatial information is ignored when classifying pixels. Isolated noise or minor colour fluctuations within an object may cause points inside an object to be improperly classified. Our example where a red pixel inside a leaf region is labelled as 'apple' is a result of this problem. A more serious problem arises when the colours in an object vary gradually across the object. Here, a statistical model consisting of a mean colour vector u_k and covariance matrix V_k may be insufficient to describe the object. Segmentation algorithms which focus on regions of pixels can avoid these problems.

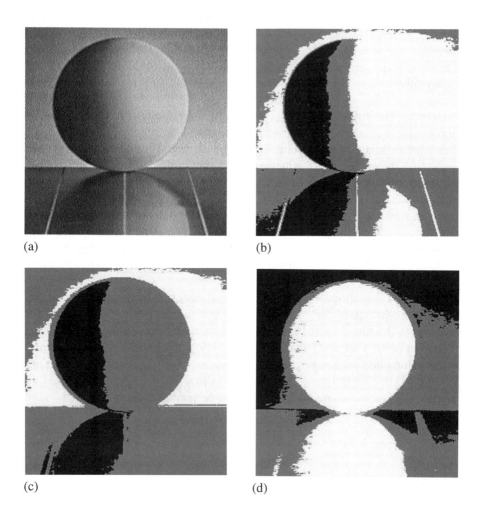

Fig. 9.1 (a) A simple colour image. See Plate VIII for a colour version. (b) Regions generated by the maximum likelihood (ML) algorithm applied to image intensity. (c) Regions generated by a colour ML algorithm in RGB space. (d) Regions generated by a colour ML algorithm in CIELAB space. The number of output classes for each case is three. Visually CIELAB space provides the best results.

9.2.1 Homogeneity measurement

The first step in developing a region-based segmentation algorithm is to decide which properties you wish regions in the segmented image to possess.
One natural choice is to have the pixels within each region be **homogeneous**.
For example, we expect most pixels within a leaf region to be a similar shade
of green. We can also expect a small percentage of points within the leaf to
have different colours as a result of natural variations. If we locate a region of
pixels which meet this criterion we can assign the label 'leaf' to the whole region, including the isolated points within the region which are not leaf green.
A successful region-based segmentation algorithm is one which partitions the
input image into disjoint regions all of which meet an appropriate homogeneity criterion.

There are several popular measurements of region homogeneity. The simplest measurement is similar to the bounding box classification described in
section 9.1.2. If all points within an image region lie within a small region
of colour space, the image region can be declared homogeneous. This can be
quantified by measuring the volume B_r of the bounding box in colour space
which contains all pixels within the region r. If the image is in RGB space
this can be calculated using:

$$B_r = (\max_r(r) - \min_r(r))(\max_r(g) - \min_r(g))(\max_r(b) - \min_r(b)).$$

A similar measurement calculates the radius R_r of the colour space sphere
which contains all of the colours of pixels in the image region. The centre
of the sphere is typically chosen to be the mean colour $\bar{c}_r$ of the region and
the radius is calculated by finding the pixel whose colour is furthest in colour
space from this mean. The Euclidean distance metric is typically used for this
purpose.

Perhaps the most common region homogeneity measurement is variance.
For colour images, we can calculate the colour covariance matrix $\mathbf{V}_r$ as described in section 9.1.3 and use either the trace $tr(\mathbf{V}_r)$ or determinant $det(\mathbf{V}_r)$
of this matrix to quantify region homogeneity. The trace is simply the sum
of the diagonal elements corresponding to the variance of each of the colour
components in the region, so it is often chosen for its computational simplicity.

If a region contains pixels from two or more objects with very different
colour ranges, the values of B_r, R_r and $tr(\mathbf{V}_r)$ will be large and the region
will be identified as not homogeneous. Each of these measures also depends
on the colour space chosen to represent the image. To devise a region homogeneity measure which corresponds well to what the human visual system

considers homogeneous, it would be wise to consider converting images from RGB space to IHS or CIELAB.

9.2.2   Seed-based region growing

Seed-based region growing is a bottom-up segmentation approach (Yakimovsky, 1976). A seed point within an object of interest is selected, and adjacent pixels are added to the region as long as the region satisfies the desired homogeneity property. The output of this procedure is a single connected region in the image. To fully partition the image into N regions, seed points in each region must be selected and the region growing process must be repeated N times.

The selection of seed points for region growing is often accomplished by manually selecting points within objects of interest. Although this is labour intensive, it does ensure that the resulting regions meet the needs of the application. An alternative is to automatically scan the image for seed points based on some expected properties of regions of interest. For example, if objects are brighter than their background, local intensity maxima could be used as seed points.

Once a seed point (x,y) has been chosen, the neighbours of that point $(x+1,y)$, $(x-1,y)$, $(x,y+1)$, and $(x,y-1)$ must be examined to see which belong in the region. If we say that all pixels whose colour is within radius R_{max} of the mean region colour $\bar{c}_r$ are part of the region, then these points should be added to the region and their neighbours considered. A new value for the mean region colour $\bar{c}_r$ can also be computed. As the region grows, the list of adjacent pixels will also grow. Eventually, the region will stop growing when all of the neighbouring pixels lie outside the colour radius R_{max}.

Although region growing is conceptually simple, there are several subtle points to consider. First, how should the list of adjacent pixels be represented and processed when adding points to the region? One approach would be to iteratively search the list of points to find the point which is **closest** to the mean colour and add this point to the region. To reduce searching time, points could be stored in a priority queue with priority based on the colour distance of the pixel to the current mean colour $\bar{c}_r$. Another approach is to perform one scan of the neighbour list and add all points to the region which meet the homogeneity criteria. For large regions, this approach will be much faster. Depending on which seed point is chosen and the order in which neighbours are examined there may be slight differences in the final region.

A second issue is the selection of the initial mean colour $\bar{c}_r$ and the colour radius R_{max}. In images where the texture of objects varies significantly, it may be desirable to base the initial value of $\bar{c}_r$ on some small neighbourhood of pixels about the seed point. This estimate will be more robust to image noise and our choice of an initial seed point. For R_{max}, it may be reasonable to base this value on the colour variance in this neighbourhood. When an object is relatively smooth, the corresponding radius will be low. When an object has significant texture, a higher value of R_{max} will be used. If this approach is taken, care must be taken to select seed points which are not near the boundary of the object of interest, since the colour variance near object boundaries is much larger than in the interior of the object. Seed-based region growing is illustrated in Fig. 9.2, which depicts a bowl of fruit on a table. The colours of the fruit in this image are visually distinct yet they have significant variations in intensity due to illumination effects. In Fig. 9.2(b) we show an example of intensity-based region growing using a seed point within the green grapes. Notice that the bright portions of the grapes are incorrectly excluded and much of one lemon is incorrectly included. Figure 9.2(c) uses the same seed point and a colour region growing algorithm in RGB space. Much more of the grapes are correctly selected and the lemon is no longer part of the region. Finally, Fig. 9.2(d) shows the grape region obtained using region growing in CIELAB colour space. The results in this case are far superior to the preceding examples, demonstrating that seed-based region growing can be very effective if the correct colour space and thresholds are chosen. A detailed evaluation of segmentation algorithms in different colour spaces should be performed for each class of images being processed (Gauch and Hsia, 1992).

9.2.3 Recursive split-merge segmentation

Recursive split-merge segmentation is a top-down approach (Horowitz and Pavlidis, 1976). Unlike region growing, which starts with a single pixel, the split-merge algorithm begins by considering the whole image as a single region. If this region is not homogeneous according to the selected metric, the region is **split** into subregions. If the initial region is a rectangle with $(x,y) \in [x_L .. x_H, y_L .. y_H]$ the most common approach is to divide this region into four rectangles with:

$$(x,y)_1 \in [x_L .. x_M, y_L .. y_M] \qquad (x,y)_3 \in [x_M .. x_H, y_L .. y_M]$$
$$(x,y)_2 \in [x_L .. x_M, y_M .. y_H] \qquad (x,y)_4 \in [x_M .. x_H, y_M .. y_H]$$

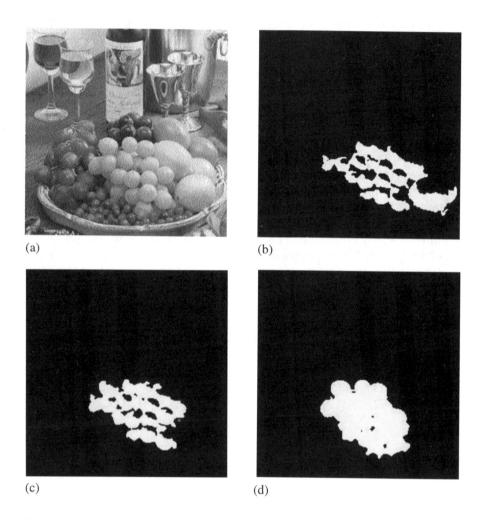

Fig. 9.2 (a) A simple colour image. See Plate IX for a colour version. (b) Region generated by seed-based region growing with an intensity distance threshold = 10. (c) Region growing in RGB space with colour distance threshold = 25. (c) Region growing in CIELAB space with colour distance threshold = 15. The same seed point was used for each method and thresholds were chosen interactively to obtain the most sensible region in each colour space.

where $(x_M, y_M) = ((x_L + x_H)/2, (y_L + y_H)/2)$. Each subregion is recursively processed in an identical manner until the image has been split into S rectangular regions which are homogeneous.

One side effect of this splitting process is that an image is typically over-segmented. This is because we split each rectangle into four equally sized subregions without checking to see if a smaller number of regions would suffice. For example, if the initial region really consists of two homogeneous objects $(x, y)_A \in [x_L \cdots x_H, y_L \cdots y_M]$ and $(x, y)_B \in [x_L \cdots x_H, y_M \cdots y_H]$, then the four regions above should be combined in the final segmented image.

Rather than trying to control the splitting process to avoid this problem, it is easier to perform a second pass over the initially segmented image to **merge** all adjacent regions which remain homogeneous when combined. In some sense this is similar to the process used when performing region growing, except that the building blocks in the merging process are much larger and can be of any shape.

The most critical aspect of region merging is keeping track of which regions are adjacent to each other. If unique region identifiers are assigned to each homogeneous region during the splitting process, a list of adjacent region pairs can be obtained with a single scan over the image. Once adjacent regions have been identified, the homogeneity function for each potential merger must be computed. For example, if we have region A with mean $\bar{c}_A$ and variance $\mathbf{V}_A$ and region B with mean $\bar{c}_B$ and variance $\mathbf{V}_B$, we need to calculate the combined mean $\bar{c}_{A \cup B}$ and variance $\mathbf{V}_{A \cup B}$ and compare $tr(\mathbf{V}_{A \cup B})$ to the merging threshold to see whether $A \cup B$ is homogeneous and thus whether the regions should be merged. When two regions are merged, region identifiers in the region adjacency list need to be updated accordingly.

The order in which we perform region merging can have a significant effect on the final segmentation results. As with region growing, there are two alternatives. We could search the entire region adjacency list for the two regions which are most homogeneous, and merge these regions first. After each merger, the entire list is searched to find the next pair of regions to combine. The computational expense of this approach is its most serious drawback. A faster alternative is to perform a limited number of passes over the adjacency list and merge any two regions which meet the merger criteria. The disadvantage of this approach is that we are no longer guaranteed to merge the most similar regions first. On the other hand, the quality of the M final regions produced is sufficient for most applications.

The homogeneity measure used for region splitting does not necessarily need to be the same as the criterion used for merging. For example, variance-based splitting can be combined with variance based merging using a low threshold for splitting and a slightly higher threshold for merging. Another alternative is to combine regions based on the distance in colour space between their mean colours $\bar{c}_A$ and $\bar{c}_B$. If this distance is below a specified threshold, the regions are merged, otherwise they are not. For both region growing and split-merge the colour space used to measure homogeneity may play a role in the quality of the segmentation results.

The application of recursive split-merge segmentation to our red ball image is shown in Fig. 9.3. The regions in Fig. 9.3(b) were calculated using an intensity-based algorithm and a variance threshold selected interactively to obtain the most visually sensible results. Notice that the ball is divided into many small regions which have very square boundaries. Slight improvement is observed in Fig. 9.3(c) where the colour segmentation algorithm described above is applied in RGB space. The background is divided into fewer regions, but the ball itself is still over-segmented. In Fig. 9.3(d) the image was segmented in CIELAB space. Although this provides the best split-merge results, there are still significant artefacts along the boundary of the ball. To avoid these problems, different schemes for splitting the image and merging regions have been explored. Watershed-based segmentation is one approach which has been successful for this purpose (Beucher, 1982; Gauch, 1997).

9.3   EDGE DETECTION AND BOUNDARY TRACKING

In many image analysis applications, boundaries of objects are of particular interest. For example, the size or shape of the outline of an object can often be used to recognize an object or detect abnormalities. For this reason, a large number of edge detection methods have been devised to locate the boundaries of objects in images based on local pixel properties. The main challenge all methods must face is how to distinguish edges from non-edges. Discovering how edge points are connected together is an equally important problem which is often overlooked. Although the methods described below make use of images in RGB space, they can be applied in any colour space with reasonable results.

9.3.1   Gradient-based edge detection

If two objects with different colours are adjacent to each other in an image, there will be a large change in colour as we move from one object to the

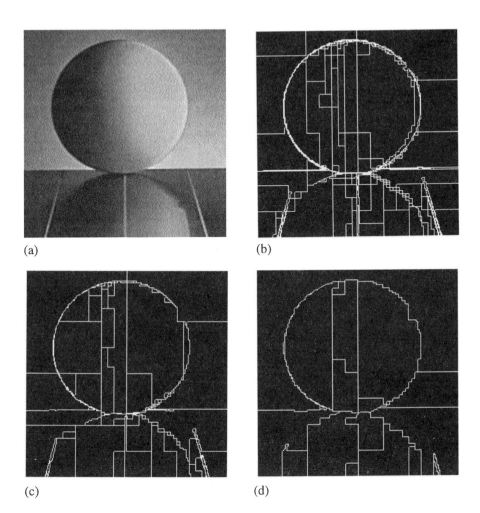

(a)   (b)

(c)   (d)

Fig. 9.3   (a) A simple colour image. See Plate VIII for a colour version. (b) Region boundaries generated using recursive split merge segmentation in RGB space with region variance threshold = 500. (c) Region boundaries using variance threshold = 1000. (d) Region boundaries using variance threshold = 2000. The number of regions decreases as the variance threshold increases. In all three cases square region artefacts can be seen.

next. Hence, boundary detection can be accomplished by searching for colour discontinuities in an image.

For a one-dimensional function $I(x)$ the derivative dI/dx measures the rate of change in intensity. When this value is above a given threshold, we can say that the function is discontinuous. Similarly, for a two-dimensional function $I(x,y)$ the rate of change in brightness is given by the gradient $\nabla I(x,y) = (\partial I/\partial x, \partial I/\partial y) = (I_x, I_y)$. When the gradient magnitude $|\nabla I(x,y)| = (I_x^2 + I_y^2)^{\frac{1}{2}}$ is above a given threshold, the function can be considered discontinuous and the point (x,y) can be labelled as an edge point (Roberts, 1965).

For colour images, the calculation of the gradient magnitude is somewhat more complex since the image is vector-valued. If we consider each channel of an RGB image separately, we can calculate $\nabla R(x,y) = (R_x, R_y)$, $\nabla G(x,y) = (G_x, G_y)$, and $\nabla B(x,y) = (B_x, B_y)$. The rate of colour change at (x,y) can then be estimated by summing the magnitudes of these gradients. Hence, for a colour image $I(x,y)$ we can evaluate:

$$\left(R_x^2 + R_y^2\right)^{\frac{1}{2}} + \left(G_x^2 + G_y^2\right)^{\frac{1}{2}} + \left(B_x^2 + B_y^2\right)^{\frac{1}{2}} > T$$

to determine whether (x,y) is an edge point or not.

An alternative is to calculate the distance in colour space between each pixel (x,y) and its neighbours using the Euclidean metric. For example, the colour difference between $I(x,y)$ and $I(x+1,y)$ is $C_x = \left(R_x^2 + G_x^2 + B_x^2\right)^{\frac{1}{2}}$. Similarly the difference in the y direction is $C_y = \left(R_y^2 + G_y^2 + B_y^2\right)^{\frac{1}{2}}$. If we let $\nabla I(x,y) = (C_x, C_y)$ for a colour image, then the gradient magnitude is $|\nabla I(x,y)| = \left(R_x^2 + G_x^2 + B_x^2 + R_y^2 + G_y^2 + B_y^2\right)^{\frac{1}{2}}$. This value can be compared to the threshold T to select edge points.

Gradient-based edge detection works well whenever object boundaries have high contrast. On the other hand, gradual colour transitions across an object boundary may result in poorly localized edges or no edges at all depending on the value of the threshold T. When the value of T is too high, we may fail to detect visually important edges. If T is too low, too many points may be identified as edges, causing edges to appear thickened. One solution to this problem is to manually select an appropriate T for each image. Another possibility is to automatically select the value of T based on the statistical properties of the colour gradient magnitude for an image. For example, T can be chosen so that only $N\%$ of pixels in the image are classified as edges.

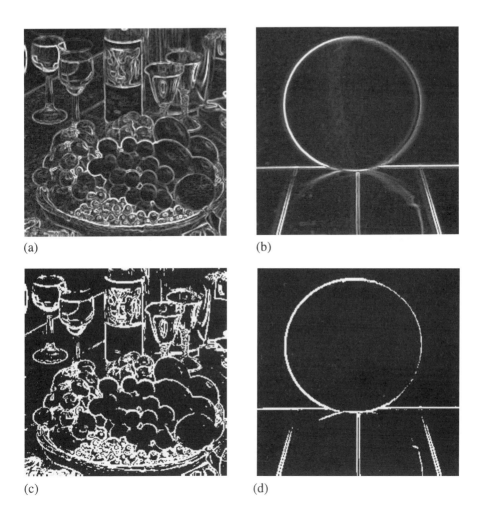

(a)   (b)

(c)   (d)

Fig. 9.4   (a) Gradient magnitude of fruit image (Plate IX) in CIELAB space. (b) Gradient magnitude of ball image (Plate VIII) in RGB space. (c) Edge points whose gradient magnitudes are in the top 20% for fruit image. (d) Edge points whose gradient magnitudes are in the top 20% for ball image. Notice that the edges are thick in some places and broken in others.

Figure 9.4 illustrates gradient-based edge detection for both the ball (Plate VIII) and fruit (Plate IX) images. Gradient magnitudes have been calculated using the Euclidean metric and scaled so that zero is shown in black and the maximum value appears white in Fig. 9.4(a) and (b). Edges in these images appear slightly blurred because the partial derivatives for each colour channel have been estimated using the Sobel operator, which corresponds to convolution with the following 3×3 masks:

$$
I_x = \begin{array}{|c|c|c|} \hline -1 & 0 & 1 \\ \hline -2 & 0 & 2 \\ \hline -1 & 0 & 1 \\ \hline \end{array}
\qquad
I_y = \begin{array}{|c|c|c|} \hline -1 & -2 & -1 \\ \hline 0 & 0 & 0 \\ \hline 1 & 2 & 1 \\ \hline \end{array}
$$

When derivatives are estimated using simple finite differences, edges are sharper but the effects of image noise are more visible. In Fig. 9.4(c) and (d) we identify edge pixels by selecting gradient magnitude thresholds such that the brightest 20% of the pixels are labelled as edges. Although the outlines of major objects have been found, the edges are thicker than one pixel wide in many places and discontinuous in others. Adjusting thresholds manually may reduce these artifacts, but this is a fundamental shortcoming of this edge detection approach.

9.3.2   Laplacian zero crossings

A problem with gradient-based edge detection is that not all object boundaries have high gradient magnitudes. For example, if an object is slightly out of focus the colour transition between an object and the background may be gradual. Edges in such cases can be found by locating local maxima in the gradient magnitude function. Since the gradient is a first-order derivative, one way to find extrema in this function is to detect the zeros in an appropriate second-order image derivative. The Laplacian operator $\nabla^2 I(x,y) = I_{xx} + I_{yy}$ has been used extensively for this purpose since it is relatively simple to calculate and is rotationally invariant like the gradient magnitude.

If finite differences are used to approximate partial derivatives, the gradient of an image $I(x,y)$ can be computed using: $I_x \cong I(x+1,y) - I(x,y)$ and $I_y \cong I(x,y+1) - I(x,y)$. Similarly, the Laplacian can be calculated using: $I_{xx} \cong I(x+1,y) - 2I(x,y) + I(x-1,y)$ and $I_{yy} \cong I(x,y+1) - 2I(x,y) + I(x,y-1)$. By creating a temporary image $T(x,y) = I_{xx} + I_{yy}$, the locations of extrema in the gradient magnitude can be found by scanning $T(x,y)$ in the x

and y directions for changes in sign. This approach for locating edges is very successful, but it does have two shortcomings which need to be addressed.

First, zero crossings in a second-order derivative indicate an extremum in the first-order derivative, but not necessarily a local maximum. Hence, many Laplacian zero crossings are **phantom edges** which really correspond to local minima in gradient magnitude. In theory, this can be resolved by examining the sign of the third derivative, but in practice images contain too much noise for this to be successful. Instead, it is common to use a low threshold on the gradient magnitude to distinguish edge points from phantom edges.

Image noise presents a second problem. Differentiation amplifies noise much like high pass filtering, so estimates of the Laplacian can be very noisy. Hence, the Laplacian zero crossings for an image may have numerous false edges caused by noise. One way to avoid this problem is to perform Gaussian smoothing prior to calculating the Laplacian zero crossings (Marr and Hildreth, 1980). The degree of smoothing applied can be adjusted by varying the width of the Gaussian filter. Other smoothing operators can also be applied with equal success.

Our discussion so far has dealt with scalar images. Since colour images are vector-valued, the Laplacian is also vector-valued. For example, the Laplacian for an image $I(x,y)$ in RGB space would be given by

$$\begin{aligned} \nabla^2 I(x,y) &= \left(\nabla^2 R(x,y), \nabla^2 G(x,y), \nabla^2 B(x,y)\right) \\ &= \left(R_{xx}+R_{yy}, G_{xx}+G_{yy}, B_{xx}+B_{yy}\right). \end{aligned}$$

The magnitude of the Laplacian can then be estimated by summing the Laplacians for each colour component. If a zero crossing of $T(x,y) = R_{xx}+R_{yy}+G_{xx}+G_{yy}+B_{xx}+B_{yy}$ is detected and the gradient magnitude is above a minimum threshold, we can say that the point (x,y) lies on an edge. This approach produces visually sensible edges for most images.

The application of Laplacian zero crossings to our fruit image is shown in Fig. 9.5. The Laplacian magnitude image in Fig. 9.5(b) was computed using the formula for $T(x,y)$ above and second derivative estimates based on a cubic spline interpolation of points in the neighbourhood of (x,y). Hence, the Laplacian for a single channel can be implemented *via* convolution with the following 3×3 mask:

$$I_{xx}+I_{yy} = \begin{array}{|c|c|c|} \hline 1 & -2 & 1 \\ \hline 4 & -8 & 4 \\ \hline 1 & -2 & 1 \\ \hline \end{array} + \begin{array}{|c|c|c|} \hline 1 & 4 & 1 \\ \hline -2 & -8 & -2 \\ \hline 1 & 2 & 1 \\ \hline \end{array} = \frac{1}{2} \begin{array}{|c|c|c|} \hline 1 & 1 & 1 \\ \hline 1 & -8 & 1 \\ \hline 1 & 1 & 1 \\ \hline \end{array}$$

(a)   (b)

(c)   (d)

Fig. 9.5 (a) A colour image of fruit. See Plate IX for a colour version. (b) The Laplacian for this image calculated in RGB space. (c) Laplacian zero crossings showing edges and phantom edges. (d) Removal of phantom edges by selecting zero crossings whose gradient magnitudes are in the top 20% for the image. Although the final edges are only one pixel wide, they are seldom closed.

The zero crossings of this function are shown in Fig. 9.5(c). Notice that there are numerous zero crossings which do not correspond to strong edges. In fact, many of these points are phantom edges. Fig. 9.5(d) shows Laplacian zero crossing pixels whose corresponding gradient magnitudes are in the top 20% for this image. This removes phantom edges but this process may also break the boundaries of objects if the threshold is not chosen carefully.

9.3.3 Directional derivative extrema

When we take a cross-section of the gradient magnitude image in the gradient direction, we obtain an edge strength profile which is perpendicular to the object boundary. The local maximum of this function can be found by locating the zeros of the derivative of this edge strength profile. Hence, edges in an image can be found by detecting the zero crossings of the directional derivative of the gradient magnitude in the gradient direction. For a scalar image $I(x,y)$ the directional derivative described above is given by (Haralick, 1984):

$$D_{\nabla I}|\nabla I| = (I_x, I_y) \left(|\nabla I|_x, |\nabla I|_y \right)$$

where

$$|\nabla I|_x = (I_x I_{xx} + I_y I_{yx}) / \sqrt{I_x^2 + I_y^2}$$

and

$$|\nabla I|_y = (I_x I_{xy} + I_y I_{yy}) / \sqrt{I_x^2 + I_y^2}.$$

We can again use finite differences to estimate first and second order partial derivatives and create a temporary image:

$$T(x,y) = \left(I_x^2 I_{xx} + 2 I_x I_y I_{xy} + I_y^2 I_{yy} \right) / \sqrt{I_x^2 + I_y^2}$$

which represents the directional derivative expanded from above. The locations of edges can be found by scanning $T(x,y)$ in the x and y directions for changes in sign. Since the denominator in this expression is always positive, it can be ignored without affecting the results. This edge detection approach yields results which are very similar to Laplacian zero crossings. It also suffers from phantom edges and noise sensitivity, but these problems can be solved using the techniques similar to those introduced for Laplacian zero crossings (Canny, 1986).

The main advantage of using directional derivatives comes when we consider vector-valued images $I(x,y)$. If we consider the colour difference between adjacent pixels when calculating the gradient magnitude we have $\nabla I(x,y) = (C_x, C_y)$ where $C_x = \left(R_x^2 + G_x^2 + B_x^2\right)^{\frac{1}{2}}$ and $C_y = \left(R_y^2 + G_y^2 + B_y^2\right)^{\frac{1}{2}}$. Substituting into the directional derivative formula we obtain:

$$D_{\nabla I}|\nabla I| = (C_x, C_y)\left(|\nabla I|_x, |\nabla I|_y\right)$$

where

$$|\nabla I|_x = (C_x C_{xx} + C_y C_{yx})/\sqrt{C_x^2 + C_y^2}$$

and

$$|\nabla I|_y = (C_x C_{xy} + C_y C_{yy})/\sqrt{C_x^2 + C_y^2}.$$

Once the values for C_x and C_y are calculated, we can use finite differences to estimate second order partial derivatives and create a temporary image:

$$T(x,y) = \left(C_x^2 C_{xx} + 2C_x C_y C_{xy} + C_y^2 C_{yy}\right)/\sqrt{C_x^2 + C_y^2}$$

Edge locations can be found by scanning $T(x,y)$ for zero crossings and correcting for phantom edges. Since the values of C_x and C_y represent the distance in colour space between adjacent pixels, these values will change if colour spaces other than RGB are used for edge detection. For example, if pixels are converted to IHS or a perceptually uniform colour space such as CIELAB, edge locations may improve visually for certain images.

In Fig. 9.6 we demonstrate directional derivative edge detection. In Fig. 9.6(a) and Plate X the input fruit image has been convolved with a Gaussian with standard deviation $\sigma = 2$ to remove noise and fine details in the image. The colour directional derivative in Fig. 9.6(b) was calculated using the formula for $T(x,y)$ above. Partial derivatives for each colour channel have been estimated using the Sobel operator. As a result of this blurring, the zero crossings shown in Fig. 9.6(c) have fewer phantom edges than the Laplacian zero crossings for the unsmoothed image in Fig. 9.5(c). Weak edges can again be removed in Fig. 9.6(d) by selecting points whose gradient magnitudes are in the top 20%. Notice, however, that smoothing has altered the shape of some objects, particularly those with sharp corners. For this reason, excessive Gaussian smoothing should be avoided, or smoothing algorithms which preserve edge locations should be applied.

(a)   (b)

(c)   (d)

Fig. 9.6 (a) A Gaussian smoothed image of fruit ($\sigma = 2$). See Plate X for a colour version. (b) The directional derivative of this image calculated in RGB space. (c) Zero crossings showing edges and phantom edges. (d) Removal of phantom edges by selecting zero crossings whose gradient magnitudes are in the top 20% for the image. Smoothing reduces the number of small object boundaries detected and also changes the shape of large object boundaries slightly.

9.3.4 Edge tracking

Once the set of pixels which lie on the boundary of an object have been identified using one of the edge detection methods described above, it is often useful to link these points together to form closed object boundaries. This can be accomplished by walking one pixel at a time clockwise (or counter clockwise) around the perimeter of the object moving from one edge point to the next. Although this sounds trivial, the decision of which point to move to at each step can be quite complex.

For example, assume that we are currently at position (x,y) and the two neighbouring points $(x+1,y)$ and $(x-1,y+1)$ are also edge points. Where should we move to? The answer depends on the coordinates of the previous point we visited. If we just moved from $(x+1,y)$ to (x,y), then clearly we should go to $(x-1,y+1)$ next. If an object is particularly narrow at one point, there may be three or more edge points adjacent to the point (x,y). This makes it difficult to base an edge tracking algorithm on an enumeration of all possible edge configurations.

A solution to this problem is to select location L_{n+1} given previous locations L_n and L_{n-1} on the edge using a clockwise (or counter clockwise) search of the 3×3 neighbourhood of L_n, starting at location L_{n-1} and stopping when the first edge point is encountered. For example, in the case where L_{n-1} is directly to the left of L_n, the seven remaining neighbours are searched in the following order:

$$\begin{bmatrix} 1 & 2 & 3 \\ L_{n-1} & L_n & 4 \\ 7 & 6 & 5 \end{bmatrix}$$

Similar searches are used for different $L_n - 1$ positions. This edge tracking procedure terminates when the entire boundary has been traversed and L_{n+1} has returned to the starting point L_0. Edge tracking using this approach will always succeed if the edge points on the object boundary are fully connected. If there are gaps in the edge, the tracking algorithm must be modified to 'jump over' the gaps or terminate with a partial object boundary.

9.4 SEGMENTATION AND EDGE DETECTION QUALITY METRICS

The quality of an image segmentation or edge detection algorithm depends on a wide range of factors including the type of images being processed and the needs of the application using the resulting regions or edges. Speed may also be an important consideration for real-time applications. For this reason, it is important to evaluate the quality of several approaches to find the 'optimal' method for a given application.

Visual inspection is often the simplest approach to measure quality. For example, if you wish to compare M segmentation algorithms using N typical images, you can compute the NM output images and display segmentation results for each input image separately. To obtain a numerical rating of each segmentation algorithm, you can have users rank the segmentation outputs on a scale from 1 to 10 and calculate the average quality ranking for each segmentation method. To reduce experimental bias, data should be collected from several individuals, making this approach very labour intensive.

It is possible to automate the algorithm assessment process if the 'true' object regions and/or object boundaries are known. One way to obtain this information is to generate synthetic images where the positions of typical objects in the image are known *a priori*. Another approach is to have an expert specify the true regions or boundaries by hand for a collection of typical images. The quality of segmentation or edge detection algorithms can then be evaluated by counting the number of correctly classified pixels C and the number of incorrectly classified pixels E and calculating the ratio $(C-E)/(C+E)$ for each of the N test images and each of the M algorithms. The method with the highest average ratio can be declared 'optimal' for this image analysis application.

10

Vector filtering

Konstantinos N. Plataniotis and
Anastasios N. Venetsanopoulos

Processing of colour image data has received increased attention lately due to the introduction of the various vector processing filters. These ranked-order type filters utilize the direction or the magnitude of the colour vectors (that is, pixel values within colour space) to enhance, restore and segment colour images. The objective of this chapter is to present and analyze the different vector processing filters. In addition to ranked-order filters, fuzzy, as well as nearest-neighbour vector filters are discussed in detail. Comparative studies involving colour images are used to assess the performance of the different vector processing filters. Results indicate that the adaptive vector processing designs reported here are computationally attractive and have excellent performance.

It is widely accepted that colour conveys information about the objects in a scene and this information can be used to further refine the performance of an imaging system. However, colour image processing has not had the same growth and development as other areas of digital signal processing. The multichannel nature of the colour image can be considered as the main reason for this slow development. The basis of the trichromatic theory of colour vision discussed in section 2.3 is that it is possible to match an arbitrary colour by superimposing appropriate amounts of three primary colours. Thus, in the various different colour spaces discussed in Chapter 4, each pixel of an image is represented by three values which can be considered as a vector. Thus the colour image can be thought of as a vector field in which the direction and length of each vector in colour space is related to the chromatic properties of the pixel (Trahanias, Pitas and Venetsanopoulos, 1994). Being a two-dimensional, three-channel signal, a colour image requires increased computation and storage during processing, as compared to a grey-scale image.

One of the most common image processing tasks is image filtering, that is the process of noise reduction in an image. Filtering is an important part of any image processing system whether the final image is utilized for visual interpretation or for automatic analysis (Venetsanopoulos and Plataniotis, 1995). Numerous filtering techniques have been proposed to date for colour image processing. Nonlinear filters applied to colour images are required to preserve edges and details, and remove impulsive and Gaussian noise. Edge information is very important for human perception. Therefore, its preservation and possibly its enhancement are very important subjective features of the performance of nonlinear image filters.

The early approaches to colour image processing usually comprise extensions of the scalar filters to colour images. In these approaches each channel of a colour image was considered to be a monochrome image itself. Thus, techniques developed for monochrome images were applied on each channel separately. Scalar order statistics techniques have played an important role in the design of robust filters for grey-scale (monochrome) image processing. This is due to the fact that, after ordering, any noise-corrupted input pixels (outliers) will be located in the extreme ranks of the sorted data array. Consequently, these abnormal values can be isolated and filtered out before the input signal is further processed. Ordering of scalar data, such as grey-scale images is well defined and has been extensively studied. Assuming that n random variables $x_1, x_2, \ldots, x_n$ are available they can be arranged in ascending order of magnitude as $x_{(1)}, x_{(2)}, \ldots, x_{(n)}$. Element $x_{(i)}$ is defined as the ith order statistic. The minimum $x_{(1)}$, the maximum $x_{(n)}$ and the median $x_{n/2}$ (n being assumed odd) are among the most important order statistics, resulting in the **min, max** and **median** filters respectively (Pitas and Venetsanopoulos, 1992).

However, the concept of input ordering cannot be easily extended to multichannel (multivariate) data such as colour image pixels, since there is no universal way to define ordering in vector spaces. A number of different ways to order multivariate data has been proposed. These techniques are generally classified into **marginal ordering** (M-ordering), where the multivariate samples are ordered along each one of their dimensions independently, **reduced** or **aggregate ordering** (R-ordering), where each multivariate observation is reduced to a scalar value according to a distance metric, **partial ordering** (P-ordering), where the input data are partitioned into smaller groups which are then ordered, and **conditional ordering** (C-ordering), where multivariate samples are ordered conditional on one of the marginal sets of observations (Barnett, 1976). In spite of this, most of the filters utilized today for colour image processing are based on the concept of multivariate ordering (vector order statistics) since it has been recognized by many researchers that vector processing is an effective way to filter out noise and to enhance colour images (Venetsanopoulos and Plataniotis, 1995).

10.1 THE VECTOR MEDIAN FILTER

The best known vector order statistics filter is the so called **Vector Median Filter** (VMF) (Astola, Haavisto and Neuvo, 1990). The VMF is a vector processing operator, which has been introduced as an extension of the scalar **Median Filter** (Pitas and Venetsanopoulos, 1992). The VMF can be derived either as a maximum likelihood estimate when the underlying probability densities are double-exponential or by using vector order statistics techniques.

In the former case, consider an m-dimensional distribution (DeGroot, 1989):

$$f(\mathbf{x}) = \gamma \exp(-\alpha|\mathbf{x} - \beta|) \tag{10.1}$$

where γ and α are scaling factors, and β is the location parameter of the distribution. The maximum likelihood estimate $\hat{\beta}$ based on the random samples $\mathbf{x}_1, \mathbf{x}_2, \ldots, \mathbf{x}_n$ from equation (10.1) can be obtained by maximizing the likelihood function:

$$L(\beta) = \prod_{i=1}^{n} \gamma \exp(-\alpha|\mathbf{x}_i - \beta|) \tag{10.2}$$

There is no closed form solution to equation (10.2). A suboptimal estimate can be found only under the assumption that $\hat{\beta}$ is one of the $\mathbf{x}_i$ if the additional requirement that $\hat{\beta}$ be one of the sample vectors is included. This leads to the definition of the vector median as:

$$\mathbf{x}_{VMF} \in (\mathbf{x}_i, i = 1, 2, \ldots, n) \tag{10.3}$$

with

$$\sum_{i=1}^{n} |\mathbf{x}_{VMF} - \mathbf{x}_i| \leq \sum_{i=1}^{n} |\mathbf{x}_j - \mathbf{x}_i|$$

The vector median of a population can also be defined as the minimal vector according to the aggregate, reduced ordering technique (R-ordering) of Barnett (1976), which leads again to equation (10.3). Let us assume that a window W of finite size n is available. The noisy image vectors inside the window W are denoted as $\mathbf{x}_j$, $j = 1, 2, \ldots, n$. If $D(\mathbf{x}_i, \mathbf{x}_j)$ is a measure of dissimilarity (distance) between vectors $\mathbf{x}_i$ and $\mathbf{x}_j$ the scalar quantity:

$$d_i = \sum_{j=1}^{n} D(\mathbf{x}_i, \mathbf{x}_j) \tag{10.4}$$

is the distance associated with the noisy vector $\mathbf{x}_i$ inside the processing window of length n.

Assuming that an ordering of the d_i's

$$d_{(1)} \leq d_{(2)} \leq \cdots \leq d_{(n)} \tag{10.5}$$

implies the same ordering to the corresponding $\mathbf{x}_i$'s:

$$\mathbf{x}_{(1)} \leq \mathbf{x}_{(2)} \leq \cdots \leq \mathbf{x}_{(n)}$$

then the VMF defines the vector $\mathbf{x}_{(1)}$ as the filter output. This selection is due to the fact that vectors which diverge greatly from the data population usually appear in higher indexed locations in the ordered sequence (Astola, Haavisto and Neuvo, 1990).

The VMF orders the colour input vectors according to their relative magnitude differences. The L_1 norm defined as:

$$d_p(i, j) = \sum_{k=1}^{m} |x_i^k - x_j^k|$$

where m is the dimension of the vector $\mathbf{x}_i$ and x_i^k is the kth element of $\mathbf{x}_i$, is utilized in most cases. However, according to the original definition proposed in Astola, Haavisto and Neuvo (1990) the L_2 norm (or **Euclidean** distance) defined as:

$$d_2(i, j) = \sqrt{\sum_{k=1}^{m} (x_i^k - x_j^k)^2}$$

can also be used to order the input vectors inside the processing window. Vector median filters can utilize either odd or even window lengths. However, in most cases we want to associate the output value with the centre value of the filter window, therefore the window size n is usually chosen to be an odd number of pixels, for example 3×3 or 5×5.

The vector ordering scheme employed by the VMF has a number of advantages. First, by utilizing a distance metric (e.g. the L_1 norm in this case) we can get an accurate and natural description of outliers (extreme values) by simply observing the values of the aggregate distance in equation (10.5). Colour vectors appearing in low ranks in the ordered sequence are vectors centrally located in the population, whereas colour vectors appearing in the higher ranks (outliers) diverge most from the data population. The ordering scheme can be used to determine the positions of the different input vectors without any *a priori* information regarding the signal or noise distributions. The ordering is entirely based on the data and is independent of an origin or fixed point in space. From a robust estimation point of view, this fact is very important, since for non-stationary signals, such as images, it is rarely possible to determine an appropriate fixed point.

It has been observed through experimentation that the VMF discards impulses and preserves edges and details in the image (Astola, Haavisto and Neuvo, 1990). However, its performance in the suppression of additive white Gaussian noise, which is frequently encountered in image processing, is inferior to that of the Arithmetic (linear) Mean Filter (AMF). If a colour image is corrupted by both additive Gaussian noise and impulsive noise, an effective filtering scheme should make an appropriate compromise between the AMF and the VMF. The so called α-trimmed VMF exemplifies this tradeoff. In this filter the $n(1-2a)$ samples closest to the vector median value are selected as inputs to an averaging-type filter. The output of the α-trimmed VMF can be defined as follows:

$$\mathbf{x}_{\alpha VM} = \sum_{i=1}^{n(1-2\alpha)} \frac{1}{n(1-2\alpha)} \mathbf{x}_{(i)}$$

with $\mathbf{x}_{(1)} \leq \mathbf{x}_{(2)} \leq \ldots \leq \mathbf{x}_{(n)}$. The trimming operation guarantees good performance in the presence of long-tailed or impulsive noise and helps in the preservation of sharp edges. On the other hand, the averaging operation causes the filter to perform well in the presence of short-tailed noise. Readers requiring a text on noise should see Pitas and Venetsanopoulos (1990). Interested readers might also read the paper by Viero, Vistamo and Neuvo (1994).

The class of filters based on order statistics is very rich. In addition to the two filters discussed above, it includes other filters such as the max/min vector filters or the *L*-Vector estimators. The *L*-Vector filter family is an important generalization of the VMF (Nikolaidis and Pitas, 1995) and is closely related to a large class of robust scalar estimators called the *L*-estimators. These robust filters have coefficients which can be chosen optimally according to the input noise probability function (Pitas, 1995).

The main reason behind the popularity and widespread use of vector order statistics based filters, such as the VMF is their simplicity. The computations involved however in the evaluation of the aggregated distances during ordering are extensive. Given a $n \times n$ processing window, $(n^2(n^2 - 1))$ distances must be computed at each window location. It is evident that the run time of the VMF heavily depends on the distance metric adopted to compute distances among the colour samples. The evaluation of the Euclidean distance requires the use of floating point operations with considerable computational overhead. On the other hand distance based on the first norm L_1 can be computed using integer arithetic with considerable computational savings (Paeth, 1995). Recently, fast approximations of the Euclidean distance have been proposed to speed up the calculations. According to Barni, Cappellini and Mecocci (1994) a linear approximation of the Euclidean norm can be used in equation (10.4) to calculate distances among the colour vectors. The new L_{2a} norm approximates the Euclidean norm by means of a linear combination of components. For vectors with positive components, as in the case of colour images, the new distance measure can be defined as follows:

$$d_p(i,j) = \sum_{k=1}^{3} \alpha_k \left| x_i^k - x_j^k \right|$$

where $\alpha_k = 1/2^{(k-1)}$, and x_i^k is the kth element of $\mathbf{x}_i$. From its definition, it is evident that the proposed approximated distance measure is more computationally effective than the classical Euclidean distance, and can considerably reduce the run time of the VMF. Vector median filter implementations based on the L_1 norm are considerably faster, although their run time is still too high for many practical applications. To speed up the filtering procedures, appropriate fast algorithms, such as running median algorithms can also be used (Pitas, 1993). In such an approach, distances which have already been calculated are not recomputed at each step. Thus, the number of distances to be evaluated can be reduced to $n(n^2 - n) + 0.5n(n - 1)$, resulting in a computational complexity of $O(n^3)$ which is significantly lower than the $O(n^4)$ of the original VMF implementation.

10.2 VECTOR DIRECTIONAL FILTERS

The vector median filter is not the only filter based on order statistics which can be used for processing of vector signals. A new type of vector processing filter was proposed recently (Trahanias, Pitas and Venetsanopoulos,

1994; Trahanias and Venetsanopoulos, 1993; Trahanias, Karakos and Venet-sanopoulos, 1996). The so-called Vector Directional Filter (VDF) family operates on the direction of the image vectors, aiming at eliminating vectors with atypical directions in the colour space. To achieve its objective the VDF utilizes the angle between the colour vectors (angular distance) to order the input vectors inside a processing window. As a result of this process a set of input vectors with approximately the same direction in the vector space is produced as the output set. Since the vectors in this set are approximately co-linear, a magnitude processing operation can be applied in a second step to produce the required filtered output.

10.2.1   The Basic Vector Directional Filter (BVDF)

The BVDF is a ranked-order, nonlinear filter which parallelizes the VMF operation. However, it employs a distance criterion, different from the L_1 norm used in the VMF. The distance criterion used is the angle between the two vectors (angular distance). This so called **vector angle criterion** defines the scalar measure:

$$a_i = \sum_{j=1}^{n} A(\mathbf{x}_i, \mathbf{x}_j) \tag{10.6}$$

with

$$A(\mathbf{x}_i, \mathbf{x}_j) = \cos^{-1}\left(\frac{\mathbf{x}_i \mathbf{x}_j^t}{|\mathbf{x}_i||\mathbf{x}_j|}\right), \tag{10.7}$$

as the distance associated with the noisy vector $\mathbf{x}_i$ inside the processing window of n pixels.

The output of the BVDF is that vector from the input set which minimizes the sum of the angles with the other vectors. In other words, the BVDF chooses the vector most centrally located without considering the magnitudes of the input vectors. The BVDF may perform well when the vector magnitudes are of no importance and the direction of the vectors is the dominant factor. However, this is usually not the case. In most colour image processing applications, the magnitudes of the vectors should also be considered. To improve the performance of the BVDF a generalized filter structure was proposed (Trahanias and Venetsanopoulos, 1993; Trahanias, Karakos and Venetsanopoulos, 1996). The new filter, called, appropriately, the Generalized Vector Directional Filter (GVDF) generalizes the BVDF in the sense that its output is a superset of the single BVDF output. Instead of a single output, the GVDF outputs the set of vectors whose angle from all other vectors is small as opposed to the BVDF which outputs the vector whose angle from all the other vectors is minimum. Thus, the output of the GVDF initially derives from a set of k input vectors with approximately the same direction in the colour space. Then, in the magnitude processing module, a final single output vector is produced by considering only the magnitude information. If prior information about the noise corruption is available, the designer may select the most appropriate magnitude processing method to maximize the performance of the filter. However, if information on the actual noise corruption is not available the applicability of the GVDF is questionable.

10.2.2   The Directional Distance Filter

To overcome the deficiencies of the GVDF a new directional filter was proposed (Karakos and Trahanias, 1995). The so-called Directional-Distance Filter (DDF) retains the structure of the BVDF, but utilizes a new distance criterion to order the vectors inside the processing window. Based on the observation that the BVDF and the VMF differ only in the quantity that is minimized, a new distance criterion was utilized by the designers of the DDF in the hope of deriving a filter which combines the properties of both these filters. Specifically, in the case of the DDF the distance inside the processing window W is defined as:

$$\beta_i = \left(\sum_{j=1}^{n} A(\mathbf{x}_i, \mathbf{x}_j) \right) \left(\sum_{j=1}^{n} ||\mathbf{x}_i, \mathbf{x}_j|| \right) \tag{10.8}$$

where $A(\mathbf{x}_i, \mathbf{x}_j)$ is the directional (angular) distance defined in equation (10.7) with the second term in equation (10.8) to account for the differences in magnitude in terms of the L_1 metric. As for any other ranked-order, multichannel, nonlinear filter, it is assumed that an ordering of the β_i 'distance':

$$\beta_{(1)} \leq \beta_{(2)} \leq \cdots \leq \beta_{(n)}$$

implies the same ordering to the corresponding input vectors $\mathbf{x}_i$:

$$\mathbf{x}_{(1)} \leq \mathbf{x}_{(2)} \leq \cdots \leq \mathbf{x}_{(n)}$$

Thus, the DDF defines the minimum order vector as its output: $\mathbf{x}_{DDF} = \mathbf{x}_{(1)}$.

10.2.3   Hybrid directional filters

The simultaneous minimization of the distance functions used in the designs of the VMF and the BVDF was attempted in the design of the DDF in order to obtain a filter that can smooth long-tailed distribution noise, such as Cauchy or Laplacian noise and preserve at the same time the chromaticity component of the image vectors (Pitas and Venetsanopoulos, 1990; Viero, Vistamo and Neuvo, 1994). Although the concept is appealing, and the resulting vector processing structure is simple, fast, and without the additional module of the GVDF there are a number of problems. Most notably, the 'distance measure' defined in equation (10.8) is heuristic, window-dependent, and it has no ties to the characteristics of the individual colour vectors. Furthermore, there is no analysis of the relative importance of the two components in the suggested 'distance function'. Thus, although the DDF can provide, in some cases, better results than those obtained by the BVDF or the GVDF, it cannot be considered as an effective, general purpose, nonlinear filter. However, its introduction inspired a new set of heuristic vector processing filters which tried to capitalize on the same appealing principle, namely the simultaneous minimization of the distance functions used in the VMF and the BVDF. Such a filter is the hybrid directional filter introduced by Gabbouj and Cheickh (1996). This filter operates on the direction and the magnitude of the colour vectors independently and then combines them to produce a unique final output. This hybrid filter, which can be viewed as a nonlinear combination of the VMF and BVDF filters, produces an output according to the following rule:

$$\mathbf{x}_{HyF} = \begin{cases} \mathbf{x}_{VMF} & \text{if } \mathbf{x}_{VMF} = \mathbf{x}_{BVDF} \\ \left(\dfrac{||\mathbf{x}_{VMF}||}{||\mathbf{x}_{BVDF}||} \right) \mathbf{x}_{BVDF} & \text{otherwise} \end{cases}$$

where $\mathbf{x}_{BVDF}$ is the output of the BVDF filter, $\mathbf{x}_{VMF}$ is the output of the VMF and $||.||$ denotes the magnitude of the vector.

Another more complex hybrid filter, which involves the utilization of the AMF, has also been proposed (Gabbouj and Cheickh, 1996). The structure of this so-called adaptive hybrid filter is as follows:

$$\mathbf{x}_{aHyF} = \begin{cases} \mathbf{x}_{VMF} & \text{if } \mathbf{x}_{VMF} = \mathbf{x}_{BVDF} \\ \mathbf{x}_{out1} & \text{if } \displaystyle\sum_{i=1}^{n} ||\mathbf{x}_i - \mathbf{x}_{out1}|| < \sum_{i=1}^{n} ||\mathbf{x}_i - \mathbf{x}_{out2}|| \\ \mathbf{x}_{out2} & \text{otherwise} \end{cases}$$

where

$$\mathbf{x}_{out1} = \left(\frac{||\mathbf{x}_{VMF}||}{||\mathbf{x}_{BVDF}||} \right) \mathbf{x}_{BVDF}$$

and

$$\mathbf{x}_{out2} = \left(\frac{||\mathbf{x}_{AMF}||}{||\mathbf{x}_{BVDF}||} \right) \mathbf{x}_{BVDF}$$

and $\mathbf{x}_{AMF}$ denotes the output of an AMF operating inside the same processing window. According to its designers, the magnitude of the output vector will be that of the mean vector in smooth regions and that of the median operator near edges.

Although these two hybrid filters attempt to parallelize the operation of the DDF, it is obvious from the definitions that these are only heuristic approximations without any relation to the characteristics of the individual colour vectors. The combination of magnitude and directional information is simplistic and has to be done at an algorithmic level resulting in an arbitrary structure which, contrary to the DDF, cannot be classified as a ranked-order nonlinear estimator. Apart from that, the hybrid filters (Gabbouj and Cheickh, 1996) are computationally demanding, since they require evaluation of both the VMF and VDF outputs. Given that the most computationally intensive part is that of vector sorting, the need to sort the input vectors twice with different sorting criteria makes the hybrid filters inappropriate for realistic image processing tasks.

10.3 ADAPTIVE VECTOR PROCESSING FILTERS

10.3.1 The framework

From the discussion above it is clear that the vector filters in use today originate from different points of view and thus their structure and properties vary widely. The large number of filters poses difficulties for the practitioner, since most of the filters are designed to perform well in a specific application and their performance deteriorates rapidly under inappropriate conditions. Thus, a nonlinear adaptive filter which performs equally well in a wide variety of applications would be of great importance. In particular, it is desirable to obtain a unifying nonlinear filter framework, which encompasses the different classes of existing nonlinear filters as special cases. Such a framework was introduced with the development of a new class of adaptive vector processing filters discussed in (Plataniotis, Androutsos and Venetsanopoulos, 1995; Plataniotis, Androutsos and Venetsanopoulos, 1996; Plataniotis, Androutsos and Venetsanopoulos, 1997; Bezdek and Pal, 1992; Dudani, 1977; Plataniotis, Androutsos, Sri and Venetsanopoulos, 1995; Plataniotis, Sri, Androutsos and Venetsanopoulos, 1996).

The adaptive filters proposed there utilize data-dependent coefficients to adapt to local image characteristics. The weights of the adaptive filters are determined using nonlinear functions of a distance criterion at each image position. Thus, the coefficients of the filter are not considered as constants, but are determined in an adaptive way. In this sense, the filter structure is data-dependent.

Once again, let us assume that a window W of finite size n (filter length) is available and that $\mathbf{x}_i$, $i = 1, 2, \ldots, n$ represents the input vectors inside the window. The general form of the adaptive filter is then given as a nonlinear transformation of a weighted average of the input vectors inside the window W (Plataniotis, Androutsos and Venetsanopoulos, 1995; Plataniotis, Androutsos and Venetsanopoulos, 1996; Plataniotis, Androutsos and Venetsanopoulos, 1997):

$$\hat{\mathbf{y}} = \sum_{i=1}^{n} w_i^s \mathbf{x}_i = \frac{\sum_{i=1}^{n} w_i \mathbf{x}_i}{\sum_{i=1}^{n} w_i} \tag{10.9}$$

In the adaptive design, the weights provide the degree to which an input vector contributes to the output of the filter. The relationship between the pixel under consideration (the window centre) and each pixel in the window should be reflected in the decision for the weights of the filter. Through the normalization procedure, two constraints necessary to ensure that the output is an unbiased estimator are satisfied. Namely:

1. Each weight is a positive number, $w_i^s \geq 0$,

2. The summation of all the weights is equal to unity, $\sum_{i=1}^{n} w_i^s = 1$.

The weights of the filter are determined adaptively using transformations of a distance criterion at each image position. These weighting coefficients are transformations of the aggregate distance between the centre of the window (the pixel under consideration) and all other samples inside the filter window. The transformation can be considered to be a membership function with respect to the specific window component. The adaptive algorithm assigns some membership function to a given vector inside the window and then uses the membership values within the window to calculate the final output.

The proposed structure is an adaptive, fuzzy design. It is adaptive since it utilizes local information to determine the filter's weights and fuzzy since the actual weights are calculated based on membership function strengths. Depending on the fuzzy membership and distance used, different filters can be devised.

In the framework described above, there is no requirement for fuzzy rules or local statistics estimations. Features extracted from local data, here in the form of a sum of distances, are used as inputs to the fuzzy weights. In the proposed methodology, the distance functions are not utilized to order input vectors. Instead, they provide selected features in a reduced space; features used as inputs for the fuzzy membership functions.

10.3.2   Distances and fuzzy weights

The most crucial step in the design of the adaptive filter is the determination of the membership function used for the construction of its weights. The fuzzy transformation is not unique. It usually depends on the specific distance measure that is applied to the input data. The different fuzzy functions must meet some desirable characteristics but mainly are required to have a smooth finite output range over the entire input range. Several candidate functions, such as triangular, trapezoidal, piecewise linear and Gaussian-like shapes can be used for this task. These functions are chosen by the designer arbitrarily, based on experience, problem specifications and computational constraints imposed by the design. Since the choice of the membership function form is very much problem dependent, the only applicable *a priori* rule is that the designer must confine himself to those functions which are continuous and monotonic.

In vector processing of colour images the main design objective is to select an appropriate fuzzy transformation, so that the pixel with the minimum distance, inside the window W, will be assigned the maximum weight. Although a number of different shapes can be used, the sigmoidal transformation, which is usually associated with inner product type of distances, is used as the membership function when the angular distance (sum of angles) is selected to calculate the distance among the vectors inside the processing window.

The fuzzy weight w_i in equation (10.9) has therefore, the following form:

$$w_i = \frac{\beta}{(1 + \exp(a_i))^r} \tag{10.10}$$

where β and r are parameters to be determined and a_i is the angular distance defined in equation (10.6). The value of r is used to adjust the weighting effect of the membership function, and β is a weight scale threshold. Since, by definition, the vector angle distance criterion delivers a positive number in the interval $[0, n\pi]$ (Trahanias, Karakos and Venetsanopoulos, 1996), the output of the fuzzy transformation introduced above produces a membership value in the interval

$$\left[\frac{\beta}{(1 + \exp(n\pi))^r}, \frac{\beta}{2} \right]$$

However, even for a moderate sized window, such as a 3×3 or 5×5 window, the lower limit of the above interval should safely be considered zero. As an example, for a modest 3×3 window and with $r = 1$ and $\beta = 2$ the corresponding interval is $[1.4 \times 10^{-12}, 1]$ and for a 5×5 window the interval becomes $[1.5 \times 10^{-35}, 1]$. Therefore, we can consider the above membership function as having values in the interval $(0, 1]$. It can easily be seen through simple calculations that the above transformation satisfies the design objective.

The angular distance utilized for the development of the vector directional filters is of course only one of the possible distance measures that can be used to measure dissimilarity between colour vectors. Another commonly used distance measure for vector ordering is the generic L_p metric (**Minkowski metric**), which is defined as follows:

$$d_p(i, j) = \left(\sum_{k=1}^{m} |x_i^k - x_j^k|^p \right)^{\frac{1}{p}} \tag{10.11}$$

where m is the dimension of the vector $\mathbf{x}_i$ and x_i^k is the kth element of $\mathbf{x}_i$. Three special cases of the L_p metric are of particular interest, namely, the **city-block distance**, the **Euclidean distance** and the **chessboard distance** (defined as $\max((x_i^1 - x_j^1), (x_i^2 - x_j^2), \ldots, (x_i^k - x_j^k)))$ that correspond to $p = 1, 2, \infty$ respectively. For the window centre (the vector under consideration) the sum of the distances to all the other vectors inside the window is the distance criterion. Let $d_p(i)$ correspond to the vector $\mathbf{x}_i$ defined as:

$$d_p(i) = \sum_{j=1}^{n} d_p(i, j)$$

where $d_p(i, j)$ is as in equation (10.11) and n is the number of input vectors inside the window. If input ordering is required, then an ordering of the $d_p(i)$'s as

$$d_{p(1)} \leq d_{p(2)} \leq \ldots \leq d_{p(n)}$$

implies the same ordering to the corresponding $\mathbf{x}_i$'s, namely:

$$\mathbf{x}_{(1)} \leq \mathbf{x}_{(2)} \leq \ldots \leq \mathbf{x}_{(n)}$$

If a generalized norm (L_p metric) is selected as the distance function, an exponential membership function can be used for the evaluation of the weights of the filter. The form of the fuzzy transformation can be defined as follows:

$$w_i = \exp\left[-\frac{d_p(i)^r}{\beta}\right] \tag{10.12}$$

where r is a positive constant, and β is a distance threshold. The parameters r and β are design parameters. The actual values of the parameters vary with the application. The above parameters correspond to the denominational and exponential fuzzy generators (Bezdek and Pal, 1992) controlling the amount of **fuzziness** in the fuzzy weight. It is obvious that since the distance measure is always a positive number the output of this fuzzy membership function lies in the interval $[0, 1]$. The fuzzy transformation is such that the higher the distance value, the lower the fuzzy weight. It can easily be seen that the membership function is 1 (maximum value) when the distance value is 0 and 0 (minimum value) when the distance value is infinite.

The fuzzy transformations of equation (10.10) and equation (10.12) are not the only way in which the adaptive weights of equation (10.9) can be constructed. In addition to fuzzy membership functions other design concepts can be utilized for the task. One such design already discussed by Plataniotis, Androutsos, Sri and Venetsanopoulos (1995) is the nearest-neighbour rule, in which the value of the weight w_i in equation (10.9) is calculated according to the following formula:

$$w_i = \frac{d_{(n)} - d_{(i)}}{d_{(n)} - d_{(1)}} \tag{10.13}$$

where $d_{(n)}$ is the maximum distance in the filtering window, measured using an appropriate distance criterion, and $d_{(1)}$ is the minimum distance, which is associated with the centremost vector inside the window.

As in the case of the fuzzy membership function, the value of the weight in equation (10.13) expresses the degree to which the vector at point i is close to the ideal, centremost vector, and far away from the worst value, the outer rank. Both the optimal rank position $d_{(1)}$ and the worst rank $d_{(n)}$ are occupied by at least one of the vectors under consideration. Plataniotis, Androutsos, Sri and Venetsanopoulos (1995) devised an adaptive vector processing filter named Adaptive Nearest-Neighbour Filter (ANNF) utilizing the general framework of equation (10.9). The weights in the ANNF were calculated by using the formula of equation (10.13) with the angular distance as a measure of dissimilarity between the colour vectors.

It is evident that the outcome of such an adaptive vector processing filter depends on the choice of the distance criterion selected as the measure of dissimilarity between vectors. As before, the (L_1) norm or the angular distance (sum of angles) between the colour vectors can be used to remove vector signals with atypical directions. However, both these distance metrics utilize only part of the information carried by the image vector. As in the case of the DDF it is anticipated that an adaptive vector processing filter based on an ordering criterion which utilizes both vector features, namely magnitude and direction, will provide a robust solution whenever the noise characteristics are unknown. To this end, a new distance measure was introduced (Plataniotis, Sri, Androutsos and Venetsanopoulos, 1996). The proposed distance measure for the noisy vector x_i inside the processing window of length n is defined as:

$$d_i = \sum_{j=1}^{n} (1 - S(\mathbf{x}_i, \mathbf{x}_j)) \tag{10.14}$$

with

$$S(\mathbf{x}_i, \mathbf{x}_j) = \left(\frac{\mathbf{x}_i \mathbf{x}_j^t}{|\mathbf{x}_i||\mathbf{x}_j|} \right) \left(1 - \frac{||\mathbf{x}_i| - |\mathbf{x}_j||}{\max(|\mathbf{x}_i|, |\mathbf{x}_j|)} \right)$$

As can be seen the similarity measure of equation (10.14) takes into consideration both the direction and the magnitude of the vector inputs. The first part of the measure in equation (10.14) is equivalent to the angular distance (vector angle criterion) of equation (10.7) and the second part is related to the normalized difference in magnitude. Thus, if the two vectors under consideration have the same length the second part of equation (10.14) equates to unity, and only the directional information is used. On the other hand, if the vectors under consideration have the same direction in the vector space (co-linear vectors) the first part (the directional information) equates to unity and the similarity measure of equation (10.14) is based only on the magnitude difference part.

Utilizing this similarity measure an adaptive vector processing filter based on the general framework of equation (10.9) and the weighting formula of equation (10.14) was devised by Plataniotis, Sri, Androutsos and Venetsanopoulos (1996). The so called Adaptive Nearest-Neighbour Multichannel Filter (ANNMF) belongs to the adaptive vector processing filter family defined through equation (10.9). However, the ANNMF combines the weighting formula of equation (10.13) with the new distance measure of equation (10.14) to evaluate its weights.

10.3.3   Comments

Although a number of different adaptive designs have been discussed above, all of them have some common design characteristics and exhibit similar behaviour. We summarize a number of them in a series of comments.

- All the adaptive vector processing filters discussed above perform smoothing of all vectors which are from the same region as the vector at the window centre. It is reasonable to make their weights proportional to the difference, in terms of a distance measure, between a given vector and its neighbours inside the operational window. At edges, or in areas with high details, the filter only smooths pixels on the same side of the edge as the centremost vector, since vectors with relatively large distance values will be assigned smaller weights and will contribute less to the final filter output. Thus, edge or line detection operations, prior to filtering, can be avoided with considerable savings in terms of computational effort. The proposed adaptive framework combines elements from almost all known classes of image filters. Namely, it combines Minkowski type distances or the angular distance function used in ranked type estimators, such as the VMF or the BVDF with averaging outputs used in linear filtering, and with data dependent coefficients used in adaptive designs.

- In the framework described above, there is no requirement for fuzzy rules or local statistics estimates. Features extracted from local data, here in the form of a sum of distances, are used as inputs to determine the form of the weights. The vector filters discussed in this section do not utilize the distance measures to order the colour input vectors.

Instead, they are used to provide selected features in a reduced space: features used as inputs for the adaptive weights.

- Unlike the BVDF an adaptive directional filter based on equation (10.9) will not act as a pure chromaticity filter since it can use both the directional filtering information, through the angle distances, for its weights as well as the magnitude component of each of the input (colour) vectors to generate the filtered result. This is a feature that differentiates our adaptive design from the chromaticity filters with grey-level processing components, such as the GVDF or the *ad hoc* hybrid structures of Gabbouj and Cheickh (1996). The GVDF selects a subset of the colour vectors and then applies grey-scale techniques only to the selected group of vectors. However, if important colour information was eliminated due to errors in the chromaticity-based decision part, the GVDF is unable to compensate using its grey-scale processing module. That is not the case in our adaptive design. The adaptive filter introduced in equation (10.9) and equation (10.10) does not discard any magnitude information based on chromaticity analysis. All the vectors inside the operational window contribute to the final output. Simply stated, the filter assigns weights to the magnitude component of each colour vector modifying in this way its contribution to the output. This natural blending of chromaticity-based weights with magnitude-based input contributions differentiates our design from the heuristic solutions used in the DDF or the hybrid filters, making the adaptive directional filter the natural setting for real-time colour image processing.

- The adaptive designs discussed here differ in their computational complexity. It should be noted at this point that the computational complexity of a given filter is a realistic measure of its practicality and usefulness, since it determines the required computing power and the associated processing time required for its implementation. The computational complexity analysis of the adaptive designs requires knowledge of the function used to calculate the weights, evaluation of this function and the exact form of the ordering process (the distance measure used). The computationally intensive part of the adaptive scheme is the distance calculation. This part, however is common to all vector processing designs. Thus, from a practical standpoint, the proposed adaptive framework yields realizations of different filters that may have reduced complexity. The remarkably flexible structure of equation (10.9) allows a wide variety of combinations that can meet a number of computational and hardware constraints, especially in real-time.

10.4   APPLICATION TO COLOUR IMAGES

In this section the performance of the different vector processing filters is evaluated in the most important area of vector processing, namely colour image filtering. Nine different filters have been used in the studies reported in this section.

Table 10.1 summarizes the filters used. Two RGB colour images have been selected for our tests. The colour images 'Lena' and 'Peppers' [1] have been contaminated using various noise source models in order to assess the performance of the filters under different noise distributions as listed in Table 10.2.

The normalized mean square error (NMSE) has been used as a quantitative measure for evaluation purposes. It is computed as:

$$\text{NMSE} = \frac{\sum\limits_{i=0}^{N}\sum\limits_{j=0}^{M} \|\mathbf{y}(i,j) - \hat{\mathbf{y}}(i,j)\|^2}{\sum\limits_{i=0}^{N}\sum\limits_{j=0}^{M} \|\mathbf{y}(i,j)\|^2}$$

where N and M are the image dimensions, and $\mathbf{y}(i,j)$ and $\hat{\mathbf{y}}(i,j)$ denote the original image vector and the estimation at pixel (i,j) respectively. Table 10.3 summarizes the results obtained for the test image 'Lena' for a 3×3 filter window. The results obtained using a 5×5 filter window are given in Table 10.4. The results for the colour image 'Peppers' are summarized in Table 10.5 and Table 10.6 respectively.

In addition to the quantitative evaluation presented above, a qualitative evaluation is necessary since the visual assessment of the processed images is, ultimately, the best subjective measure of the efficiency of any method. Therefore, we present sample processing results in Plate XI.

From the results listed in the tables, it can easily be seen that adaptive designs provide consistently good results in all types of noise, outperforming the other vector processing filters under consideration. The adaptive designs discussed here attenuate both impulsive and Gaussian noise with or without outliers present in the test image. The versatile design of equation (10.9) allows for a number of different adaptive/fuzzy filters, which can provide solutions to many types of different filtering problems. Simple adaptive designs, such as the ANNF can preserve edges and smooth noise under differ-

---

[1]The 'Peppers' image is part of the USC-SIPI Image Data Base and is reproduced with permission.

ent conditions, outperforming other widely used vector processing filters. If knowledge of the noise characteristics is available, the designer can tune the parameters of the adaptive filter to obtain better results. Finally, considering the number of computations, the computationally intensive part of the adaptive algorithm is the distance calculation. However, this step is common in all vector processing filters considered here. In summary, our design is simple, does not increase the numerical complexity of the multichannel algorithm and delivers excellent results for complicated multichannel signals, such as real colour images.

CONCLUSION

Vector processing filters have been the subject of this chapter. A number of different designs, adaptive and non-adaptive, have been discussed in detail. Adaptive vector filters which combine in a novel way data-dependent weights, average filters, and distance measures were introduced and analyzed. Experimental analysis with colour images has demonstrated the efficiency of the adaptive designs in smoothing out noise. The adaptive vector processing filters discussed here outperform other nonlinear filters, such as the Vector Median Filter (VMF), the Directional-Distance filter (DDF) and the heuristic directional filters of Gabbouj and Cheickh (1996). Moreover, the adaptive vector processing filters preserve the chromaticity component, which is very important in the visual perception of colour images. Future work in this area should address the development of double-window vector processing filters where an outlier rejection scheme will be utilized prior to the adaptive filter.

Table 10.1 Vector filters compared.

Name	Description and Reference(s)
VMF	Vector median filter (Astola, Haavisto and Neuvo, 1990)
BVDF	Basic vector directional filter (Trahanias, Pitas and Venetsanopoulos, 1994) and (Trahanias and Venetsanopoulos, 1993)
GVDF	Generalized vector directional filter (Trahanias and Venetsanopoulos, 1993) and (Trahanias, Karakos and Venetsanopoulos, 1996)
DDF	Directional-distance filter (Karakos and Trahanias, 1995)
HF	Hybrid directional filter (Gabbouj and Cheickh, 1996)
AHF	Adaptive hybrid directional filter (Gabbouj and Cheickh, 1996)
FVDF	Fuzzy vector directional filter, equation (10.10), $\beta = 2$, $r = 1$ (Plataniotis, Androutsos and Venetsanopoulos, 1996)
ANNF	Adaptive nearest-neighbour filter, equation (10.13) (Plataniotis, Androutsos, Sri and Venetsanopoulos, 1995)
ANNMF	Adaptive nearest-neighbour multichannel filter, equation (10.13), equation (10.14) (Plataniotis, Sri, Androutsos and Venetsanopoulos, 1996)

Table 10.2 Noise distributions.

Number	Noise model
1	Gaussian ($\sigma = 30$)
2	impulsive (4%)
3	Gaussian ($\sigma = 15$) impulsive (2%)
4	Gaussian ($\sigma = 30$) impulsive (4%)

Table 10.3 NMSE $(\times 10^{-2})$ for the 'Lena' image, 3×3 window.

Filter	Noise			
	1	2	3	4
None	4.2083	5.1694	3.6600	9.0724
BVDF	2.8962	0.3848	0.4630	1.1354
GVDF	1.4600	0.3000	0.6334	1.9820
DDF	1.5240	0.3255	0.6483	1.6791
VMF	1.6000	0.1900	0.5404	1.6791
FVDF	0.7335	0.2481	0.4010	1.0390
ANNF	0.8510	0.2610	0.3837	1.0860
ANNMF	0.6591	0.1930	0.3264	0.7988
HF	1.3192	0.2182	0.5158	1.6912
AHF	1.0585	0.2017	0.4636	1.4355

Table 10.4 NMSE $(\times 10^{-2})$ for the 'Lena' image, 5×5 window.

Filter	Noise			
	1	2	3	4
None	4.2083	5.1694	3.6600	9.0724
BVDF	2.8819	0.7318	0.6850	1.3557
GVDF	1.0800	0.5400	0.4590	1.1044
DDF	1.0242	0.5126	0.6913	1.3048
VMF	1.1700	0.5800	0.5172	1.0377
FVDF	0.7549	0.3087	0.4076	0.9550
ANNF	0.6260	0.4210	0.4360	0.7528
ANNMF	0.5445	0.2505	0.3426	0.6211
HF	0.7700	0.3841	0.4890	1.1417
AHF	0.6762	0.3772	0.4367	0.7528

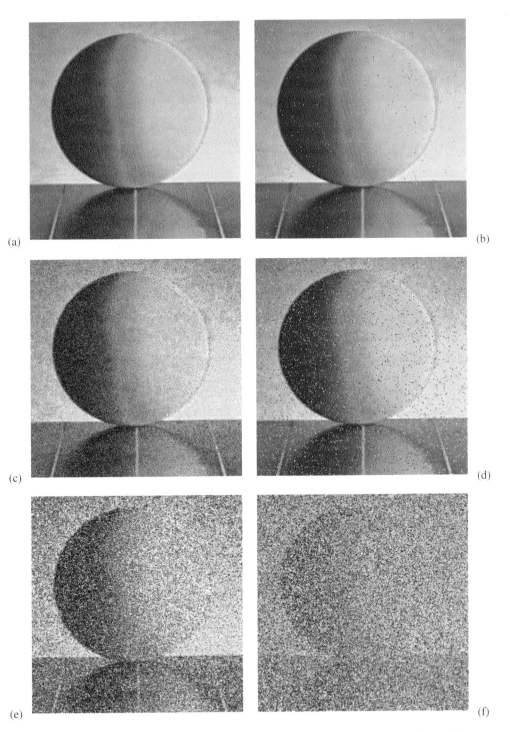

(a)

(b)

(c)

(d)

(e)

(f)

Plate I: A colour image corrupted with different levels of Gaussian noise (a), (c), and (e) and impulse noise (b), (d), and (f). Images (a) and (b) have variance 100, (c) and (d) have variance 1000, and (e) and (f) have variance 10 000. (Page 151)

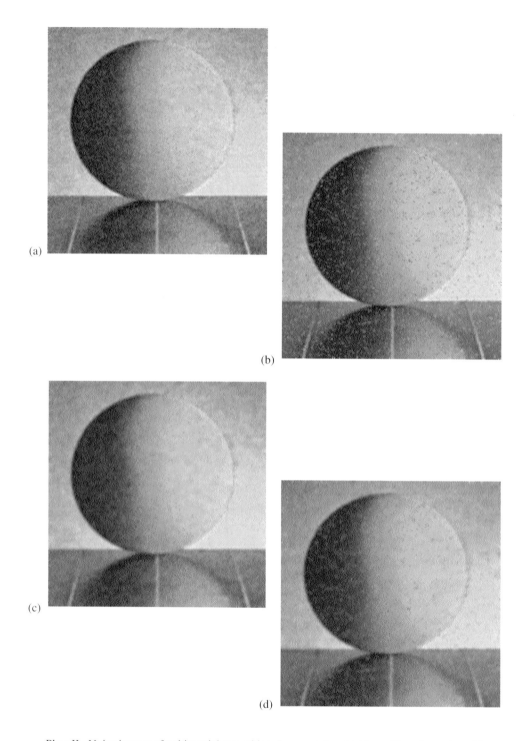

(a)

(b)

(c)

(d)

Plate II: Noisy images after binomial smoothing. Images (a) and (c) have Gaussian noise with variance 1000, (b) and (d) have impulse noise with variance 1000. One iteration of smoothing has been applied to images (a) and (b), three to (c) and (d). (Page 152)

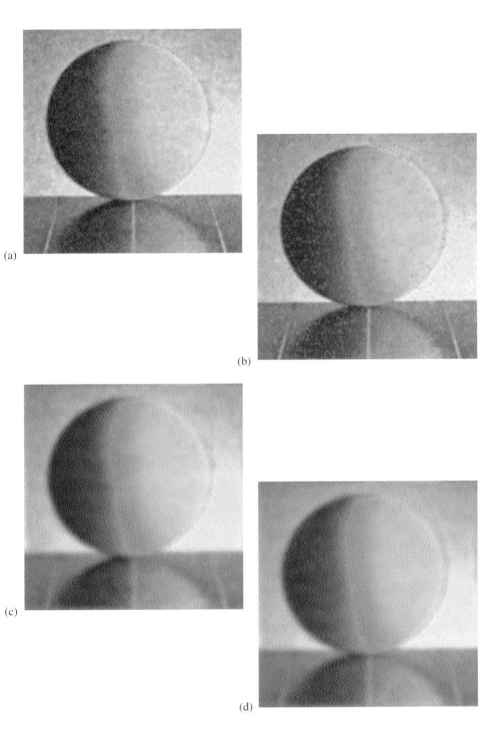

(a)

(b)

(c)

(d)

Plate III: Noisy images after Gaussian smoothing. Images (a) and (c) have Gaussian noise with variance 1000, (b) and (d) have impulse noise with variance 1000. The standard deviation of the Gaussian convolution mask for images (a) and (b) is one, for (c) and (d) it is three. (Page 154)

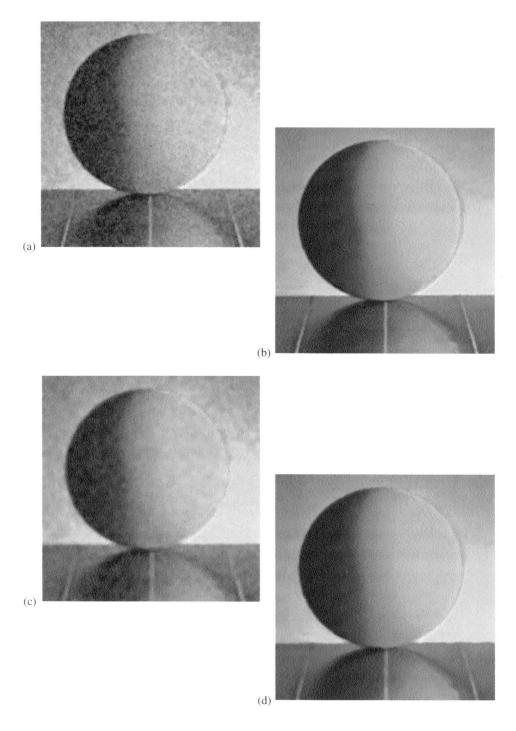

(a)

(b)

(c)

(d)

Plate IV: Noisy images after median smoothing. Images (a) and (c) have Gaussian noise with variance 1000, (b) and (d) have impulse noise with variance 1000. A 3 x 3 sorting neighbourhood was used for (a) and (b), while a 6 x 6 neighbourhood was used for (c) and (d). (Page 155)

(a)

(b)

(c)

(d)

Plate V: Image enhancement using (a) histogram equalization, (b) adaptive histogram equalization (AHE) after dividing the image into 3 x 3 regions, (c) AHE after dividing the image into 5 x 5 regions, and (d) AHE after dividing the image into 7 x 7 regions. (Page 159)

(a)

(b)

(c)

(d)

Plate VI: Image enhancement using unsharp masking with: (a) 3, (b) 6, (c) 12, and (d) 24 blurring iterations. In each case 80% of the blurred image is subtracted from the original. (Page 160)

(a)

(b)

(c)

(d)

Plate VII: Image enhancement using constant variance enhancement using a: (a) 3 x 3, (b) 6 x 6, (c) 12 x 12, and (d) 24 x 24 statistics neighbourhood. (Page 161)

Plate VIII: A simple colour image. (Page 169)

Plate IX: Fruit. (Page 173)

Plate X: A Gaussian smoothed image of Plate IX ($\sigma = 2$). (Page 184)

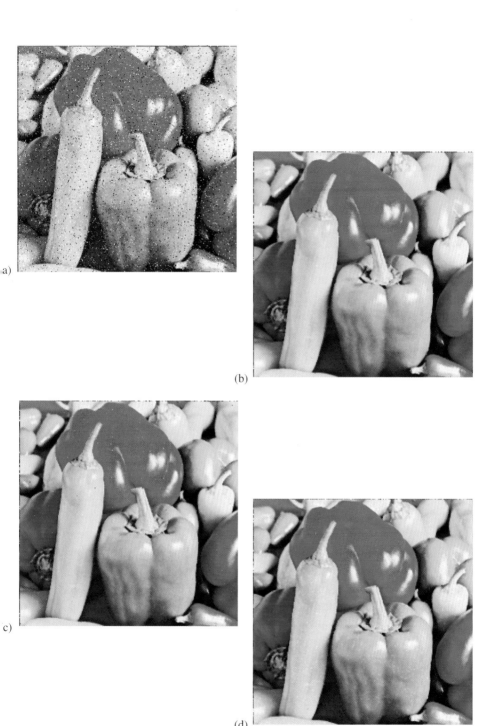

(a)

(b)

(c)

(d)

Plate XI: (a) 'Peppers' image corrupted with 4% impulses, (b) VMF filtered version of (a), (c) HF filtered version of (a), (d) ANNMF filtered version of (a); all using a 3 x 3 window. (The original image is part of the USC-SIPI Image Data Base and is reproduced with permission.) (Page 205)

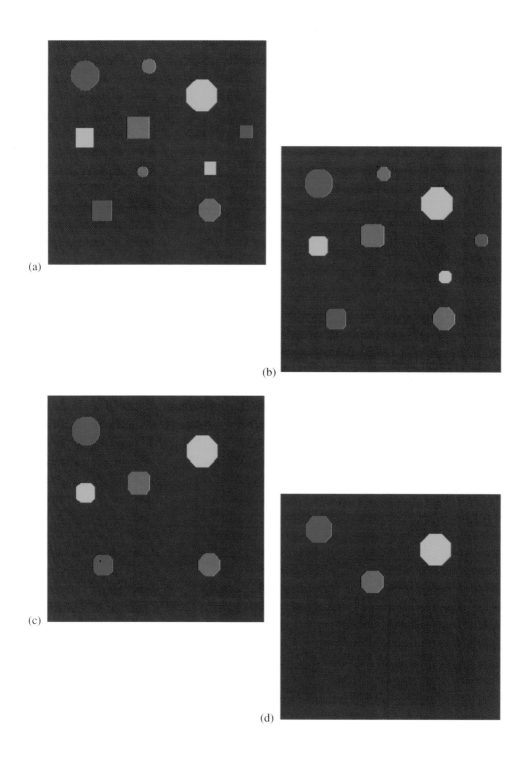

Plate XII: Vector multiscale opening with $d(\mathbf{f}(x,y)) = f_Y(x,y)$: (a) original image. Results of vector opening with (b) $n = 3$, (c) $n = 4$, (d) $n = 6$. (Page 221)

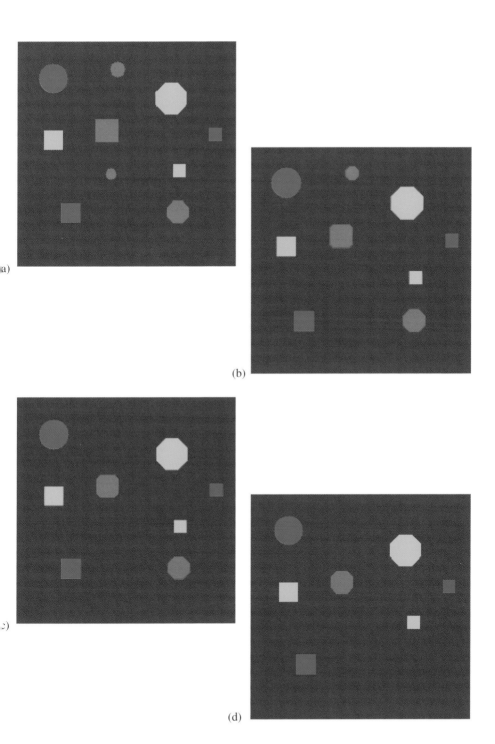

(a)

(b)

(c)

(d)

Plate XIII: Vector multiscale opening with $d(\mathbf{f}(x,y)) = f_R(x,y)$. (a) original image. Results of vector opening with (b) $n = 3$, (c) $n = 4$, (d) $n = 6$. (Page 224)

Plate XIV: Original image used for noise suppression experiments. (Page 226)

(a)

Plate XV: (a) Noisy version of original image in Plate XIV, high spectral correlation; (b) result of component-wise filter applied in RGB space; (c) result of vector morphological filter. (Page 226)

(b)

(c)

a)

(b)

Plate XVI: (a) Noisy version of original image in Plate XIV, low spectral correlation; (b) result of component-wise filter applied in RGB space; (c) result of vector filter. (Page 226)

(c)

Plate XVII: Original image 'Boats'. (Reproduced from the USC-SIPI Image Data Base with permission from the USC Signal and Image Processing Institute.) (Page 260)

Plate XVIII: Reconstructed 'Boats' image of Plate XVII after JPEG compression to 0.6 bpp. (Page 260)

Plate XIX: Reconstructed 'Boats' image of Plate XVII after JPEG compression to 0.26 bpp. (Page 260)

Plate XX: Reconstructed 'Boats' image of Plate XVII after sub-band/wavelet compression to 0.6 bpp. (Page 263)

Plate XXI: Reconstructed 'Boats' image of Plate XVII after sub-band/wavelet compression to 0.18 bpp. (Page 263)

Plate XXII: Original image 'Lena'. (Reproduced from the USC-SIPI Image Data Base with permission from the USC Signal and Image Processing Institute.) (Page 270)

Plate XXIII: The image of Plate XXII quantized to 16 colours using the median-cut algorithm. (Page 284)

Plate XXIV: The image of Plate XXII quantized to 16 colours using the variance minimization quantization algorithm. (Page 284)

Plate XXV: The image of Plate XXII quantized to 16 colours using the octree algorithm. (Page 286)

Plate XXVI: Plate XXII dithered using Floyd-Steinberg dithering and a 16-colour palette desined using the octree algorithm. (Page 289)

(a)

(b)

(c)

Plate XXVII: (a) Frame from original colour video sequence 'Salesman'; (b) frame from the same sequence coded in *intraframe* mode of a standard H.263 coder at a target bitrate of 64 kbps; (c) a 'P' frame (coded in *interframe*, or predictive mode) showing distortions after rapid motion. (Page 298)

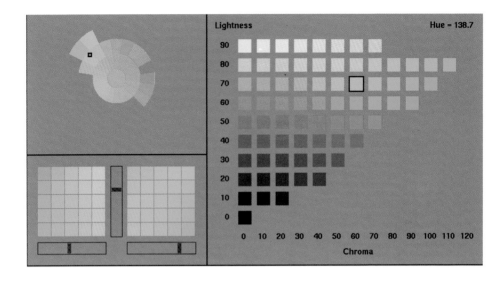

Plate XXVIII: The CIELAB colour picker. (Page 321)

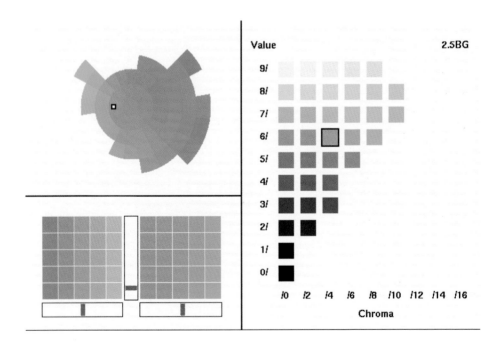

Plate XXIX: The Munsell colour picker. (Page 323)

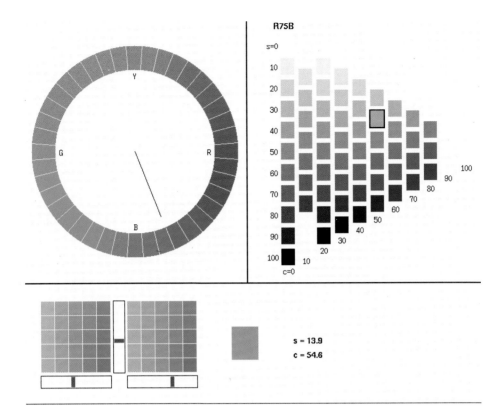

R75B

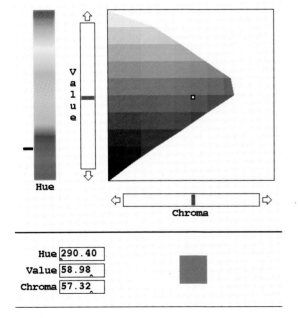

Plate XXX: The Natural Colour System picker. (Page 325)

Hue 290.40
Value 58.98
Chroma 57.32

Plate XXXI: The TekColor
colour picker. (Page 325)

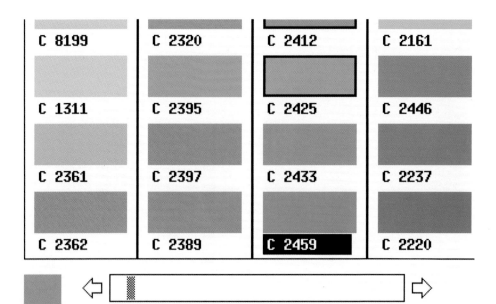

C 8199	C 2320	C 2412	C 2161
C 1311	C 2395	C 2425	C 2446
C 2361	C 2397	C 2433	C 2237
C 2362	C 2389	C 2459	C 2220

ΔE=5.87

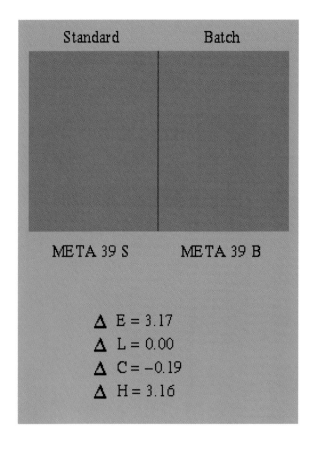

Standard   Batch

META 39 S   META 39 B

Δ E = 3.17
Δ L = 0.00
Δ C = −0.19
Δ H = 3.16

Plate XXXII: Locating a shade from an arbitrary colour specifier. (Page 325)

Plate XXXIII: Examining colour difference. (Page 326)

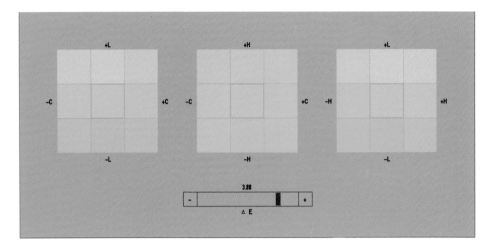

Plate XXXIV: Specifying colour tolerance. (Page 326)

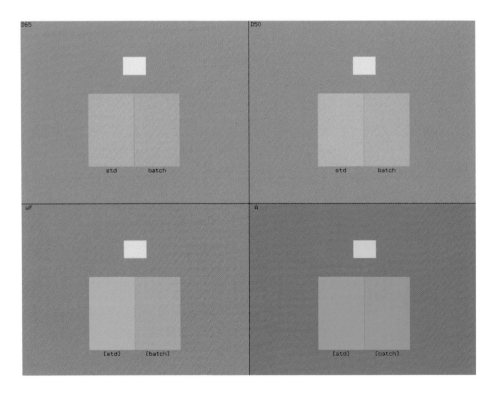

Plate XXXV: Visualizing metamerism. (Page 327)

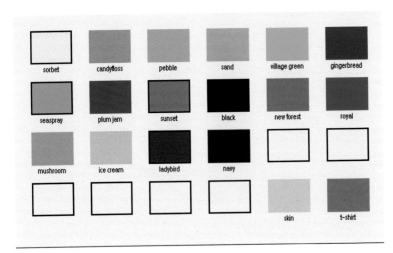

Plate XXXVI:
A sample
palette.
(Page 329)

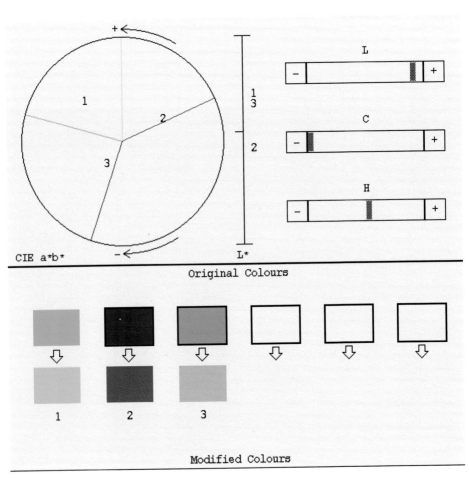

Plate XXXVII: Colour adjustment. (Page 329)

Table 10.5 NMSE ($\times 10^{-2}$) for the 'Peppers' image, 3×3 window.

Filter	Noise			
	1	2	3	4
None	5.0264	6.5257	3.2890	6.5076
BVDF	3.9267	1.5070	0.8600	1.4911
GVDF	1.8640	0.4550	0.3613	0.4562
DDF	3.5090	0.5886	0.5336	0.5893
VMF	1.8440	0.3763	0.3260	0.3786
FVDF	1.4550	0.4246	0.3412	0.4046
ANNF	1.1230	0.5110	0.3150	0.5180
ANNMF	0.9080	0.3550	0.3005	0.3347
HF	1.5892	0.4690	0.3592	0.4781
AHF	1.4278	0.4246	0.3566	0.4692

Table 10.6 NMSE ($\times 10^{-2}$) for the 'Peppers' image, 5×5 window.

Filter	Noise			
	1	2	3	4
None	5.0264	6.5257	3.2890	6.5076
BVDF	4.2698	2.7920	1.6499	4.1350
GVDF	1.2534	0.6977	0.6600	0.7030
DDF	2.1440	0.7636	0.7397	0.7612
VMF	1.3390	0.6740	0.6563	0.6812
FVDF	2.1120	0.7310	0.6971	0.7178
ANNF	1.0027	0.5230	0.5200	0.6210
ANNMF	0.8050	0.4471	0.4047	0.4458
HF	1.0040	0.9970	0.7684	0.9970
AHF	1.1167	0.9841	0.7632	0.9841

11

Morphological operations

Mary L. Comer and Edward J. Delp

Mathematical morphology has been shown to be useful for the processing and analysis of binary and greyscale images (Serra, 1982; Haralick, Sternberg and Zhuang, 1987). Morphology has been used to perform noise suppression, texture analysis, shape analysis, edge detection, skeletonization, and multiscale filtering for applications such as medical imaging, geological image processing, automated industrial inspection, image compression, and ECG signal analysis (Serra, 1982; Maragos, 1990; Schonfeld and Goutsias, 1991; Overturf, Comer and Delp, 1995; Chu and Delp, 1989). This chapter describes the extension of mathematical morphology to colour images.

Many techniques developed for use with monochrome images can be extended to colour images by applying the algorithm to each of the colour component images separately. An important question arises: is component-wise spatial filtering sufficient? Algorithms that exploit spectral correlations could provide better performance and be more computationally efficient. It is widely accepted that for nonlinear filtering techniques, such as mathematical morphology and median filtering, an alternative to the component-wise approach is needed. For this reason, a number of nonlinear filtering algorithms developed specifically for multivariate image processing have been proposed (Astola, Haavisto and Neuvo, 1990; Hardie and Arce, 1991; Comer and Delp, 1992; Goutsias, 1992; Trahanias and Venetsanopoulos, 1993). In Astola, Haavisto and Neuvo (1990) the vector median filter was developed as an extension of the median filter for scalar-valued signals. The vector median of a collection of vectors is the vector from the collection which has the minimum aggregate distance from all other vectors in the collection (Astola, Haavisto and Neuvo, 1990; Barnett, 1976). The output of the vector median filter at a given pixel of a colour image is the vector median of the colour vectors inside the filter window. Vector filtering is discussed in detail in Chapter 10. In Hardie and Arce (1991) a class of ranked-order based filters for multichannel images was presented. At each pixel, a region of confidence is determined based on the order statistics of the vectors at neighbouring pixels. If the vector at the given pixel does not lie inside this region of confidence, then it is assumed to be an outlier, and it is replaced by the vector in the region of confidence which is at a minimum distance. Vector directional filters for multichannel image processing were introduced by Trahanias and Venetsanopoulos (1993). These filters process vector-valued signals in two steps. First, vectors are processed based on direction, or angle, resulting in the removal of vectors with atypical directions. Then, magnitude processing is performed using any classical greyscale image processing filter.

As with these other nonlinear filters, the techniques which have been proposed for the extension of mathematical morphology to colour images have been based on the concept of ranking multivariate data (Comer and Delp, 1992; Goutsias, 1992). The following sections describe morphological filters which use the vector ranking concepts discussed in Barnett (1976). After a brief review of binary and greyscale mathematical morphology, the vector morphological filtering operations are defined, a set-theoretic analysis relative to these vector operations is presented, and the use of vector morphology for the applications of multiscale image analysis and noise suppression is illustrated.

11.1 MATHEMATICAL MORPHOLOGY

Mathematical morphology is based on **set theory**. A morphological operation defined on a binary image is referred to as **binary morphology**. This involves representing the image as a set $X \subseteq \mathbb{R}^2$ or $\mathbb{Z}^2$ (depending on whether the image is defined on a discrete or continuous lattice — an image defined on a discrete lattice will be referred to as a discrete-space image and an image defined on a continuous lattice will be referred to as a continuous-space image), where $\mathbb{R}$ is the set of real numbers and $\mathbb{Z}$ is the set of integers. Points in the image foreground are members of X and points in the background are members of the complement of X, designated X^c. The image is transformed by another set, known as the **structuring element**. The shape and size of the structuring element determine the resultant image (Serra, 1982; Haralick, Sternberg and Zhuang, 1987; Maragos and Schafer, 1987; Stevenson and Arce, 1987; Song and Delp, 1990; Song and Delp, 1991).

11.1.1   Binary morphology

There are four basic binary morphological operations: **dilation, erosion, opening**, and **closing**, represented by the symbols $\oplus$, $\ominus$, $\circ$, and $\bullet$, respectively. Binary dilation and erosion are defined as follows:

$$
\begin{aligned}
X \oplus H &= \{(x,y) : H_{(x,y)} \cap X \neq \emptyset\} \\
X \ominus H &= \{(x,y) : H_{(x,y)} \subseteq X\}
\end{aligned}
$$

where X is the original image, $H \subseteq \mathbb{R}^2$ or $\mathbb{Z}^2$ is the structuring element, and $H_{(x,y)}$ is the translate of the set H by the vector $(x,y) \in \mathbb{R}^2$ or $\mathbb{Z}^2$. Opening is defined as erosion followed by dilation, and closing is defined as dilation followed by erosion.

11.1.2   Greyscale morphology

Greyscale morphological operations are an extension of binary morphological operations to greyscale images. For greyscale operations, the image will be represented by the function $f(x,y)$, where $(x,y) \in \mathbb{R}^2$ or $\mathbb{Z}^2$, or simply f, and the structuring element will be the function $h(x,y)$, or h. Greyscale dilation and erosion are defined as follows:

$$
\begin{aligned}
(f \oplus h)(x,y) &= \sup_{(r,s)\in H} \{f(x-r,y-s)+h(r,s)\} \\
(f \ominus h)(x,y) &= \inf_{(r,s)\in H} \{f(x+r,y+s)-h(r,s)\}
\end{aligned}
$$

where $\sup\{\}$ and $\inf\{\}$ denote the supremum and infimum operators, respectively, and $H \subseteq \mathbb{R}^2$ or $\mathbf{Z}^2$ is the **support** of $h(x,y)$ (the region over which the structuring element is defined). A special class of greyscale morphological filters, referred to as function-and-set-processing (FSP) filters (Maragos and Schafer, 1987), results when $h(x,y) = 0$ for every $(x,y) \in H$. The resulting operations of dilation, erosion, opening, and closing are then written as $f \oplus H$, $f \ominus H$, $f \circ H$, and $f \bullet H$. Detailed descriptions of the operations defined in this section can be found in (Serra, 1982; Haralick, Sternberg and Zhuang, 1987).

11.2 COLOUR MORPHOLOGY

The extension of mathematical morphology to colour images is not straight-forward. Serra (1988) discusses the generalization of morphology to its most basic elements, and concludes that the axioms can be reduced to three key ideas: an order relationship (for example, set inclusion for binary morphology), a supremum or an infimum pertaining to that order, and the possibility of admitting an infinity of operands. The first two of these, the order relationship and the supremum (or infimum), are missing in colour images, because there is no unambiguous way to order two or more colours. The fact that these fundamental concepts of morphology do not apply to colour images makes it difficult to define 'colour morphology'. However, it is possible that some of the techniques can be extended to colour images.

The problem of ordering multivariate data is not unique to mathematical morphology. Although there is no natural means for total ordering of multivariate samples, much work has been done to define concepts such as median, range, and extremes in multivariate analysis. Barnett (1976) proposed the classification of these sub-ordering principles into four groups: marginal ordering, reduced ordering, partial ordering, and conditional ordering.

11.2.1 Component-wise morphology

In **marginal ordering** ranking takes place within one or more of the marginal sets of samples (Barnett, 1976), i.e. scalar ranking is performed within each channel. Thus, to order a collection of colour vectors using marginal ordering the components in each spectral band are ordered independently of the components in other spectral bands. Morphological operations which are defined using marginal ordering are referred to as component-wise operations. Because the component images are filtered separately with the component-wise filter, there is a possibility of altering the spectral composition of the image, e.g. the colour balance and object boundaries. For example, there is a possibility with this approach that an object could be removed or enhanced in one or two of the R, G, and B components, but not in all of them. This effect near spatial edges in an image is referred to as 'edge jitter' (Astola, Haavisto and Neuvo, 1990). This effect would be unacceptable for many applications. For example, in object recognition, colour and shape both play important roles. If the component-wise filter is applied to a colour image as a step in object recognition, then a change in the spectral composition of image objects may produce errors in further processing of the colour image obtained from the filter.

11.2.2   Vector morphology

A different way to examine the problem of colour morphology is to treat the colour at each pixel as a vector. To motivate this approach, consider a colour image with only two colours, representing an object and a background region. Let $\mathbf{f}$ be a colour image which consists of the two colours $\mathbf{f}_1 = [R_1, G_1, B_1]^T$ and $\mathbf{f}_2 = [R_2, G_2, B_2]^T$. One way to analyze geometrical features of this image, which is perhaps more natural than the component-wise approach, is to view the pixels which represent the object as a set, and the pixels which represent the background as the complement of this set, as is done with a binary image. The justification for this approach is that both binary images and two-colour colour images represent scenes with two regions, and the geometrical information in both cases is determined by which pixels belong to the object and which belong to the background. Thus, we define the sets X and X^c as

$$X = \{(x,y) : \mathbf{f}(x,y) = \mathbf{f}_1\}$$
$$X^c = \{(x,y) : \mathbf{f}(x,y) = \mathbf{f}_2\}$$

In this case colour dilation, erosion, opening, and closing would be defined the same way as binary dilation, erosion, opening, and closing, with X as the image foreground and X^c as the background. In this vector approach the data at each pixel in the filtered image represent either an object pixel or a background pixel, rather than, for example, the red component from the object and the blue and green components from the background, as is possible with the component-wise approach. To extend the vector approach to colour images with more than two colours, it is necessary to define an order relation which orders the colours as vectors, rather than ordering the individual components.

This will be done using **reduced ordering**. In reduced ordering each multivariate observation is reduced to a single value, which is a function of the component values for that observation, with the multivariate samples ranked according to this single value (Barnett, 1976). To illustrate this type of ordering, let $\mathbf{x}_1, \mathbf{x}_2, \ldots, \mathbf{x}_n$ be a collection of multivariate samples, where each $\mathbf{x}_i$ is a vector in $\mathbb{R}^p$. The n samples are to be ordered using a reduced ordering scheme. The first step is to map each $\mathbf{x}_i$ to a scalar value $d_i = d(\mathbf{x}_i)$ where $d : \mathbb{R}^p \to \mathbb{R}$. After d_i has been obtained for each i, the vectors $\mathbf{x}_1, \mathbf{x}_2, \ldots, \mathbf{x}_n$ are ordered based on $d_1, d_2, \ldots, d_n$ as follows:

$$\mathbf{x}_{(1)} \leq \mathbf{x}_{(2)} \leq \cdots \leq \mathbf{x}_{(n)}$$

where $\mathbf{x}_{(r)}$ is the vector with corresponding scalar value $d_{(r)}$, and $d_{(r)}$ is the rth smallest element of the set $\{d_1, d_2, \ldots, d_n\}$.

We now use reduced ordering as described above to define vector morphological filtering operations for colour images. The structuring element for the vector morphological operations defined here is the set H, and the scalar-valued function used for the reduced ordering is $d : \mathbb{R}^3 \to \mathbb{R}$. The operation of vector dilation is represented by the symbol $\oplus_v$. The value of the vector dilation of $\mathbf{f}$ by H at the point (x, y) is defined as:

$$(\mathbf{f} \oplus_v H)(x, y) = \mathbf{a} \tag{11.1}$$

where

$$\mathbf{a} \in \{\mathbf{f}(r, s) : (r, s) \in H_{(x,y)}\} \tag{11.2}$$

and

$$d(\mathbf{a}) \geq d(\mathbf{f}(r, s)) \; \forall \, (r, s) \in H_{(x,y)} \tag{11.3}$$

Similarly, vector erosion is represented by the symbol $\ominus_v$, and the value of the vector erosion of $\mathbf{f}$ by H at the point (x, y) is defined as

$$(\mathbf{f} \ominus_v H)(x, y) = \mathbf{b} \tag{11.4}$$

where

$$\mathbf{b} \in \{\mathbf{f}(r, s) : (r, s) \in H_{(x,y)}\} \tag{11.5}$$

and

$$d(\mathbf{b}) \leq d(\mathbf{f}(r, s)) \; \forall \, (r, s) \in H_{(x,y)} \tag{11.6}$$

Vector opening is defined as the cascade of vector erosion and vector dilation, and vector closing is defined as the cascade of vector dilation and vector erosion. With the above definitions for vector morphological operations we must impose the restriction that the set H be a finite set, because if H is not finite then it is possible that no value of $\mathbf{a}$ satisfies equations (11.2) and (11.3) or no value of $\mathbf{b}$ satisfies equations (11.5) and (11.6).

With these definitions the output vector at each point in the image is, by definition, one of the vectors in the original image, so there is no possibility of introducing new colour vectors into the image. It is possible that two different colour values of $\mathbf{a}$ could satisfy equations (11.2) and (11.3) or two different colour values of $\mathbf{b}$ could satisfy equations (11.5) and (11.6). In this case the output of the vector filter can be chosen based on positions in the structuring element window.

Often the metric used to perform reduced ordering is some type of **distance metric** (Barnett, 1976). The output of the vector filter will depend not only on the input image and the structuring element, but also on the scalar-valued function used to perform the reduced ordering. For many image processing applications it might make sense to use a characteristic of the human visual system, such as luminance, as a metric for reduced ordering.

11.2.3   Analysis of vector morphology

Since mathematical morphology is based on set theory, it is important to investigate the vector morphological operators defined above in terms of set operations. Serra (1982) discusses the need to analyze greyscale morphological operations not only in terms of transformations of greyscale functions, but also in terms of set transformations on the cross sections of those greyscale functions. In fact, greyscale morphology was originally developed by representing functions as sets, using either cross-sections or the umbra representation of a function, and applying binary morphological operations to those sets (Serra, 1982). Similarly, if we define the concept of cross-sections of a colour image, then vector morphological operations can be analyzed in terms of set transformations on these cross-sections.

We first review the set-theoretic analysis of FSP morphological operations for greyscale images. For a greyscale image f (discrete-space or continuous-space) the set

$$X_t(f) = \{(x,y) \in D : f(x,y) \geq t\}, \quad t \in V$$

where $V \subseteq \mathbb{R}$ or $\mathbb{Z}$ is the range of the function f; is known as the cross-section of f at level t (Serra, 1982). If the function f is upper semi-continuous then the image can be reconstructed from its cross-sections by

$$f(x,y) = \sup\{t \in V : (x,y) \in X_t(f)\}$$

The following equations show the relationship between greyscale dilation and erosion of the image f by the set H and binary dilation and erosion of the cross sections of f by H:

$$X_t(f \oplus H) = X_t(f) \oplus H \quad \Longleftrightarrow \quad (f \oplus H)(x,y) = \sup_{(r,s) \in H_{(x,y)}} \{f(r,s)\}$$

$$X_t(f \ominus H) = X_t(f) \ominus H \quad \Longleftrightarrow \quad (f \ominus H)(x,y) = \inf_{(r,s) \in H_{(x,y)}} \{f(r,s)\}$$

Thus, the cross section at level t of $f \oplus H$ is equal to the binary dilation of the cross section at level t of f by the set H and the cross section at level t of $f \ominus H$ is equal to the binary erosion of the cross section at level t of f by the set H.

To derive analogous equations relating vector morphological operations to binary morphological operations, we propose the following definition for cross sections of a colour image: The set

$$X_t(\mathbf{f}) = \{(x,y) : d(\mathbf{f}(x,y)) \geq t\}$$

is the cross-section of the colour image $\mathbf{f}$ at level t with respect to the function $d : \mathbb{R}^3 \to \mathbb{R}$. Reconstruction of a colour image $\mathbf{f}$ from its cross-sections is possible only if each value of d has a unique colour vector associated with it. In this case, reconstruction of $\mathbf{f}$ is given by

$$d(\mathbf{f}(x,y)) = \sup\{t \in V : (x,y) \in X_t(\mathbf{f})\}$$

Then, $\mathbf{f}(x,y)$ is the colour vector corresponding to $d(\mathbf{f}(x,y))$.

The following proposition relates the vector operations of equations (11.1) to (11.6) to the operations of binary morphology.

Proposition 1 *Subject to the constraints*

$$(\mathbf{f} \oplus_v H)(x,y) \in \{\mathbf{f}(r,s) : (r,s) \in H_{(x,y)}\} \tag{11.7}$$

and

$$(\mathbf{f} \ominus_v H)(x,y) \in \{\mathbf{f}(r,s) : (r,s) \in H_{(x,y)}\} \tag{11.8}$$

the following equations provide necessary and sufficient conditions for the vector dilation and erosion of f to be equivalent to binary dilations and erosions of the cross-sections of f:

$$X_t(\mathbf{f} \oplus_v H) = X_t(\mathbf{f}) \oplus H \Leftrightarrow d((\mathbf{f} \oplus_v H)(x,y)) \geq d(\mathbf{f}(r,s)) \; \forall (r,s) \in H_{(x,y)} \tag{11.9}$$

and

$$X_t(\mathbf{f} \ominus_v H) = X_t(\mathbf{f}) \ominus H \Leftrightarrow d((\mathbf{f} \ominus_v H)(x,y)) \leq d(\mathbf{f}(r,s)) \; \forall (r,s) \in H_{(x,y)} \tag{11.10}$$

Proof. First, we assume that $X_t(\mathbf{f} \oplus_v H) = X_t(\mathbf{f}) \oplus H$ and prove the right-hand side of equation (11.9). Using this assumption, we have

$$
\begin{aligned}
\{(x,y) : d((\mathbf{f} \oplus_v H)(x,y)) \geq t\} &= \{(x,y) : H_{(x,y)} \cap X_t(\mathbf{f}) \neq \emptyset\} \\
&= \{(x,y) : \exists (r,s) \in H_{(x,y)} \ni d(\mathbf{f}(r,s)) \geq t\}
\end{aligned}
$$

Hence,

$$d((\mathbf{f}\oplus_v H)(x,y)) \geq t \Leftrightarrow \exists(r,s) \in H_{(x,y)} \ni d(\mathbf{f}(r,s)) \geq t$$

Let

$$t_{max} = \max_{(r,s)\in H_{(x,y)}} \{d(\mathbf{f}(r,s))\}$$

Then

$$d((\mathbf{f}\oplus_v H)(x,y)) \geq t_{max} \geq d(\mathbf{f}(r,s)) \ \forall(r,s) \in H_{(x,y)}$$

Thus,

$$d((\mathbf{f}\oplus_v H)(x,y)) \geq d(\mathbf{f}(r,s)) \ \forall(r,s) \in H_{(x,y)}$$

Now we assume that $d((\mathbf{f}\oplus_v H)(x,y)) \geq d(\mathbf{f}(r,s)) \ \forall(r,s) \in H_{(x,y)}$ and prove the left-hand side of equation (11.9). Let $(x,y) \in X_t(\mathbf{f}) \oplus H$. Then

$$H_{(x,y)} \cap X_t(\mathbf{f}) = H_{(x,y)} \cap \{(r,s) : d(\mathbf{f}(r,s)) \geq t\} \neq \emptyset$$

which means that $\exists(r,s) \in H_{(x,y)}$ such that $d(\mathbf{f}(r,s)) \geq t$, and hence,

$$d((\mathbf{f}\oplus_v H)(x,y)) \geq t$$

by the assumption that the right-hand side of equation (11.9) holds. Thus, $(x,y) \in X_t(\mathbf{f}\oplus_v H)$, and

$$X_t(\mathbf{f}) \oplus H \subseteq X_t(\mathbf{f}\oplus_v H)$$

Now, let $(x,y) \in X_t(\mathbf{f}\oplus_v H)$. Then

$$d((\mathbf{f}\oplus_v H)(x,y)) \geq t$$

which means that

$$\exists(r,s) \in H_{(x,y)} \ni d(\mathbf{f}(r,s)) \geq t$$

because of the constraint of equation (11.7). Hence,

$$H_{(x,y)} \cap X_t(\mathbf{f}) \neq \emptyset$$

which implies that

$$(x,y) \in \{(r,s) : H_{(r,s)} \cap X_t(\mathbf{f}) \neq \emptyset\}$$

Thus $(x,y) \in X_t(\mathbf{f}) \oplus H$, and

$$X_t(\mathbf{f} \oplus_v H) \subseteq X_t(\mathbf{f}) \oplus H$$

Next, we assume that $X_t(\mathbf{f} \ominus_v H) = X_t(\mathbf{f}) \ominus H$ and prove the right-hand side of equation (11.10). We have

$$\begin{aligned}\{(x,y) : d((\mathbf{f} \ominus_v H)(x,y)) \geq t\} &= \{(x,y) : H_{(x,y)} \subseteq X_t(\mathbf{f})\} \\ &= \{(x,y) : d(\mathbf{f}(r,s)) \geq t \forall (r,s) \in H_{(x,y)}\}\end{aligned}$$

Hence,

$$d((\mathbf{f} \ominus_v H)(x,y)) \geq t \Leftrightarrow d(\mathbf{f}(r,s)) \geq t \forall (r,s) \in H_{(x,y)}$$

Then, for any $t_0 \in \mathbb{R}$,

$$d((\mathbf{f} \ominus_v H)(x,y)) = t_0 \Rightarrow d(\mathbf{f}(r,s)) \geq t_0 \forall (r,s) \in H_{(x,y)}$$

and, thus,

$$d((\mathbf{f} \ominus_v H)(x,y)) \leq d(\mathbf{f}(r,s)) \; \forall (r,s) \in H_{(x,y)}$$

Now we assume that $d((\mathbf{f} \ominus_v H)(x,y)) \leq d(\mathbf{f}(r,s)) \; \forall (r,s) \in H_{(x,y)}$ and prove the left-hand side of equation (11.10). Let $(x,y) \in X_t(\mathbf{f}) \ominus H$. Then $H_{(x,y)} \subseteq X_t(\mathbf{f})$, which implies that for every $(r,s) \in H_{(x,y)}$, $d(\mathbf{f}(r,s)) \geq t$, and hence, using the constraint of equation (11.8),

$$d((\mathbf{f} \ominus_v H)(x,y)) \geq t$$

Thus $(x,y) \in X_t(\mathbf{f} \ominus_v H)$. Now, let $(x,y) \in X_t(\mathbf{f} \ominus_v H)$. Then

$$d((\mathbf{f} \ominus_v H)(x,y)) \geq t$$

and hence,

$$\forall (r,s) \in H_{(x,y)} d(\mathbf{f}(r,s)) \geq t$$

which means that

$$H_{(x,y)} \subseteq X_t(\mathbf{f})$$

and thus, $(x,y) \in X_t(\mathbf{f}) \ominus H$. □

Proposition 1 is important because it provides the following interpretation of the vector morphological operations defined in equations (11.1) to (11.6): vector dilation of $\mathbf{f}$ by H is equivalent to thresholding $\mathbf{f}$ at each level $t \in \mathbb{R}$ to obtain a set containing all pixels representing objects with d value greater than or equal to t and performing a binary dilation by H on this set, and then recombining the dilated sets to form the output colour image.

In the following sections we describe the use of colour morphology for multiscale image analysis and noise suppression.

11.3   MULTISCALE IMAGE ANALYSIS

The representation of image objects at multiple scales is important in many computer vision and image processing applications. A multiscale representation of an image consists of a set of images which are derived by filtering the original image with a family of filters of varying scale, or spatial extent. Multiscale image analysis using morphological filtering has been suggested for applications such as shape-size distributions, image compression, and edge enhancement (Maragos, 1990; Maragos, 1989; Overturf, Comer and Delp, 1995).

The type of filtering used to obtain a multiscale representation of an image determines the properties of the resultant multiscale representation (and thus the multiresolution representation, also). Linear filters, such as wavelet filters, quadrature mirror filters, and Gaussian filters, are often used. Multiscale image analysis using morphological filtering has been suggested for applications such as shape-size distributions, image compression, and edge enhancement (Maragos, 1990; Maragos, 1989; Chen and Yan, 1989; Overturf, Comer and Delp, 1995; Sun and Maragos, 1989; Toet, 1989). Although the linear filtering approach has the advantage that a multiscale representation obtained using linear filtering can be viewed as a space-frequency representation and its frequency content can be studied using Fourier analysis, the morphological filtering approach is more appropriate for quantifying shape information at different scales (Maragos, 1989; Chen and Yan, 1989).

In this section we investigate the use of the vector and component-wise approaches for obtaining multiscale morphological representations of colour images. We concentrate in particular on multiscale FSP operations on discrete-space images. The multiscale FSP opening and closing of a discrete-space greyscale image f by the finite, connected set $H \subseteq \mathbb{Z}^2$ at scale n are defined as (Maragos, 1989):

$$f \circ nH = (f \ominus nH) \oplus nH \tag{11.11}$$

and

$$f \bullet nH = (f \oplus nH) \ominus nH \tag{11.12}$$

for $n = 0, 1, 2, \ldots$, where

$$nH = H \oplus H \oplus \cdots \oplus H \quad (n-1 \ dilations)$$

We can perform a multiscale opening or closing on a colour image by replacing the greyscale erosions and dilations of equations (11.11) and (11.12) by either component-wise colour dilations and erosions or vector dilations and erosions. It is clear that the component-wise filter can have some unexpected effects in terms of spectral filtering. This raises the question of how to include spectral information in a multiscale colour image representation. The vector filter has an advantage over the component-wise filter in addressing this issue because it will not alter the colour composition of image objects. To illustrate the importance of this advantage for multiscale image analysis, Fig. 11.1 shows the component-wise multiscale opening in RGB space of a colour image, and Fig. 11.2 shows the vector multiscale opening of the same image, using luminance as the scalar-valued function for the reduced ordering. The structuring element H used to create these multiscale representations is shown in Fig. 11.3. The vector filter provides a more appropriate multiscale representation of the image than the component-wise filter. The component-wise multiscale representation would be particularly unsatisfactory if the image was to be represented by its multiscale edges (Mallat and Zhong, 1992). In this case a colour edge detector applied at each scale would not produce an accurate representation of the image structure. Although the component-wise filter could be applied in a colour space other than RGB, the undesirable effect illustrated in Fig. 11.1 could still occur, since the individual component images would still be filtered independently.

Since the output of the vector filter depends on the scalar-valued function used for reduced ordering, the selection of this function provides flexibility in incorporating spectral information into the multiscale image representation. For example, certain linear combinations of the tristimulus values can be used. This would be written as

$$d(\mathbf{f}(x,y)) = a_R f_R(x,y) + a_G f_G(x,y) + a_B f_B(x,y) \qquad (11.13)$$

if the image is filtered in the RGB colour space. For the case $a_R = 0.299$, $a_G = 0.587$, and $a_B = 0.114$, $d(\mathbf{f}(x,y))$ becomes the luminance image. Multiscale opening in this case would suppress bright objects at each scale. A multiscale representation obtained using the luminance image as the scalar-valued function and the structuring element shown in Fig. 11.3 is shown in Plate XII.

The values of a_R, a_G, and a_B can also be selected to enhance or suppress specific colours. For example, if $a_R = 1$, $a_G = 0$, and $a_B = 0$, then the effect of a multiscale opening would be to suppress objects with high red content. Similarly, if $a_R = 0$, $a_G = 1$, and $a_B = 0$ then green objects would be suppressed by a multiscale opening and if $a_R = 0$, $a_G = 0$, and $a_B = 1$ then blue objects would be suppressed. This would lead to a family of images parameterized by shape, size, and 'colour', which could be useful for an application

(a)   (b)

(c)   (d)

Fig. 11.1 Component-wise multiscale opening: (a) original image; (b) result of component-wise opening, $n = 3$; (c) result of component-wise opening, $n = 4$; (d) result of component-wise opening, $n = 5$.

(a)   (b)

(c)   (d)

Fig. 11.2   Vector multiscale opening with $d(\mathbf{f}(x,y)) = f_Y(x,y)$ (luminance image): (a) original image; (b) result of vector opening, $n = 3$; (c) result of vector opening, $n = 4$; (d) result of vector opening, $n = 5$.

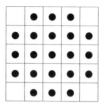

Fig. 11.3 Structuring element used for multiscale smoothing experiments.

such as object recognition. Plate XIII shows a multiscale representation (using the same original image and structuring element as those used for Plate XII) obtained using the red image as the scalar-valued function. It can be seen that the difference in the values of a_R, a_G, and a_B used for the two multiscale representations shown in Plate XII and Plate XIII strongly influences the resultant representation.

11.4   IMAGE ENHANCEMENT

In this section we investigate the use of colour morphology for noise suppression. To simulate noisy colour images with spectrally correlated noise, the classical process used to whiten correlated random variables is reversed (Fukunaga, 1990). At each pixel, a vector $\mathbf{Z}$, of three uncorrelated, unit-variance, random samples is generated. These samples are then mapped, through a linear transformation, to a vector $\mathbf{N}$ with three samples with covariance matrix:

$$\Sigma = \begin{bmatrix} \sigma_R^2 & \rho_{RG}\sigma_R\sigma_G & \rho_{RB}\sigma_R\sigma_B \\ \rho_{RG}\sigma_R\sigma_G & \sigma_G^2 & \rho_{GB}\sigma_G\sigma_B \\ \rho_{RB}\sigma_R\sigma_B & \rho_{GB}\sigma_G\sigma_B & \sigma_B^2 \end{bmatrix}$$

where σ_R^2, σ_G^2, and σ_B^2 are the variances of the red, green, and blue noise components, respectively, and ρ_{RG}, ρ_{RB}, and ρ_{GB} are the spectral correlation coefficients for the red and green, red and blue, and green and blue noise components, respectively.

To simulate the noise spatially, an ε-mixture of Gaussian noise is added to the R, G, and B components of a colour image (Chu and Delp, 1989). The

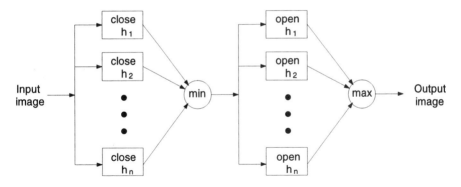

Fig. 11.4 Block diagram of 2DCO filter for greyscale images.

probability density function of this noise is given by:

$$
\begin{aligned}
f_{\mathbf{N}}(\mathbf{n}) \quad = \quad & \varepsilon \frac{1}{(2\pi)^{3/2}|\Sigma_1|^{1/2}} \exp\left(-\tfrac{1}{2}\mathbf{n}^{\mathrm{T}}\Sigma_1^{-1}\mathbf{n}\right) \\
& +(1-\varepsilon)\frac{1}{(2\pi)^{3/2}|\Sigma_2|^{1/2}} \exp\left(-\tfrac{1}{2}\mathbf{n}^{\mathrm{T}}\Sigma_2^{-1}\mathbf{n}\right)
\end{aligned}
$$

where $\mathbf{N} = [N_1, N_2, N_3]^{\mathrm{T}}$ is the random noise vector, and Σ_1 and Σ_2 are covariance matrices corresponding to impulsive and non-impulsive noise, respectively. Thus, with probability ε, a noise vector with covariance matrix Σ_1 is added at a given pixel. The variances $\sigma_{R_1}^2, \sigma_{G_1}^2$, and $\sigma_{B_1}^2$ for this noise are large. This allows spatially impulsive noise to be simulated. With probability $1 - \varepsilon$, a noise vector with covariance matrix Σ_2 is added. The variances $\sigma_{R_2}^2, \sigma_{G_2}^2$, and $\sigma_{B_2}^2$ for this noise are small. This allows for simulation of non-impulsive noise.

11.4.1 Experimental results for noise suppression

This section presents results of experiments involving noise suppression in colour images. The greyscale morphological filter used for all of these experiments was the 2DCO filter introduced by Stevenson and Arce (1987) and extended by Song and Delp (1991). A block diagram of this filter is shown in Fig. 11.4. It consists of a cascade of two stages each of which consists of multiple morphological operators of one type (opening or closing). In each stage multiple operators are applied to the input image using different structuring elements. The output of each stage is selected according to the morphological operator used: maximum for the output of the opening stage, minimum for the output of the closing stage. If the first stage consists of closing operators and the second stage consists of opening operators, then for a greyscale input image f and a set of structuring elments $\{h_1, h_2, \ldots, h_n\}$, the output y of the first stage is given by:

$$y = \min\{f \bullet h_1, f \bullet h_2, \ldots, f \bullet h_n\} \tag{11.14}$$

The output z of the second stage is given by

$$z = \max\{y \circ h_1, y \circ h_2, \ldots, y \circ h_n\} \tag{11.15}$$

The purpose of each stage is to preserve the geometrical features in the image that match any one of the given structuring elements. For the experiments presented in this section, four structuring elements consisting of lines of length 3 with different directions were used (Stevenson and Arce, 1987).

The 2DCO component-wise filter consists of applying the greyscale 2DCO filter to each of the three component images independently, whereas the 2DCO vector filter is defined by replacing the greyscale morphological operations of equations (11.14) and (11.15) by vector operations.

Noise was added to a colour image in RGB space for two different values of spectral correlation between component images. Plate XIV shows the original image used for the noise suppression experiments. The noisy image shown in Plate XV(a) has parameters $\varepsilon = 0.05$; $\sigma_{R_1} = \sigma_{G_1} = \sigma_{B_1} = 100$, for the impulsive noise and $\sigma_{R_2} = \sigma_{G_2} = \sigma_{B_2} = 10$ for the non-impulsive noise; and $\rho_{RG} = \rho_{RB} = \rho_{GB} = 0.95$ for both impulsive and non-impulsive noise. The noise between component planes in this image is highly correlated.

The result of a component-wise 2DCO filter applied in RGB space is shown in Plate XV(b).[1] Also shown, in Plate XV(c), is the result from the application of the vector 2DCO filter. The Euclidean norm:

$$d(\mathbf{f}(x,y)) = \sqrt{(f_R(x,y))^2 + (f_G(x,y))^2 + (f_B(x,y))^2}$$

was used as the metric for reduced ordering. This filtering method has results similar to the component-wise filtering in RGB space.

The second noisy image used in the experiments is shown in Plate XVI(a). In this image, the values of ε and the variances are the same as those used for

---

[1] We are not aware of any generally accepted metric for measuring distortion in noisy colour images. Therefore, evaluation of our results is subjective. The actual images shown in this chapter are available at: http://www.ece.purdue.edu/~ace

Plate XV. However, this image has $\rho_{RG} = \rho_{RB} = \rho_{GB} = 0.45$ for both types of noise. Thus, the spectral correlation of the noise is lower.

The result of component-wise filtering in RGB space is shown in Plate XVI(b). It can be seen that this filter has the same level of performance for this image as it did for the image of Plate XV. Also shown, in Plate XVI(c) is the result of the application of the vector morphological filter to the image with lower noise spectral correlation. This method does not perform as well as the component-wise RGB filter for noise with lower spectral correlation. With spectrally uncorrelated noise the vector approach will not perform as well as the component-wise approach due to the restriction that the output of the vector filter must be one of the input vectors inside the filter window.

ACKNOWLEDGEMENT

The work on which this chapter is based was partially supported by the AT&T Foundation and a National Science Foundation Graduate Fellowship.

12

Frequency domain methods

Stephen J. Sangwine and Amy L. Thornton

Image processing, in common with other branches of signal processing, has a well-developed literature covering image manipulation in the **frequency domain**. An image is a two-dimensional array of pixels and is referred to as being in the **spatial domain**. (In signal processing, the corresponding domain is usually the time domain, the signals, such as audio, being functions of time.) It is possible, however, to transform an image into the **spatial frequency domain** using, for example, a Fourier transform, and manipulate the frequency domain representation of the image. In the spatial domain, the image is represented as the variation of luminance and/or chrominance with position in the pixel array (corresponding to the imaging plane in the camera or other acquisition device), whereas in the frequency domain, the image is represented by its spatial frequency components, each having a magnitude and phase. This representation has it roots in the mathematical technique of Fourier Series analysis, whereby a periodic function or signal may be decomposed into a series of sinusoidal components, or alternatively, viewed as a superposition or summation of the sinusoidal components. The frequency domain representation of the image is usually referred to as its 'spectrum'.

It is conventional in image processing to talk of the 'frequency domain', omitting the adjective 'spatial', it being understood that when we talk of frequencies, we mean spatial frequencies expressed in cycles per unit length, and not frequencies in cycles per second (Hz); the unit length being either the width or height of the image, or the width or height of a pixel (thus we can talk of picture cycles or cycles per pixel). We adopt this convention in the rest of this chapter.

Frequency domain methods include many types of image manipulation and we now discuss some of them briefly. Many of these are discussed in books on image processing, for example by Castleman (1996) and Sonka,

Hlavac and Boyle (1993) as applied to greyscale images and the separate colour components of colour images.

Classic linear filtering, which can be implemented by convolution (discussed on page 152) in the spatial domain, can also be implemented in the frequency domain by computing the Fourier transforms of the image and the convolution mask, and multiplying together the image and mask spectra pixel-by-pixel. The resulting spectrum is the spectrum of the filtered image, which is obtained using an inverse Fourier transform. The process of transformation to the frequency domain takes time, of course, but for large convolution masks, this time and the time taken to compute the pixel-by-pixel products is less than that taken to perform the convolution. However, frequency domain methods are more important where more than one image is involved. For example, in object recognition it may be possible to have a reference image of a known object which will be seen in another image or images at some unknown position, scale, and possibly rotation. Because the object appears in both images, its frequency content is also in both images, although not directly comparable, and it is possible to find the object by correlating the sets of frequency domain information. This sort of technique can find objects without iteration; producing the coordinates and rotation and scaling parameters in a single-pass process.

There are other transforms which could be applied, and Castleman (1996) describes many of them. However, they will not be discussed here, because the problem in applying any of these transforms to colour images is essentially the same as the problem in applying the Fourier transform: colour image pixels are **triples** of colour or colour and luminance components, and the conventional Fourier transform handles real or complex data, that is, at most, **pairs** of pixel components. Often, those who use Fourier transforms in signal and image processing think of the input from the time or spatial domains as being real-valued (for example, sampled voltages or greyscale pixel values) because there is no meaning to an imaginary voltage. It is important to realize, however, that the Fourier transform is defined for *complex* input data: indeed many digital implementations of the transform operate on complex input data, the imaginary part of which is set to zero.

In the rest of this chapter, therefore, recent work on colour pixel representations for complex Fourier transforms and on 4-tuple (or quaternion) Fourier transforms will be presented, and the future of this field discussed.

12.1   REVIEW OF THE 2D DISCRETE FOURIER TRANSFORM

The two-dimensional discrete Fourier transform and its inverse are given by:

$$F(u,v) = S \sum_{m=0}^{M-1} \sum_{n=0}^{N-1} f(m,n) \exp[-j2\pi(\frac{mu}{M} + \frac{nv}{N})] \qquad (12.1)$$

$$f(m,n) = S \sum_{u=0}^{M-1} \sum_{v=0}^{N-1} F(u,v) \exp[j2\pi(\frac{mu}{M} + \frac{nv}{N})] \qquad (12.2)$$

where $S = 1/\sqrt{MN}$ is a scaling factor (for a square image, of course $N = M$). Some formulations of the transform have different scaling factors for the forward and inverse transforms: the advantage of using the same factor for each is that the same computer code can compute both forward and inverse transforms, only the sign of the complex exponentials needing to be changed.

In equations (12.1) and (12.2), $f(m,n)$ represents the spatial domain image, and $F(u,v)$ represents its Fourier transform or spectrum. If $f(m,n)$ is a greyscale image it will have a real (i.e. not complex) value at each coordinate (m,n). The values in the spectrum $F(u,v)$ however, are complex. If $f(m,n)$ is real, certain symmetry conditions apply to $F(u,v)$ which are discussed by Castleman (1996).

Direct evaluation of the discrete Fourier transform is possible only for small images because the double summation makes the transform complexity of $O(N^2 M^2)$: every point in the output array is calculated by summing all points in the input array, each multiplied by a complex exponential. Fortunately, there are much faster algorithms for computing the DFT, discovered in the 1960s and known as Fast Fourier Transform algorithms or FFTs. The details are outside the scope of this chapter and can be found in Castleman (1996) or Blahut (1985) or Brigham (1988).

It is obvious that colour images can be transformed by computing separate Fourier transforms for each colour component (for example, R, G and B) and this is what many commercial image manipulation packages do: the user must first separate the colour image into three component images which can be treated as greyscale images. Linear filtering operations on these component images applied in the frequency domain will work, and the component images can then be reassembled to make a colour image. However, this approach has significant limitations which motivated the work described in this chapter. Primarily, component-wise transformation does not allow the image spectrum to be handled **holistically**: you get three separate spectra. If you want to compute correlations of images in, for example, object recognition, where the same object may appear in two images, but *not with the same (precise) colouration,* you need a single spectrum for each image, not three separate spectra. A change in colouration between one image and another may be caused by illumination of a different colour temperature, as would happen in outdoor image capture under varying conditions of daylight (noon *versus* dusk, for example) or artificial light. What is needed is some way to handle colour image pixels as **vectors.** Similar issues arise in other developing areas of colour image processing discussed in this book, for example, in non-linear filtering discussed in Chapter 9 and Chapter 10 and in morphological filtering discussed in Chapter 11.

12.2 COMPLEX CHROMATICITY

Thornton and Sangwine (1995) partially achieved a single spectrum for a colour image by working with the chromaticity information alone, discarding the luminance. In Chapter 4 various different colour spaces are discussed, and in some of these spaces, one of the coordinate axes corresponds to luminance alone, the other two axes representing the colour, or chromaticity information. Examples of colour spaces where this applies are HSI, CIELAB, and YUV. Thornton and Sangwine used a coordinate scheme based on geometrical projection from RGB space onto a complex plane denoted by Z. The origin of this plane coincides with the point $(0,0,0)$ in RGB space, and the plane is perpendicular to the greyline in RGB space which extends from $(0,0,0)$ to $(255,255,255)$ (assuming an 8-bit discrete RGB space). Fig. 12.1 shows the relationship of the Z plane to the projections of the R, G and B coordinate axes of RGB space: R' is the projection of the R axis of RGB space onto the plane, and similarly for the other axes. The equation to convert from RGB coordinates to Z coordinates is as follows, and is derived by straightforward geometry:

$$Z = \cos\theta \left[\frac{1}{2} r(g+b) + j \frac{\sqrt{3}}{2} g - b \right]$$

$\cos\theta$ is a scale factor which represents the cosine of the angle between the RGB axes and the Z plane. It may be omitted. Intensity is defined as $I = (r+g+b)/3$ as in a typical formulation of HSI space (section 4.6 on page 4.6). To convert from I and Z back to RGB the following equations are used, derived by inversion of the equations just given.

$$r = I + \frac{2}{3}\Re(Z)$$

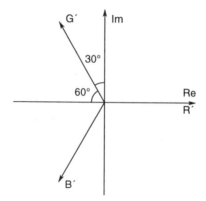

Fig. 12.1 The IZ plane [after Thornton and Sangwine (1995)]

$$g = I + \frac{1}{\sqrt{3}}\Im(Z) - \frac{1}{3}\Re(Z)$$

$$b = I - \frac{1}{\sqrt{3}}\Im(Z) - \frac{1}{3}\Re(Z)$$

where $\Re(Z)$ and $\Im(Z)$ are the real and imaginary parts of Z respectively. These equations may produce values outside the valid range of RGB coordinates, and a clipping algorithm may be needed to adjust such values.

12.2.1 Phase correlation

Using the above representation (Thornton and Sangwine, 1996; Thornton and Sangwine, 1997) it has been shown possible to apply **phase correlation** (Kuglin and Hines, 1975) to chromaticity images (that is colour images with no luminance or intensity component), otherwise known as **isoluminant** images (McCabe, Caelli, West and Reeves, 1997). Phase correlation operates as follows. A reference image (g_1) contains a view of an object to be located in a second image g_2. The reference image could show a single object, or multiple objects, best seen against a plain background. The second image could contain other objects as well as background clutter. Phase correlation produces a spatial domain image P, whose magnitude (P being complex), will contain a peak corresponding to the translation of the object between the two images:

$$P = \left| \mathcal{F}^{-1} \left\{ \frac{G_1 G_2^*}{|G_1 G_2^*|} \right\} \right| \qquad (12.3)$$

where G_1 is the Fourier transform of g_1, that is $G_1 = \mathcal{F}(g_1)$; $\mathcal{F}$ denotes a Fourier transform; and $\mathcal{F}^{-1}$ denotes an inverse Fourier transform, and $*$ denotes a complex conjugate. Note that G_1 and G_2 are multiplied point-by-point, not by using matrix multiplication, and the G_2^* means that each pixel in G_2 is replaced by its complex conjugate. Phase correlation can be explained in relatively non-mathematical terms as follows. The Fourier transform of each image contains information about the frequency content of the image, including the frequency content due to the object of interest. The position of the object in the image affects only the phase of the frequency points, that is the arguments of the complex numbers in the spectrum. Any given frequency always appears at the same position in the spectrum. In other words, the Fourier transform is translation invariant. When the two spectra are multiplied, point-by-point, the difference in phase of each frequency value is computed: this is easily understood by considering the complex values in exponential form:

$$A_1 \exp(j\theta_1) A_2 \exp(j\theta_2) = A_1 A_2 \exp(j(\theta_1 + \theta_2))$$

Now, because the complex conjugate of G_2 was used in equation (12.3), all its phase values were negated, and therefore they are subtracted from those of G_1. The denominator of equation (12.3) divides by the magnitudes of the frequency values, thus removing all but the phase differences. The result is known as the **normalized cross-power spectrum** . Finally, the inverse Fourier transform produces a surface in the spatial domain which is largely flat because the uncorrelated phases due to the background do not add coherently. The correlated differences due to the object of interest, however, add to produce a peak, which can be found by classical signal processing methods, including thresholding if the peak is of sufficient amplitude.

In Thornton and Sangwine (1996) the complex values in g_1 and g_2 represent the chromaticity of pixels in the two images. In particular, the argument of each pixel represents hue, and the modulus represents, essentially, saturation; concepts which are explained in section 4.6. The phase correlation surface therefore (which is complex) should have zero phase at the peak if the object seen in the two images had the same hue, otherwise the phase at the peak reveals the difference in hue between the two images. Figure 12.2 shows an example phase correlation surface computed from two colour images. The difference between the two images is the position and colour of

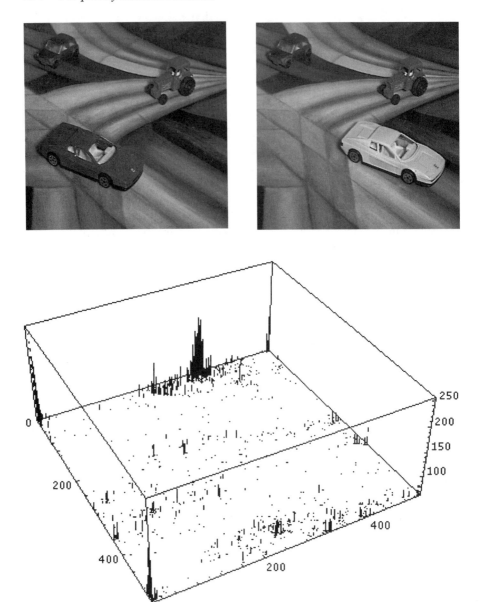

Fig. 12.2   Monochrome versions of two colour images and the phase correlation surface computed from the original colour images.

the car. In one image the car is red, in the other yellow. The principal peak in the phase correlation surface reveals the displacement of the car between the two images, and its phase reveals the difference in hue. A clearer result is obtained by using an image of the reference object against a plain background. When a reference image with a multi-coloured background is used there is more noise in the phase correlation surface, and the hue value of the peak may be obscured.

12.2.2   The spatial chromatic Fourier transform

McCabe, Caelli, West and Reeves (1997) have also employed a complex Fourier transform to handle chromatic pixel information. They employed a form of YUV space, similar to that described in section 4.3 and in section 5.3.2. Ignoring Y, they then employed a complex Fourier transform with the real and imaginary parts of the complex pixels corresponding to red-green and blue-yellow axes respectively.

The significance of this work lies in the interpretation of the spectral domain information: they show how points in the frequency domain represent sinusoidally varying chromatic information in the spatial domain, the spatial frequency and orientation of the sinusoid being determined by position in the spectral domain (as with Fourier transforms of greyscale images); and the chromatic sequence of the sinusoid by the values of the complex spectral points, which determine a *path* through colourspace. This is a very significant insight which deserves to be extended to the work described in the next section.

This interpretation of the chromatic properties of the spectral domain data has lead McCabe, Caelli, West and Reeves (1997) to a possible approach to chromatically sensitive *linear* filters for the first time. Undoubtedly there will be applications for such filters in time, but there is still a great deal of fundamental research needed to understand how they may be designed, and what their detailed properties are.

12.3   THE QUATERNION FOURIER TRANSFORM

The work described so far in this chapter was a first step towards true Fourier transforms of colour images, but it suffers from the limitation that only two components of the colour space can be handled because complex numbers have only two components. A natural question to ask is whether there are 'higher order' complex numbers which could be used, and the answer is that there are. The **quaternion numbers** were discovered by Hamilton (1866) in the 1860s. They are not widely known other than by mathematicians, some quantum physicists and a few aerospace and robotics engineers. Sangwine (1996) showed that it is possible to define a Fourier transform using these numbers with four components per pixel, and that it is therefore possible to compute holistic Fourier transforms of colour images. Much remains to be done, however, to establish the properties of the transform, and it has been established by Thornton, for example, that the phase correlation technique is not trivially extensible to quaternion Fourier transforms, there being four symmetrically positioned peaks in the phase correlation surface rather than one.

12.3.1   The quaternion numbers

A quaternion number has a real part and three imaginary parts and it can be represented in Cartesian form as:

$$a + ib + jc + kd \tag{12.4}$$

where a, b, c and d are real numbers, and i, j and k are complex operators (generalizations of the complex operator j — usually denoted by i outside engineering). The properties of the three operators are as follows:

$$i^2 = j^2 = k^2 = ijk = -1$$

$$\begin{aligned} ij &= k & jk &= i & ki &= j \\ ji &= -k & kj &= -i & ik &= -j \end{aligned} \tag{12.5}$$

Note that within the ring of quaternions, there are three complex sub-fields. A quaternion-space (which is a 4-space) may be sectioned to give three complex sub-spaces or planes.

Given a quaternion $Q = a + ib + jc + kd$, its quaternion conjugate is $Q^* = a - ib - jc - kd$, and its modulus is given by:

$$|Q| = \sqrt{a^2 + b^2 + c^2 + d^2} \tag{12.6}$$

Euler's theorem generalises to the complex sub-fields with operators i, j and k, thus:

$$e^{(\pm i\theta)} = \cos\theta \pm i\sin\theta$$

$$e^{(\pm j\theta)} = \cos\theta \pm j\sin\theta$$
$$e^{(\pm k\theta)} = \cos\theta \pm k\sin\theta$$

Multiplication of quaternions is *not commutative*, and there are left and right quotients from division. This makes use of the quaternions less straightforward than use of the complex numbers and one must exercise care in algebra and in computer coding to ensure that products are written and coded in the correct order. The fundamental reason for this problem is that quaternion multiplication is essentially a vector cross-product whose direction (in 4-space in this case) is reversed by interchanging the operands. For a mathematical, but accessible, modern account of quaternion numbers see Artmann (1988).

12.3.2   The quaternion Fourier integral

The discrete quaternion Fourier transform (DQFT) described in this chapter was developed from the work of Ell (1992),(1993) who studied the analysis of two-dimensional linear systems, particularly with regard to their stability criterion. Ell introduced the quaternion Fourier integral of a two-dimensional function: (using two time variables):

$$H[j\omega, kv] = \int_{-\infty}^{\infty} \int_{-\infty}^{\infty} e^{-j\omega t} h(t,\tau) e^{-kv\tau} \, dt \, d\tau$$

with inverse defined as:

$$h(t,\tau) = \frac{1}{4\pi^2} \int_{-\infty}^{\infty} \int_{-\infty}^{\infty} e^{j\omega t} H[j\omega, kv] e^{kv\tau} \, dv \, d\omega$$

This transform appears very similar to the two-dimensional Fourier integral (Castleman, 1996):

$$F(u,v) = \int_{-\infty}^{\infty} \int_{-\infty}^{\infty} f(x,y) e^{-i(ux+vy)} \, dx \, dy$$

Indeed, the only significant difference is the separation of the two complex exponentials and the use of two different complex operators j and k. These small differences should not lead the reader into thinking that the transform is a trivial development, however. The use of quaternions here makes a fundamental difference to the two-dimensional nature of the transform, which is what Ell needed in his work.

Ell did not discuss the possibility of applying this transform to image processing, neither did he discuss the meaning of his two-dimensions of time. He also seems to have assumed that the time-domain function to be transformed ($h(t, \tau)$) is real-valued, but there is no reason to limit the function to be real or even complex-valued, and this makes possible the application of this transform to handle colour images, as discussed in the next section.

12.3.3   The discrete quaternion Fourier transform

The DQFT and its inverse, deduced by Sangwine (1996), is:

$$F(u,v) = S \sum_{m=0}^{M-1} \sum_{n=0}^{N-1} e^{-j2\pi\frac{mu}{M}} f(m,n) e^{-k2\pi\frac{nv}{N}} \tag{12.7}$$

$$f(m,n) = S \sum_{u=0}^{M-1} \sum_{v=0}^{N-1} e^{j2\pi\frac{mu}{M}} F(u,v) e^{k2\pi\frac{nv}{N}} \tag{12.8}$$

where the discrete array $f(m,n)$ is of dimension $M \times N$, and S is as in equations (12.1) and (12.2), with which equations (12.7) and (12.8) should be compared. The ordering of the exponential factors here is important because of the non-commutative multiplication of quaternions. As discussed at the end of the last section, the function to be transformed ($f(m,n)$) is assumed to be quaternion-valued. The spectrum, $F(u,v)$ is also quaternion-valued.

The two-dimensional quaternion-valued functions which are transformed can be colour images (in RGB format, for example). These can be handled by placing the three colour components into the three imaginary parts, leaving the real part zero. (Symmetry suggests this arrangement, but there is no mathematical reason known why some other arrangement should not be used.) It has been shown, and reported in Sangwine (1996), that a colour image in RGB format may be transformed into the spatial frequency domain and then inverse transformed back into the spatial domain to recover the image exactly, and that multiplication of the spectral domain image with the two-dimensional frequency response of a simple filter produces a low-pass or high-pass filtered image.

12.3.4 Basis functions

The combination of two complex exponentials from different complex sub-fields gives rise to basis functions which are combinations of cosines and sines (cosinusoidal basis functions) as are found in the two-dimensional DFT. In the DQFT however, the orthogonality of the two complex exponentials creates a separation of information in the spectral domain that does not occur with the DFT. This may be seen as follows. Consider a quaternion-valued pixel $a + ib + jc + kd$ multiplied by the two complex exponentials (which forms one contribution to a spectral domain point):

$$(\cos\alpha - j\sin\alpha)(a + ib + jc + kd)(\cos\beta - k\sin\beta) \tag{12.9}$$

where $\alpha = 2\pi mu/M$ and $\beta = 2\pi nv/N$. This may be seen to multiply out as shown in equation (12.10) (taking care with ordering of any combinations of i, j and k, and using the rules given in equation (12.5) above to simplify terms as required).

$$
\begin{aligned}
&+ a\cos\alpha\cos\beta - b\,\sin\alpha\,\sin\beta + c\,\sin\alpha\cos\beta + d\cos\alpha\,\sin\beta \\
+i\,(&+ a\,\sin\alpha\,\sin\beta + b\cos\alpha\cos\beta + c\cos\alpha\,\sin\beta - d\,\sin\alpha\cos\beta\,) \\
+j\,(&- a\,\sin\alpha\cos\beta - b\cos\alpha\,\sin\beta + c\cos\alpha\cos\beta - d\,\sin\alpha\,\sin\beta\,) \\
+k\,(&- a\cos\alpha\,\sin\beta + b\,\sin\alpha\cos\beta + c\,\sin\alpha\,\sin\beta + d\cos\alpha\cos\beta\,)
\end{aligned}
\tag{12.10}
$$

For comparison, we now consider the same situation in the DFT as given in equation (12.1): a single complex-valued pixel is multiplied by the complex exponential and forms one contribution to a spectral domain point:

$$(a + ib)(\cos\alpha - i\sin\alpha)(\cos\beta - i\sin\beta) \tag{12.11}$$

where α and β are as above. This may be seen to multiply out as shown in equation (12.12).

$$
\begin{aligned}
&+ a\cos\alpha\cos\beta - a\,\sin\alpha\,\sin\beta + b\,\sin\alpha\cos\beta + b\cos\alpha\,\sin\beta \\
+i\,(&- a\,\sin\alpha\cos\beta - a\cos\alpha\,\sin\beta + b\cos\alpha\cos\beta - b\,\sin\alpha\,\sin\beta\,)
\end{aligned}
\tag{12.12}
$$

It is apparent from equations (12.10) and (12.12) that the two-dimensional DFT and the DQFT differ fundamentally in the way that they represent the spectral information, although the positioning of points in the spectral domain as a function of horizontal and vertical spatial frequency is the same. The DQFT separates the four possible combinations of horizontal and vertical cosine and sine components into the four components of the quaternion-valued spectral point, whereas the DFT merges some of these together ($a \cos \alpha \cos \beta$ and $a \sin \alpha \sin \beta$ for example) into a single real or imaginary component.

Notice that, in the DQFT, as in the DFT, the combinations of any given cosine and sine product are separated into different real or imaginary parts. For example, the combination $\cos \alpha \cos \beta$ appears in the real part of the DFT spectrum with coefficient a, and in the imaginary part with coefficient b; and in the four parts of the DQFT spectrum with each of the coefficients a, b, c, or d. Thus, any given spatial variation of a colour component from the set $\{r, g, b\}$ is separated into different real/imaginary parts of the spectral point.

12.4   DISCUSSION

Application of frequency domain methods to colour images depends crucially on formulation of a holistic transform that can handle all three components of a colour pixel and produce a single spectrum. This can be done in a limited way using the complex Fourier transform, and the work of McCabe, Caelli, West and Reeves (1997) has indicated an approach towards the interpretation of the spectral information.

Sangwine's (1996) work with quaternions shows a possible way forward, but much remains to be done with interpretation of the spectral domain information, and in establishing the detailed properties of the transform, such as symmetry. Thornton and Sangwine's (1996) work on phase correlation must be extended to quaternion-valued transforms: if this cannot be done, then it is likely that other approaches for correlating images will also not work correctly. This particular problem is a relatively simple one to understand and offers the most likely way to a better understanding of the quaternion transform.

Finally, the work of McCabe, Caelli, West and Reeves (1997) shows a possible approach to the understanding of the spectral information in the quaternion-valued spectrum as well as possible approaches to **colour sensitive filtering** , in which linear filters may be definable to have both spatial and chromatic sensitivity. This requires new developments in the theory of 2D linear systems to extend the concepts of impulse response and frequency response to a 4-space: a spatial filter has an impulse response whose Fourier transform is the spatial frequency response of the filter. A colour sensitive filter would have an impulse response with different spatial frequency responses for different colours. This may be represented by cross-sections or paths through quaternion-valued spectral space, but at the time of writing, this is mere speculation.

ACKNOWLEDGEMENT

The work of Amy L. Thornton was supported by a PhD studentship awarded by the Research Endowment Trust Fund of The University of Reading.

13

Compression

Marek Domański and Maciej Bartkowiak

13.1 IMAGE AND VIDEO COMPRESSION

The digital data representing still images fills large files with size increasing with picture resolution. Introduction of colour increases the memory requirements by a factor of 1.5 to 4 as compared to monochrome images. For example, a file containing row data representing a medium resolution 640×480 colour picture with 8-bit representation of its RGB components exhibits a size of 0.92 MB. Storage of such uncompressed pictures leads to unreasonable demands for disk space and would cause visual databases to be inefficient. The need for compression is even more apparent for digital video. For example, an uncompressed digitized video signal of a television channel according to the ITU-T Recommendation BT.601 needs 216 Mbps (ITU, 1994b). Transmission of such a large bitstream needs unacceptably large bandwidth even using advanced efficient modulation techniques. Therefore digital image and video compression techniques are necessary for applications in communications, multimedia, medical systems, consumer electronics, remote surveillance *etc.* Exemplary present and future areas of application include:

- videophones, videoconferencing;

- digital television, digital high-definition television (HDTV);

- interactive video services: video on demand, teleshopping, tele-education *etc.*;

- visual databases;

- telemedicine;

- interactive games;

- remote control and surveillance.

There are large prospective markets for all of the above and therefore much money has been invested in research in these areas and a lot of research has been done on still image and digital video data compression. Video data compression has been shown to be one of the most challenging tasks in research on information and communication technology. Many hundreds of technical papers have been published in journals and proceedings of international conferences, symposia and workshops. The exciting results obtained inspired definition of entirely new applications, e.g. new multimedia services, which again stimulated further research on compression for channels with various bitrates.

In general, compression is possible because of the redundancy in uncompressed images and video data. There are two kinds of information which can be removed during compression:

redundant information – the information that can be obtained from other pieces of information existing in a picture or video sequence;

irrelevant information – the information which is not relevant for the viewer, e.g. information which is impossible or difficult for a viewer to perceive or even information which has been classified as less important.

Removal of redundant information leads to **lossless encoding** where the original image can be retrieved during the decoding (decompression) process without any loss of information. On the other hand, classification of information as irrelevant is somewhat subjective. This source of compression is exploited in **lossy coding** in which some degradation of the image occurs: the original image cannot be perfectly reconstructed from its compressed data. Although the degradation may not be perceivable, it may be annoying after several cycles of compression and decompression. Nevertheless significant reduction in the size of an image/video file usually leads to substantial degradation of the quality of decompressed pictures as shown on Plate XIX and Plate XXI.

In order to store or transmit compressed image/video data, the original data representing images or frames from a video sequence are fed into a device which performs compression, often called **source coding** in textbooks related to image communication. Such a device is called a **coder** (or **source**

coder). The coder produces a compressed stream of digital data which is decompressed by a **decoder** after transmission or when the information is re-trieved from storage. For bidirectional communication there must be both a coder and a decoder installed at each end of a link. These two devices are often implemented as one device called a **codec**. Codecs for still images are mostly implemented as computer software. Video codecs may also be imple-mented in this way, mostly for testing or low-cost applications, but practical applications need high-speed implementations based on sets of very sophis-ticated dedicated VLSI chips.

There are many coding techniques applicable to colour images. These techniques can be classified into groups according to the way in which they deal with colour.

The most common approach to colour image compression consists of sep-arate encoding of the three image components. Lossy coding mostly operates on the luminance-chrominance representation with subsampled chrominance. The rationale for this is that the chrominance frequency band is narrower than that for luminance (see also section 13.2.9). The total number of samples in both chrominance components is often only 50% of the number of luminance samples. Moreover the chrominance components are often more compressed than luminance, at least in high-compression applications. Thus the amount of chrominance data in the compressed bitstream can be 20% less than the amount of compressed luminance data. This figure explains why most ef-fort is focused on luminance compression. Selected issues related to separate encoding of components from still images are reported briefly in section 13.2.

A more sophisticated approach is based on **vector processing** of the three components where their mutual dependencies are exploited (see sec-tion 13.3). A special case of this approach is related to calculation of the palette representation of an image (which is itself a compression technique) and possibly further compression of the data related to the palette representa-tion.

13.1.1   Compression efficiency

The degree of data reduction achieved by a compression process or algorithm is called **compression ratio**:

$$compression\ ratio = \frac{N_i}{N_o}$$

where N_i is the length of the data before compression (bits) and N_o is the length of the data after compression (bits). The data size after compression is often expressed also as an average number of bits per pixel (mainly for still images) or as the bitstream rate needed to transmit a video sequence, in kbps or Mbps.

By purely lossless compression, the compression ratio obtained for a given image or a sequence of images depends only on the technique applied. Obtainable compression ratios using lossless and nearly lossless coding are small and usually do not exceed 4 to 5. Such coding is needed in applications where visual data must remain unchanged over many consecutive cycles of compression and decompression. Medical and cartographic applications as well as digital television sequences of contribution quality are good examples of this.

Higher compression ratios can be achieved by lossy coding in which various techniques give different degrees of compression and different quality of the decompressed pictures. Of course, quality decreases as compression ratio increases. Therefore measures of compression ratio *versus* quality measured for various test images or sequences define the efficiency characteristics of a lossy technique, rather than a single compression ratio as for the lossless techniques. Thus the often asked popular question: 'What is the highest possible compression?' has little meaning for lossy coding. For example, digital television video signals compressed to bitrates over 6 to 8 Mbps can be retrieved with no perceivable degradation while low-resolution videophone pictures compressed for transmission through very low bitrate channels (even under 30 kbps) like analogue subscriber loops can only be decompressed with much annoying degradation.

In order to evaluate codecs the quality of the decompressed images must be measured. There are two basic ways of doing this:

subjective evaluation by assessment of the quality of the decompressed images or assessment of impairments between the original and decompressed images made by a panel of viewers who vote on a numeric scale;

objective evaluation by pixel-wise calculation of an error measure between the original and the decompressed image.

Subjective evaluation consists in averaging of scores given by some number of viewers (at least 15). The result is called mean opinion score (MOS). Assessment of television images is normalized by the International Telecommunication Union (ITU). The respective Recommendation BT.500 (ITU,

1994a) defines experimental conditions like the minimum number of participants, lighting conditions, timing *etc.* Moreover it defines a scale from 1 (the lowest score) to 5 (the highest score) for both quality and impairment assessment. The technique recommended for evaluation of compression errors is called the **double-stimulus continuous quality-scale method** (DSCQS). The experiment participants are asked to assess the quality of two pictures or sequences displayed consecutively for a short time of about 10–15 s each. The observers mark their notes on a continuous scale from 1 to 5. Usually, one image or sequence is the unimpaired reference picture or sequence while the other is that obtained after decompression. A variant of the method mentioned above is the **double stimulus impairment scale method**. In this method the observers are asked to vote on the impaired pictures, keeping in mind the reference unimpaired pictures. Moreover, **single stimulus methods** are in use. In these methods a single image or sequence of images is presented and the observers make their assessments on the presentation.

The usual method of objective evaluation is to calculate the peak signal-to-noise ratio (PSNR) on the basis of the normalized mean square error (NMSE) between the original (uncompressed) image and the decompressed image:

$$\text{PSNR (dB)} = -10 \log \text{NMSE},$$

where for a monochrome image or the luminance component of a colour image NMSE is defined as:

$$\text{NMSE} = \frac{\sum\limits_{i=1}^{N} e_i^2}{N \cdot 255^2} \tag{13.1}$$

where e_i is the error between the reconstructed and the original image calculated at the ith pixel, N is the number of pixels, and 255 is the dynamic range assuming an 8-bit representation (the most common). The PSNR measure defined above is a modification of the signal-to-noise ratio measure already discussed in section 8.1. Here, PSNR is calculated relative to the dynamic range rather than to signal variation while in section 8.1 signal-to-noise ratio was calculated relative to the signal variance. The values of PSNR are about 6 dB above those of signal-to-noise ratio. Therefore, PSNR is not only easier to calculate than signal-to-noise ratio but it also gives more optimistic results, and this may account for its popularity in reporting the results of compression.

Some authors use simple, mean square error (MSE) or its root (RMSE):

$$\text{MSE} = \frac{\sum_{i=1}^{N} e_i^2}{N}$$

Note that many papers on colour image compression report results calculated in this way for the luminance component only. Of course, a more relevant way to calculate PSNR is to calculate e_i using the colour difference formulae (see section 3.4). Such cases may need to replace the number 255 in the denominator of equation (13.1) by the dynamic range of the respective colour difference. For example, Westerink, Biemond and Boekee (1988) found experimentally that perceptually proper division of bits between luminance and chrominance in the compressed bitstream is obtained by application of the Euclidean distance in the YUV space when the squared distances along the U and V axes are weighted with the factor 0.3.

Nevertheless, there is no commonly accepted method for calculation of the PSNR of colour images. Moreover, the results obtained using PSNR often do not coincide with those obtained in subjective tests. It is well known that small errors randomly distributed over the whole image reflect badly in objective measures, i.e. small values of PSNR, but they often correspond to invisible degradation of image quality. On the other hand, relatively high-valued errors concentrated in a particular part of an image can permit the PSNR to be relatively high (i.e. good) while the subjective assessment is very low because of an annoying corruption of a particular portion of the picture.

This observation calls into question the usefulness of this objective measure which, however, does exhibit a very important advantage – it is easy to calculate. This fundamental advantage makes PSNR very popular. Nevertheless its disadvantages have motivated many researchers to search for better objective quality measures. Unfortunately they have not gained much success so far (*Proceedings International Conference on Image Processing*, 1996, Session 17A1: Image quality evaluation, pp.869-940). Recently, some measures that coincide relatively well with the subjective methods have been described. These methods try to model more general properties of the human visual system rather than to perform the calculations at the pixel level, e.g. (Tan, Ghanbari and Pearson, 1997).

There are several other issues related to evaluation of compression methods. Resilience to transmission errors, complexity aspects, interworking with other systems, and scalability, are examples of the related issues discussed in the references on image and video coding given in this chapter.

13.1.2   Digital representation of images for compression

Images for compression may be in various formats which are defined by:

- colour space, like RGB, YC_bC_r, CMY/CMYK *etc.* (discussed in Chapter 4 and Chapter 5);

- the number of bits per component sample;

- spatial resolution;

- temporal resolution (for video).

The RGB space is not an efficient representation for compression because there is significant correlation between the colour components in typical RGB images. For compression, a luminance-chrominance representation (such as YC_bC_r, YUV, YIQ, CIELAB etc) is considered superior to the RGB representation because of lower inter-component correlation. RGB images acquired from colour cameras and scanners are therefore transformed to one of the luminance-chrominance spaces prior to compression.

Nevertheless colour space conversion introduces some errors caused by rounding of the results of arithmetic operations. Therefore, the RGB representation may be used for strictly lossless compression with a low compression ratio.

Because of the system complexity requirements the **numbers of bits per sample** of each component should be low, but on the other hand, these numbers need to be high enough to avoid visible impairments. Reduction of the number of bits per sample is equivalent to an increase in the size of the quantization steps between quantization levels. When the quantization steps exceed the colour discrimination thresholds of the human visual system smooth changes of colour become step-like. So called **false contours** appear in an image with too low a number of quantization levels. This effect is very annoying in many natural pictures, for example, of human faces.

Since the number of levels of visual stimuli recognized by the human eye is of the order of 100 (Wyszecki and Stiles, 1982), it is assumed that the 256 levels offered by an 8-bit representation are 'safe enough'. Representation of each component by 8-bit samples leads to so-called **true colour** 24-bit representation of colour image samples which is the most common format in today's high-quality computer display and storage devices. The popular

Table 13.1 Minimum number of quantization levels necessary to keep the quantization error unperceivable (i.e. below the discrimination threshold) for the RGB and Munsell spaces (after Gan, Kotani and Miyahara (1994)).

	Gamma-correction	
Representation	*None*	$(\gamma = 3)$
R	2^{14}	2^{10}
G	2^{16}	2^{12}
B	2^{12}	2^{9}
H	8 times the value of C	
V	2^{8}	
C	2^{6}	

display hardware used in cheaper computer video adapters utilizes fewer bits by application of 6-bit video digital-to-analog converters. Even worse limitations apply to the display mode called **high colour**, where only 15 or 16 bits for the (R, G, B) triple are assigned to a pixel. This limitation does not seem to be annoying for everyday usage of computer displays. However, precise analysis of quantizing errors (Ikeda, Dai and Higaki, 1992a; Ikeda, Dai and Higaki, 1992b) leads to the conclusion that uniform quantization of R, G and B components to 256 levels causes in the worst case visual errors 20 times greater than the same quantization performed on the components of the CIELAB colour space. Therefore for high quality applications, like medical, chemical or physical research, the RGB-based system is extremely inefficient and would require more than 8 bits for an individual component in order to keep the error below the threshold of visual perception as shown in Table 13.1 (Gan, Kotani and Miyahara, 1994). All these results are related to serious non-uniformity of the RGB colour space, which means that the distance between some pairs of points in the RGB space corresponds to an unnoticeable colour difference while the same distance in another part of the space corresponds to a quite significant difference in colour sensation (Pratt, 1991; Wyszecki and Stiles, 1982). The same colour accuracy could be obtained within the Munsell colour space using only 9, 8 and 6 bits for the H, V

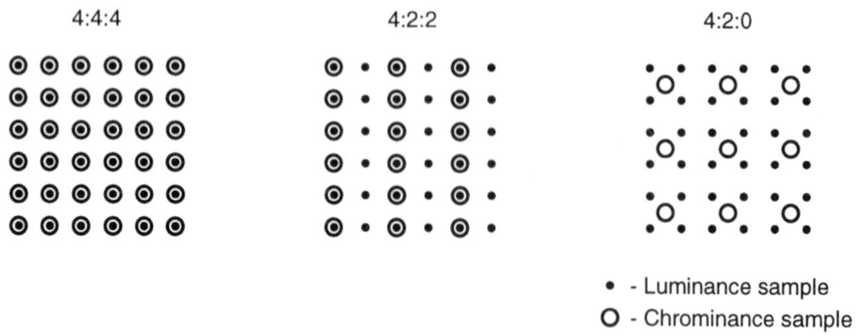

4:4:4   4:2:2   4:2:0

• - Luminance sample
O - Chrominance sample

Fig. 13.1 Basic chrominance sampling schemes: 4:4:4 (no chrominance subsampling) and 4:2:2, 4:2:0 (with chrominance subsampling).

and C coordinates, respectively (Table 13.1). Therefore the JPEG still image compression standard (ISO/IEC, 1994) specifies 12-bit component samples as an option for lossy coding while it allows up to 16 bits per sample of a component for lossless coding.

A 24-bit representation is also typical for the YC_bC_r representation of digital video. Nevertheless 10-bit representation of the digital television components is used in some studio and high-fidelity applications.

Spatial resolution is usually the same for all components of the RGB representation, while in the luminance-chrominance systems the chrominance components are mostly subsampled with respect to the luminance component. The basic video chrominance sampling schemes are shown in Fig. 13.1. The 4:2:2 sampling scheme is used in high-fidelity applications like contribution-quality television while 4:2:0 is very common in high-compression applications. Basic video formats are summarized in Table 13.2. The digital television studio format defined by the ITU-R Recommendation BT.604-1 (ITU, 1994b) is the basic digital format for 25 Hz (as used in Europe) and 30 Hz (as used in the USA and Japan). Subsampling leads to several formats such as SIF and CIF *etc.* Higher spatial and temporal resolutions are related to high-definition television (HDTV) formats (about $10^6 - 2 \times 10^6$ luminance pixels per frame) and super high-definition formats (over 4×10^6 luminance pixels per frame).

Table 13.2 Basic digital video formats.

	Digital TV D1	*SIF*	*CIF*	*QCIF*	*sub-QCIF*
Chrominance subsampling	4:2:2	4:2:0	4:2:0	4:2:0	4:2:0
Luminance format (active pixels)	720×576 720×480	352×288 352×240	352×288 or 360×288	176×144	128×96
Chrominance format (active pixels)	360×576 360×480	176×144 176×120	176×144 or 180×144	88×72	64×48
Bits per component sample	8 (10)	8	8	8	8
Frame rate (Hz)	25 or 30	25 or 30	30*	30*	30*

* or a submultiple.

13.2   COMPONENT-WISE STILL IMAGE COMPRESSION

13.2.1   Entropy coding

Consider a group of **lossless** coding techniques related to the following idea: assign shorter bit strings to those symbols which are more frequent in an image while longer bit strings will be assigned to less frequent symbols. The set of encoded symbols is a set of numbers, e.g. a set of grey levels or a set of values of a colour image component. A variable-length bit string called a **code** is assigned to each symbol. Therefore entropy coding (EC) techniques are often called **variable-length coding**.

A well-known result from information theory says that the average number of bits per symbol must not be less that the value of the **entropy**:

$$H = -\sum_{i=1}^{K} p_i \log_2 p_i \tag{13.2}$$

where p_i is the probability of occurrence of the ith symbol and K is the number of symbols. Practical methods usually result in numbers of bits per symbol being very close to this minimum value defined by the entropy. The inspection of the above formula results in a corollary that application of variable-length codes becomes more efficient when the symbol probability distribution is more nonuniform.

Useful codes must be uniquely decodable in the sense that there is only one possible input sequence that could have produced a given encoded sequence. A very well-known example of such codes are the Huffman codes (Huffman, 1952). In order to produce an optimum code, the statistics of the signal must be known. Therefore a coder using such a code reads the data twice: once to estimate the statistics and once to encode the data.

The same code table containing all the codes needed to encode a given set of symbols must be used by both the coder and the decoder. There are two basic ways to ensure this:

1. The code table is transmitted together with the data, e.g. for each image. Optimal codes can be used at the cost of transmission of the table of codes. The coder calculates the optimal code table automatically for a given set of data, e.g. for an image.

2. The coder and decoder use predefined tables which have been derived for some assumed data statistics. Neither calculation of the optimal code nor transmission of the code table is needed but the coding efficiency may decrease as the actual data statistics differ from the assumed.

The Huffman codes and their modifications are widely used in image compression. The code generation algorithm and its modifications are discussed elsewhere e.g. (Jayant and Noll, 1984; Gersho and Gray, 1992; Held and Marshall, 1996). For the sake of brevity, we finish our consideration of Huffman codes with a simple example (Table 13.3). A very useful extension is Huffman block coding where variable-length codewords are assigned to combinations of symbols instead of individual symbols. There are more advanced techniques of entropy coding often used in image data compression. In **arithmetic coding** (Pasco, 1976; Rissanen, 1976) variable-length codes are assigned to variable-length blocks of symbols. Because it does not require assignment of integer-length codes to fixed length blocks of symbols (as Huffman coding does) it can approach more closely the lower bound established by equation (13.2). Implementations of arithmetic coders are more complicated and need to overcome the precision problem; however, they tend to yield a higher compression ratio than Huffman coders. The algorithms for arithmetic coding are described in many textbooks e.g. (Gersho and Gray, 1992; Held and Marshall, 1996) and will be omitted here. The Q-coder (Mitchel and Pennebaker, 1988) is one of several variants of the arithmetic coder.

Table 13.3   A Huffman code table for differential coding of DC value (ISO/IEC, 1994).

Symbol	Code word (binary)
0	00
1	010
2	011
3	100
4	101
5	110
6	1110
7	11110
8	111110
9	1111110
10	11111110
11	111111110

Ziv-Lempel coding assigns fixed-length codes to variable-length blocks of symbols. It is named after its inventors Ziv and Lempel who published the algorithm (Ziv and Lempel, 1977; Ziv and Lempel, 1978) which was later corrected and improved by Welch (1984). An advantage of Ziv-Lempel coding is that, in contradistinction to both previously mentioned techniques, it does not need any knowledge of the input signal statistics. Variants of this technique are often used for file compression in various computer environments.

Since good-contrast images exhibit flat luminance histograms, the compression ratio achieved by entropy coding is usually small. All the above-mentioned techniques of entropy coding result in low compression ratios usually not exceeding 1.5 for luminance and something more for chrominance. Much better lossless compression results can be obtained by joint application of entropy and predictive coding (see section 13.2.3). Actually all data streams produced by modern coders are entropy-encoded.

13.2.2   Run length coding

Run-length coding (RLC) is a lossless compression process that results in a string of repeated symbols being converted into a pair (v, r), where v (value) denotes the repeated symbol and r (run) is the number of times the symbol is repeated. Run-length coding is used in bilevel image or facsimile compression. Its applications to colour images include mostly coding of transform coefficients as discussed in section 13.2.4.

13.2.3   Differential pulse code modulation

Differential pulse code modulation (DPCM) predicts the component value of the next pixel based on the values of its neighbours. The differences between the actual value u and predicted values p of pixels forms the prediction error or difference image d as illustrated in Fig. 13.2. In the simplest case of linear prediction, the predicted value p is a neighbouring pixel value stored in the predictor P. More complicated linear predictors calculate the predicted value as a linear combination of the values of the neighbouring pixels. The decoder must receive some initial value or values to start its predictor. The values of consecutive pixels are calculated as sums of the predicted values p and the differences d received from the coder. The predictor usually makes its predictions for consecutive pixels in a row or column. The coder/decoder pair is realizable if the predictor, except for the initial values, reads only the pixels whose values have already been decoded. For example, if an image is processed row-wise from left to right and from top to bottom, the value of a consecutive pixel in the jth row and ith column can be predicted from the pixels in the row $j-1$ or from the pixels in column $i-1$ as shown in Fig. 13.3. Images can be modelled as signals with high correlation which is exploited in DPCM. As the histogram of a difference image exhibits a

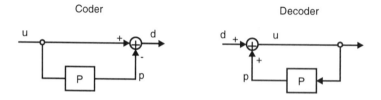

Fig. 13.2 Lossless DPCM coder and decoder.

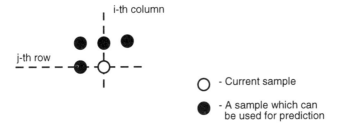

Fig. 13.3 Prediction

high peak about zero, application of entropy coding to encode the difference image is efficient. Therefore DPCM together with entropy coding is a typical lossless image coding technique. For the R, G, and B components of natural scene images compression ratios of 2 to 4 are usually achievable. Note that no quantization is used in the DPCM coder for lossless coding. Introduction of quantization of the prediction error d results in lossy coding with higher compression.

13.2.4   Transform coding

Transform coding is the basic tool of contemporary lossy coding techniques. Usually, it is performed independently on the luminance and chrominance components. Very many aspects and variants of transform coding have already been studied in a great number of papers; an interesting review of them was given by Clarke (1985). The main idea of the method is to calculate the $(N \times N)$-pixel two-dimensional transform of the $(N \times N)$-pixel blocks from each image component. Orthogonal transforms exhibit the property of energy compaction, i.e. the higher values of the transform coefficients tend to concentrate in one area of the block. Therefore rough quantization or even neglect of other coefficients has a relatively low impact on the quality of the image calculated from the quantized transform coefficients *via* the inverse transform. This operation is performed in the decoder which receives the coefficients quantized and then encoded using run-length and entropy coding.

Profound research has been carried out on the choice of the most efficient orthogonal transform, i.e. for a transform that can be implemented *via* fast algorithms and which efficiently compacts the signal energy into as few as possible coefficients. As a result of this research, image and video compression standards incorporated the **discrete cosine transform** (DCT) performed on (8 × 8)-pixel blocks. The block size of 8 × 8 pixels has been chosen as a compromise between implementation complexity and achievable energy compaction.

The coefficients $F(k,l)$ of the discrete cosine transform are calculated from the matrix of the pixel samples $f(m,n)$ from a given image component:

$$F(k,l) = \frac{1}{4}C(k)C(l) \sum_{i=0}^{7} \sum_{j=0}^{7} f(i,j) \cos\left[\frac{\pi(2i+1)k}{16}\right] \cos\left[\frac{\pi(2j+1)l}{16}\right] \quad (13.3)$$

where $i, j = 0, 1, 2, \ldots 7$ are integer coordinates within a pixel block, $k, l = 0, 1, 2, \ldots 7$ are integer frequency samples, $f(i,j)$ are the values of the image components (mostly Y, C_r or C_b) for the pixel (i,j) within the block, and

$$C(z) = \begin{cases} 1/\sqrt{2} & z = 0 \quad z \in \{k,l\} \\ 1 & z \neq 0 \end{cases}$$

Direct calculation according to equation (13.3) is inefficient because it requires 64 multiply-accumulate operations for each DCT coefficient, i.e. 4096 multiply-accumulate operations for each 8 × 8 block. There exist fast two-dimensional DCT algorithms with significantly reduced complexity. For example, the algorithm proposed by Cho and Lee (1991) needs only 96 multiplications and 466 additions for an 8 × 8 block. Unfortunately the most efficient two-dimensional algorithms are highly irregular. More regular implementations can be obtained in separable solutions, that is by successive calculation of one-dimensional 8-point DCTs for all rows and then by column-wise transformation of the result. The one-dimensional 8-point discrete cosine transform:

$$F(k) = \frac{1}{2}C(k) \sum_{l=0}^{7} f(l) \cos\left[\frac{(2l+1)k\pi}{16}\right]$$

can be implemented using 11 to 13 multiplications and 29 additions (Lee, 1984; Arai, Agui and Nakajima, 1988). In some structures some of the multiplications perform scaling of the final output to the correct range. If the output is to be quantized, as is usual, the scaling factors can be absorbed into the divisions needed to quantize the outputs. Only 5 multiplications and 29 additions are actually needed before quantization in order to implement the one-dimensional 8-point transform (Arai, Agui and Nakajima, 1988). A deeper discussion can be found in Rao and Yip (1990). The discrete cosine transformation results in an 8×8 block of transform coefficients. The coefficient $F(0,0)$ is called the DC coefficient. In practice, it always exhibits the highest value. The other coefficients $F(k,l)$ are called the AC coefficients. Their values tend to decrease as the indices k and l increase, that is, most of the signal energy is in the lower frequency components. Therefore the higher coefficients can be represented with a smaller number of bits. The usual way of doing this is to quantize the transform coefficients. This quantization is an irreversible lossy compression operation in the DCT domain. Each transform coefficient $F(k,l)$ is quantized in a uniform quantizer defined by its quantization step $Q(k,l)$. The quantized value $F_Q(k,l)$ of the transform coefficient $F(k,l)$ is an integer:

$$F_Q(k,l) = \left\langle \frac{F(k,l)}{Q(k,l)} \right\rangle$$

where $\langle \cdot \rangle$ denotes rounding to an integer value. This rounding may be performed by different means.

A matrix $\mathbf{Q}(k,l)$ defined for each colour component of an image is called a **quantization table**. The choice of quantization table strongly influences the coding efficiency and is related to the problem of optimal bit allocation which has been discussed in many papers (Clarke, 1985). A standard calculation proposed by Huang and Schultheiss (1963) based on the assumption of Gaussian distribution of the coefficient values leads to a relation between the numbers of bits and the coefficient variances. Among other approaches, one that deals with quantization steps based on the visibility of 8×8 DCT basis functions for luminance and chrominance has to be noted. Typical quantization tables proposed as an example in the JPEG still image compression standard (see section 13.2.8) are shown in Fig. 13.4 and Fig. 13.5 (ISO/IEC, 1994). The compression ratio, and hence the quality of the reconstructed image, is controlled by scaling the quantization matrix. Scaling by a factor larger than 1 results in a higher compression ratio and higher impairment of the reconstructed image. If the quantization tables are scaled by a factor 0.5, the resulting reconstructed image is usually nearly indistinguishable from the source image. A technique of perceptual optimization of the quantization tables for given images based on the visibility of the quantization noise has been developed by Watson (1994). An improvement of some dB in PSNR (by constant compression ratio) may also be gained by adaptive techniques which change the quantization tables locally within a component of an image. Among many techniques reported (Clarke, 1985), the scheme of Chen

16	11	10	16	24	40	51	61
12	12	14	19	26	58	60	55
14	13	16	24	40	57	69	56
14	17	22	29	51	87	80	62
18	22	37	56	68	109	103	77
24	35	55	64	81	104	113	92
49	64	78	87	103	121	120	101
72	92	95	98	112	100	103	99

Fig. 13.4 An example quantization table for the luminance component (after ISO/IEC (1994)).

17	18	24	47	99	99	99	99
18	21	26	66	99	99	99	99
24	26	56	99	99	99	99	99
47	66	99	99	99	99	99	99
99	99	99	99	99	99	99	99
99	99	99	99	99	99	99	99
99	99	99	99	99	99	99	99
99	99	99	99	99	99	99	99

Fig. 13.5 An example quantization table for both the chrominance components (after ISO/IEC (1994)).

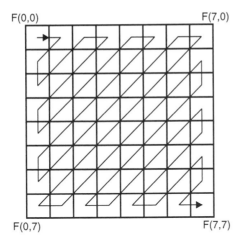

F(0,0)   F(7,0)

F(0,7)   F(7,7)

Fig. 13.6 Zig-zag scan.

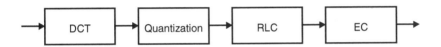

| DCT | Quantization | RLC | EC |

Fig. 13.7 Transform coding scheme.

and Smith (1977) may be regarded as being representative for adaptive algorithms. The idea is to switch the bit allocation scheme according to the activity class of the block while this activity is measured by the total AC coefficient energy of the block. The activity is measured independently for the blocks of each component.

After quantization the coefficients $F_Q(k,l)$ are ordered according to the zig-zag scan shown in Fig. 13.6. This ordering converts the matrix of transform coefficients into a sequence of coefficients which starts with the DC coefficient followed by 63 AC coefficients. There are often values repeated for some consecutive coefficients. Therefore run length coding (RLC) is usually applied to encode the sequence of AC coefficients as discussed in section 13.2.2. The pairs (v,r) are then compressed using entropy coding using the sequence of processing blocks shown in Fig. 13.7. During the entropy and run length decoding process, the quantized values of the transform coefficients are obtained. Dequantization is performed by multiplying the coefficients by the respective quantization steps. Therefore the quantization tables

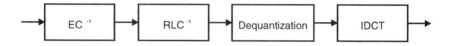

Fig. 13.8   Transform decoding scheme.

as well as the tables of entropy coding have to be sent to the decoder as **side information**. The values of the pixels in the individual image components are reconstructed via the **inverse discrete cosine transformation** (IDCT):

$$f(i,j) = \frac{1}{4} \sum_{k=0}^{7} \sum_{l=0}^{7} C(k)C(l)F(k,l) \cos\left[\frac{\pi(2i+1)k}{16}\right] \cos\left[\frac{\pi(2j+1)l}{16}\right]$$

The inverse transformation is implemented using the same fast algorithms as the forward transformation. The general scheme of transform decompression is illustrated in Fig. 13.8. Plate XVII shows an original colour image. Plate XVIII shows the reconstructed image after JPEG compression at 0.6 bpp, while Plate XIX shows the reconstructed image after JPEG compression at 0.26 bpp. For high compression ratios a phenomenon called **blocking effect** becomes very annoying as shown in Plate XIX. This effect is a characteristic artefact for block transform coding whereby the blocks of pixels which were transform coded become visible in the decompressed image.

13.2.5   Subband coding

Subband coding (SBC) sometimes also called **wavelet coding** is a well established technique to compress independently the components of colour images and video (Vetterli, 1984; Woods and O'Neil, 1986). Although it is not included in existing image compression standards, it is used in commercially available software. The basic idea is to split up the two-dimensional frequency band of an image into subsampled channels which are encoded using techniques accurately matched to the individual signal statistics and possibly to the properties of the human visual system in the individual subbands.

Practical image subband coding techniques mostly use separable decomposition, i.e. one-dimensional filters are used in order to separate the frequency bands both horizontally and vertically. The reason is that separable filter implementations are much simpler than implementations of nonseparable two-dimensional filters. On the other hand, the gain in coding efficiency

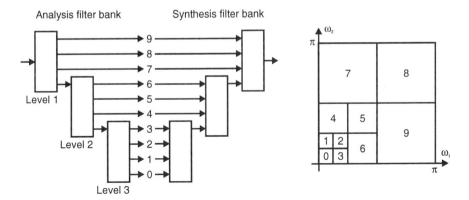

Fig. 13.9 Tree-structured decomposition of the two-dimensional frequency band (a three-level system).

obtained by application of nonseparable filters is usually small or negligible (Woods, 1991).

The published literature on image subband coding describes a great variety of separable decomposition schemes with the number of channels in the range of 4 to 64. Among them the multi-stage wavelet decomposition schemes are very popular as their spectral resolution fits image power spectra which decay with distance from the origin. Experimental results with standard test images of resolution of about 512×512 show that the 10-band three-stage (three-level) decomposition schemes (Fig. 13.9) are a good practical choice for luminance decomposition. At each level the two-dimensional frequency band is decomposed into four subbands horizontally and vertically subsampled by a factor of 2. Only the lowest frequency subband is input to the next level of decomposition. Both chrominance components are usually subsampled by a factor of 2 horizontally and vertically at the input to the coder. Therefore the decomposition scheme for chrominance needs one stage less than that for luminance. Three levels of decomposition of a real image are shown in Fig. 13.10. The images are processed in the coder by an analysis filter bank and, in the decoder, by a synthesis filter bank. The synthesis filter banks are either the same as the analysis filter banks or closely related to them.

Processing of an image by a multi-level separable filter bank is in fact a sequence of processes related horizontally and vertically by two-band filter banks. These filter banks can be implemented very efficiently together

Fig. 13.10 Three-level subband decomposition of the luminance component of the colour test image 'Boats' (Plate XVII)

with the following subsampler in a so called **polyphase structure**. The number of arithmetic operations in such a two-band filter bank is approximately half of that in a single respective filter operating at the input sampling frequency. Several filter bank design methods have been developed (Vaidyanathan, 1993; Vetterli and Kovacevic, 1995) and used in image coding. Design of filter banks influences complexity and efficiency of the coding system and therefore should be made carefully. Very well-known and still widely used filter banks are the finite impulse response (FIR) quadrature mirror filters introduced by Johnston (1980). Other interesting FIR designs (Mallat, 1989; Smith and Barnwell, 1988; Egger and Li, 1995) have also often been used in image data compression. Moreover recursive filters have been successively applied in image coding (Ramstad, 1988; Domański, 1988).

The spectral decomposition does not introduce any compression in itself. Actual compression is achieved by encoding of the subbands. The lowest frequency subband is a strongly-decimated replica of the original image as can be seen in Fig. 13.10. The quality of reconstruction of this subband has a great influence on the quality of the fully reconstructed image, therefore this subband has to be coded with relatively high fidelity. In order to achieve a high compression ratio the other subbands are coded more crudely. Earlier works used DPCM (see section 13.2.3) and vector quantization, which is discussed in the next section, to encode the subbands independently. Such techniques need a method for estimation of an optimal bit allocation scheme (Woods, 1991). An interesting variant is to use the perceptually optimal bit allocation for all three components. The techniques of hierarchical quantization (Domański and Świerczyński, 1994) and embedded zerotree wavelets (Shapiro, 1993; Said and Pearlman, 1996) exploit the mutual dependencies between subbands. In particular, the latter technique is considered as a very efficient compression method allowing compression ratios of about 100 with a PSNR of about 28 dB for the luminance component. This technique outperforms DCT-based transform coding. Plate XVIII and Plate XX compare two images reconstructed from the 0.6 bpp bitstream generated by a DCT-based transform coder (Plate XVIII) and a Shapiro subband coder (Plate XX). Strong artefacts are perceivable at high compression (Plate XIX).

The characteristic degradation caused at high compression ratios by subband coding is colourful ringing at edges and smoothing of details as shown in Plate XX and Plate XXI. These images should be compared with Plate XVIII and Plate XIX respectively which show the results of reconstruction from JPEG compressed images at identical or very similar compression ratios. Detailed reviews can be found in the books by Woods (1991), Vetterli and Kovacevic (1995) and in the paper by Ramstad (1988).

13.2.6   Vector quantization

The main idea of vector quantization (VQ) is to quantize vectors formed from groups of image pixels into so called **codevectors** from a limited set called a **codebook**. Usually, nearest neighbour quantizers are used. Such a quantizer assigns to a given vector that vector from the codebook that is its nearest neighbour as shown in Fig. 13.11 (Gersho and Gray, 1992; Jayant and Noll, 1984). The codebook is usually designed for a class of images, for example, as represented by a set of training images. The basic technique is the algorithm proposed by Linde, Buzo and Gray (1980). This algorithm is in fact designed to iteratively improve a given initial codebook. The design of a codebook with N codewords can be stated as follows:

1. In some arbitrary way, choose N codewords yielding a small distortion of the data vectors represented by them.

2. At each step clusters of vectors represented by individual codevectors are calculated.

3. The codevectors are replaced by the centroids of the clusters just obtained.

4. The procedure repeats until the process converges to a solution which is a minimum of the total representation error (but possibly only a local minimum).

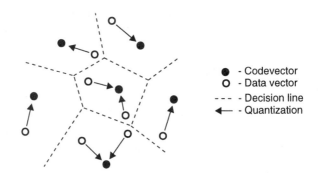

Fig. 13.11   Vector quantization to the nearest codebook vector.

Figure 13.12 illustrates this process. This iterative algorithm may be applied for still image codebook generation but it is too slow for real-time video applications. For such applications much faster noniterative but nonoptimal algorithms (e.g. binary split, pairwise nearest neighbour) are applicable (Gersho and Gray, 1992).

Colour quantization and palette representation generation (see section 13.3) are examples of vector quantization. The process allocates a label addressing the codevector to each triple of colour component values. Other approaches can use higher dimensionality vectors including also values of other pixels. The compression ratio tends to increase as the dimensionality of the quantized vectors grows.

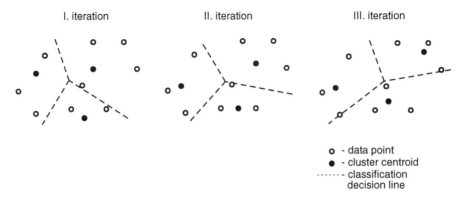

o - data point
● - cluster centroid
······ - classification
decision line

Fig. 13.12   A two-dimensional example of three consecutive iterations of the LBG algorithm.

13.2.7   Fractal coding

Another approach, closely related to vector quantization, is based on the theory of fractal objects which exhibit high visual complexity but low information content. The first practical solution was given by Jacquin (1992) who proposed fractal block coding which exploits self-similarity between pairs of blocks under a set of geometrical affine operations like translations, rotations, flipping and scaling. In his approach the coding process is based on dividing the image into blocks and then, for each such block, finding one that, when transformed, approximates it well. Encoding requires a relatively long processing time, but decoding is fast.

13.2.8   Still image compression standards

Image compression standardization has the goal of ensuring interoperation of codecs from different manufacturers. The standards define the rules required to create and decode the stream of compressed data. During the last 15 years several standards have been established by ISO (International Organisation for Standardization) , IEC (International Electrotechnical Commission) and ITU (International Telecommunication Union) for different areas of application. Some of the standards define strictly the image input format and compression algorithm while others are more generic. Standards for video compression are briefly reported in section 13.4.2. Here, formal standards for still colour images will be briefly noted.

JPEG (ISO/IEC, 1994) is the basic standard to compress component-wise colour images. The standard neither specifies the colour basis of the images nor the number of components. Moreover, the component resolution is not specified and may be different between components. For lossy coding it is reasonable to convert RGB images into the luminance-chrominance representation. However, the conversion (colour space transformation and downsampling) can consume over 30% of the encoding processing time and, similarly, the inverse conversion may need more than 30% of the computational power consumed by decompression. JPEG codecs operate in one of the following modes:

- Sequential encoding: each image component is encoded in a single scan.

- Progressive encoding: the image is encoded in multiple scans and each scan improves the quality of the reconstructed image by encoding additional information.

- Lossless encoding.

- Hierarchical encoding: the image is encoded at multiple resolutions, so that lower-resolution versions may be accessed without first having to reconstruct the full resolution image.

The first two modes are based on DCT-coding while the lossless mode uses spatial DPCM and entropy coding.

JPEG coders operate for a large range of compressions between 2 bits/pixel and 0.2 bit/pixel with the quality of the reconstructed images changing from indistinguishable from the original to moderate quality with some annoying degradations (Pennebaker and Mitchell, 1993; Wallace, 1991). The performance of the JPEG coding technique can be assessed from Plate XVIII and Plate XIX. An objective performance comparison of the JPEG codec implementation from the Independent JPEG Group (IJG, Undated) with the wavelet coder (Davis, Danskin and Heasman, 1997) is shown in Fig. 13.13. This plot has been prepared using Godlove's formula for calculation of colour error in Munsell colour space (discussed in section 4.8).

Colour facsimile standard

This uses the CIELAB colour space for image representation and the baseline sequential JPEG algorithm (Bhaskaran and Konstantinides, 1995).

13.2.9   Choice of colour space

The problem of the choice of colour space is relevant for lossy compression techniques. Since each transformation from one colour coordinate system to another is related to noninteger scaling of the component values, the finite state representation imposes subtle distortion due to numerical errors and primarily, the rounding operation.

The search for the optimal colour representation for image compression can be addressed from several points of view. In general, the choice of colour space may influence the coding efficiency either very slightly or significantly (up to 70% compression ratio improvement) (Van Dyck and Rajala, 1994; Abel, Bhaskaran and Lee, 1992; Charrier, Knoblauch and Cherifi, 1997), depending on the particular properties of the coding method employed, as well as the quality measure used to assess the coding efficiency. The latter must not be derived from a simple error measure calculated as a colour difference in one of the colour spaces considered, since it might favour its own colour space. At the least, a colour error formula which is sufficiently relevant to colorimetric measurement or better, subjective tests and mean opinion score can be recommended as the basis of ultimate judgement.

Considering various colour spaces, three aspects can be taken into account:

1. Colour images generally do not contain three times as much information as their greyscale counterparts. The individual colour components

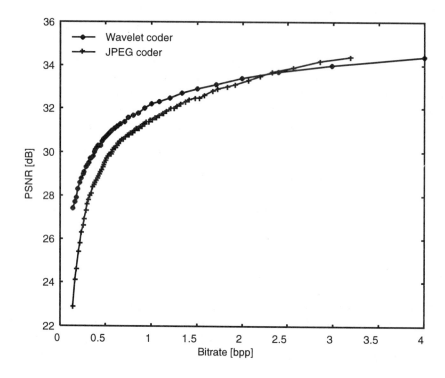

Fig. 13.13 The JPEG and wavelet encoding of the colour test image 'Lena' (Plate XXII): PSNR *versus* the number of bits per pixel.

often exhibit significant redundancy.

The common solution is to apply an orthogonal decomposition of the three-dimensional colour data in order to compact the image energy into possibly few channels (Pratt, 1971). The YC_bC_r or YIQ coordinates are often proposed as a representation which provides nearly as much energy compaction as a theoretically optimal decomposition based on the eigenvectors of the data covariance matrix, which is in fact the basis of the Karhunen-Loeve transform (or KLT). However, the energy compaction is not directly a measure of redundancy and therefore the significant gain in coding efficiency may justify the computational over-head in calculation of the KLT basis (Abel, Bhaskaran and Lee, 1992).

2. The perceptual uniformity implies constant sensitivity to coding arte-facts throughout the colour space.

   This important property is related to the efficiency of the coding sys-tems that apply uniform scalar quantization to the colour components or any of their derivatives in a general sense. Among others, the CIELAB colour space and AC_1C_2, (so called **cone space**) (Van Dyck and Rajala, 1994) can be suggested as those yielding decompressed images that are visually better at constant compression ratio than those obtained using quantization in the RGB or YIQ spaces.

3. A convenient control over coding artefacts can be achieved using a colour space that separates individual characteristics of the colour, as perceived by human beings.

   The colour spaces separating the luminance from the chrominance in-formation can be suggested to fulfil such requirements, and one from the family of HLS, IHS, HSV seems to give the best solution here. Allo-cating an optimal number of bits to the three channels related to light-ness, hue and saturation, one can easily adapt the compression ratio to one's individual sensitivity to the perceived discrepancies of various aspects of colour in the reconstructed image. Moreover, several classes of images exhibit a very limited range of certain colour features which can be highly compressed using this representation.

An important issue related to lossy (i.e. involving quantization in any general sense) coding of colour images is related to the gamut mismatch of various colour spaces. Due to the finite number of quantization steps, the original colour space of the input image must be completely covered in the colour space chosen for coding. Since the coordinate systems are usually nonparal-lel, many potential combinations of quantization levels lie outside the original gamut (Fig. 13.14). This yields reduced efficiency in a finite state representa-tion of the image in other than the original colour space. Despite its imperfec-tions, the YC_bC_r space is commonly chosen as the intermediate representation of colour images for compression purposes. The popular implementations of image compression standards, like JPEG employ this space as the default with the only alternative being the CMYK representation used mostly for prepress image manipulation due to the very high requirements for the accuracy of colour conversions between CMYK and other systems as discussed in Chap-ter 15.

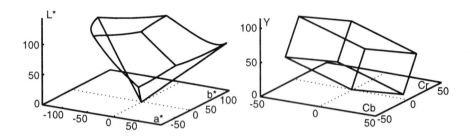

Fig. 13.14 The RGB colour gamut plotted in the CIELAB and YUV coordinate systems.

The important advantage of the luminance-chrominance colour representation which makes it especially suited for compression, is related to the specific properties of the colour information within images of natural scenes:

- While the image energy is distributed almost equally among the three R, G and B components, both spatially and spectrally, the power spectra of the individual luminance and both chrominance components exhibit unequal distribution favouring the luminance component as the one containing the bulk of fine detail high frequencies as illustrated by the power spectra of the 'Lena' image (Plate XXII) shown in Fig. 13.15.

- Whereas the sensitivity of the human visual system is relatively high for chrominance inaccuracy, it is very reduced in frequency, so the chrominance data needs only a fraction of the luminance resolution in order to impose sufficient sharpness on the perceived image. Therefore, common compression techniques employ various subsampling schemes for the chrominance channels in order to exploit this feature.

Fig. 13.15 The power spectra of the standard colour image 'Lena' (see Plate XXII): R, G, B components (upper row), Y, C_r, C_b components (lower row)

13.3　TECHNIQUES EXPLOITING MUTUAL DEPENDENCIES BETWEEN COLOUR COMPONENTS

Let us introduce a common notation. Denoting the given image $\mathbf{I}$ as a matrix of size N_1 by N_2, a pixel (i, j) is represented as a three-component vector $\underline{X}_{i,j}$:

$$\mathbf{I} = [\underline{X}_{i,j}], \qquad\qquad i = 1 \dots N_1 - 1, \quad j = 1 \dots N_2 - 1$$

$$\underline{X}_{i,j} = \{R_{i,j}, G_{i,j}, B_{i,j}\}, \text{ or}$$
$$\underline{X}_{i,j} = \{Y_{i,j}, C_{b_{i,j}}, C_{r_{i,j}}\}, \text{ or}$$
$$\underline{X}_{i,j} = \{C_{i,j}, M_{i,j}, Y_{i,j}\}, \text{ etc.}$$

13.3.1　Vector control of component-wise processing

Most of the adaptive image compression techniques utilize local spatial information to adapt their control procedures to the local image content. Thus, it is possible to obtain better efficiency by allocating more bits to those regions which require more exact representation, while visually flat and smooth regions can be treated more crudely. Since for most natural colour images the edges and similar patterns occur at approximately the same location in each component (Maragos, Mersereau and Schafer, 1984), it is reasonable to estimate the spatial activity using jointly the information from three channels. Usually, a discrete estimate of the gradient operator is used as a measure of the local activity:

$$\hat{\nabla}\mathbf{I} = \left[\frac{\Delta'\underline{X}_{i,j}}{\Delta i}, \frac{\Delta''\underline{X}_{i,j}}{\Delta j} \right]$$

where:

$$\Delta'\underline{X}_{i,j} = X_{i+1,j} - X_{i,j} \qquad\qquad (13.4)$$

or

$$\Delta'\underline{X}_{i,j} = X_{i+1,j} - X_{i-1,j} \qquad etc. \qquad (13.5)$$

and $\Delta i = 1$ or $\Delta i = 2$, respectively. Similarly

$$\Delta''\underline{X}_{i,j} = X_{i,j+1} - X_{i,j} \qquad\qquad (13.6)$$

or

$$\Delta''\underline{X}_{i,j} = X_{i,j+1} - X_{i,j-1} \qquad etc. \qquad (13.7)$$

and $\Delta j = 1$ or $\Delta j = 2$, respectively. Refer to section 9.3.1 for a detailed discussion on various methods of gradient estimation.

The conceptual difference between component-wise and vector control is related to the way the gradient is employed. While the component-wise approach controls the compression parameters of each colour component independently, using the respective components of the gradient matrix, the vector approach uses some **norm** of the gradient matrix,

$$\left\| \frac{\Delta X_{i,j}}{\Delta i} , \frac{\Delta X_{i,j}}{\Delta j} \right\|$$

to control the three channels simultaneously. Experimental results show that adaptive control of coding parameters benefits from the vector approach since the estimation of spatial activity is more precise when using whole information about colour changes at a given pixel position.

A special case of the vector quantization (VQ) technique (discussed in section 13.2.6) called address-vector quantization exhibits local smoothness and also continuity of edges across block boundaries within images coded using VQ. Thanks to local stationarity, some combinations of neighbouring blocks can be predicted with high probability (Feng and Nasrabadi, 1988). An extension of this technique to the case of colour images has been proposed by its authors (Feng and Nasrabadi, 1989). Application of an algorithm for joint prediction of the blocks of three-component vectors on the basis of an assumed 3-dimensional statistical model of inter-block coincidence leads to a technique offering bitrates as low as 0.5 to 0.6 bpp with PSNR=28 to 31 dB, which is at least twice as good a compression ratio compared to traditional vector quantization.

A combination of subband compression and vector quantization techniques gives a class of algorithms for efficient compression of image subbands (discussed in section 13.2.5) under given bitrate constraints. Li and Jain (1996) extended such a technique called **entropy constrained vector quantization** (proposed by Senoo and Girod (1992)) by the application of a so called activity map in order to control the suppression of upper subbands. Various local spatial activity estimation methods within different colour spaces have been investigated. Bitrates of 0.28 to 0.6 bpp have been reported as yielding good visual quality of the reconstructed images.

13.3.2   Joint colour and spatial compression techniques

Many compression techniques originally proposed for coding of greyscale images and extended to the case of colour images can be expressed as a sequence of linear and non-linear operations on vector-valued signals. If the colour components are coded separately, much of the processing effort is redundant and the compression is not efficient, since similar information is coded several times. In the past, a decorrelation in terms of linear transformation of the components was proposed as a remedy for the redundancy. The image data was converted to a more appropriate colour space, i.e. YUV, or YIQ space, or the coordinate system defined by the eigenvectors of the data (the same as used in the Karhunen-Loeve transform). This representation leads to the compaction of the signal energy along one of the coordinates (see section 13.2.9), however application of the global decorrelation does not yield their local independence (Abel, Bhaskaran and Lee, 1992).

The concept of joint colour and spatial compression is a natural extension of the idea behind vector control of component-wise processing. Basically, it consists of joint treatment of the vector component by application of a Euclidean distance whenever colour difference has to be calculated. In the case of quantization this leads to replacement of the scalar quantization with vector quantization in colour space. A good motivation for the application of vector methods is that the natural mutual dependency between colour components is exploited by them, whereas, due to unremovable cross-component redundancy, it is an obstacle in efficient coding of separate colour components.

As stated in section 13.2.3, introduction of quantization to coding of the prediction error within a DPCM scheme leads to a class of lossy compression techniques. Maragos, Mersereau and Schafer (1984) proposed a multichanel prediction together with joint (i.e. vector) quantization of the three colour components of the prediction error. High quality reconstructed colour images are obtained at a rate of 1 bpp, while the separate encoding of each colour component independently leads to 3 bpp bitrate at a comparable quality of the reconstructed image. This simple example shows the great potential hidden in the vector approach to processing and coding of colour images.

Block truncation coding (Delp and Mitchell, 1979) is a block-based method, where each block is compressed by preserving only the very first statistical moments of the data. Several extensions have been proposed for coding of colour images with a vector approach consisting of joint treatment of the colour data (Yang, Lin and Tsai, 1994). If each block of 4×4 pixels is coded by two representative colours, an average compression ratio of about

1:14 can be achieved with very good quality of the reconstructed images.

Compression techniques based on several variations of vector quantization using codevectors of the joint spatial and colour information have been successfully used in various applications (Boucher and Goldberg, 1984; Barilleaux, Hinkle and Wells, 1987; Oehler, Riskin and Gray, 1991; Wang and Chang, 1992). Experimental results show that simultaneous treatment of spatial and colour information within square pixel blocks of variable or fixed size 2×2 to 4×4 leads to superior performance over the results of traditional component-wise compression at the same bitrates from the range 0.5 to 2 bits/pixel, since the mean square error measure is minimized in a multispectral sense. The increase of computational costs in the joint spatial and colour codebook design is not very high though, thanks to the much faster convergence of the typical LBG (Linde, Buzo and Gray, 1980) algorithm applied here.

A vector extension of fractal-based coding (see section 13.2.7) has been proposed by Zhang and Po (1995). The approach extends the 3-dimensional affine transform traditionally employed in compression of monochrome images to a 5-dimensional affine transform in order to express the affine self-similarity possible between different components of the colour image. The results show that there is a compression ratio improvement of about 1.5 over that achieved when colour components are coded separately. Using a Euclidean distance measure and variable block size, bitrates below 1 bpp have been achieved with good image quality and a PSNR of about 31 dB.

13.3.3 Palette representation and its applications

Full representation of colour images in digital storage systems is usually based on the three-component scheme. The commonly used colour spaces are RGB, YC_bC_r or CMY/CMYK. The number of bits allocated to each colour component varies depending on the application and in typical arrangements is equal to eight, which is mostly imposed by the convenience of storage in computer memory organized in eight-bit words. This kind of representation is called **true colour** in the terminology of computer display and storage devices. A number of file formats (section 7.2.1) (e.g. TIFF, PPM, TARGA, PNG, BMP *etc.*) used for storage of images in digital media support true colour mode, despite the large size of such files.

Whether previously compressed or not, the image has to be present in its direct (i.e. true colour) form at some point in the display system in order to allow the line scan device to access each pixel as required to form the proper signal for the CRT. Considering popular personal computers, terminals and workstations, the main economy factor is the cost of the high-speed video RAM used to buffer the display data. The motivation of the search for more efficient representation of digital colour images is obvious due to the substantial amount of memory required for the storage of the three components independently. The most straightforward approach to this problem is related to applying so called **palette** (or **palettized**) **representation**. This is based on the observation that only a small subset of all possible combinations of the intensities of red, green and blue appear in a typical digital image, and therefore it is possible to code each unique pixel colour in the image as an address in a look-up table that stores the true colour representation of the pixel.

The technique applied here could involve determining the complete set of all colours $\{C_n\}$ present in the picture **I**, and building an arbitrarily ordered **palette** P containing all the colours found:

$$\mathbf{P} = [\mathbf{C}_1, \mathbf{C}_2, \ldots \mathbf{C}_M], \quad \text{where} \quad \underline{C}_n = \{R_n, G_n, B_n\}$$

(where M denotes the number of unique colours observed in the original image I). The image in the new representation, $\hat{\mathbf{I}}$, would then be described by a matrix of scalars, instead of the three-component vectors. The value stored, $\hat{X}_{i,j}$ is an index to one of the palette elements:

$$\hat{\mathbf{I}} = [\hat{X}_{i,j}], \quad \forall i, j \quad \hat{X}_{i,j} = n \Rightarrow \underline{X}_{i,j} = \underline{C}_n$$

For the great majority of images this representation is slightly more efficient than true colour, taking into account the total number of bits to be stored including the palette. This efficiency strongly depends on the 'colourfulness' of the particular picture. The number of distinct colours observed in typical images like 'Lena' and 'Boats', shown in Plate XXII and Plate XVII, is usually about 100 000 or more. Moreover, this kind of transformation is completely lossless. Its drawbacks are the computational costs related to determining the complete set of colours and further pixel indexing, and the difficulty of further data compression (except entropy coding). Also, the storage gain is not very high. Therefore this approach has not found a practical application.

A similar approach which gained great popularity in computer systems consists of application of a limited palette of colours. The representation uses a **colour palette** (or a **colourmap** in computer terminology) of arbitrarily limited size (usually a power of two, not greater than 256, for the sake of convenience):

$$\mathbf{P} = [\underline{C}_1, \underline{C}_2, \ldots \underline{C}_K], \quad \text{where} \quad \underline{C}_n = \{R_n, G_n, B_n\} \quad \text{and} \quad K < M$$

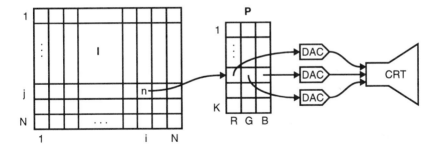

Fig. 13.16 Image display with a limited palette of colours.

Due to the size of the palette being reduced (e.g. to 256), the amount of memory needed is reduced to one third compared to that needed to store the original image. The great importance of this kind of representation of colour images is related to the low complexity of the decoding process. The additional hardware required for decoding of the actual colour of pixels on the fly consists of a simple lookup-table containing the three components for each palette element as shown in Fig. 13.16. Palette representation has been also adopted for the storage of images in digital media in order to avoid the need for additional computations which would be necessary to reduce the number of colours prior to display. Therefore many file formats (e.g. SUN Microsystems raster format, Microsoft BMP, GIF, TIFF, PNG, *etc.*) support the limited palette representation. In this case the palette is usually stored as a part of the file header. Special coding algorithms have been developed for further data compression of such data streams (see section 13.3.6). This kind of representation is called **pseudocolour** or **indexed colour** in computer terminology and in software packages for image processing and editing. Sometimes, images of this kind are generated by assigning various palette colours to grey levels of monochrome images to improve their visualization. This technique of visualization (e.g. for medical applications) is often referred as **pseudo-colouring**.

The representation of images with arbitrarily limited palette size is lossless as long as the number of colours present in the picture is no greater than the limit M, and they all appear in the palette. These conditions may be satisfied in the case of synthetic images of computer graphics, however they are hardly ever met in the real world of images of natural scenes. Therefore some loss of quality can usually be observed in images with a limited palette size.

Since the allowed number of available colours K is no longer greater than the number of all unique colours M observed in the original image $\mathbf{I}$, some kind of approximation is necessary. In this case, each colour from the palette approximates a whole set of slightly different colours from the original image. To perform this approximation, in general, the visual difference between palettized and the original image should be minimized. In practice, however, a pixelwise classification technique minimizes some error function ε in order to find the optimal representative for the colour of each point:

$$\hat{X}_{i,j} = \min_{k=1...K} \varepsilon(\underline{X}_{i,j}, \underline{C}_k) \qquad \forall i, j \tag{13.8}$$

where the function $\varepsilon(x,c)$ corresponds to the colour difference between its arguments.

The approximation of the actual colours of the pixels within the image using the colours from the limited set often leads to visible degradation of the image quality as shown on Plates XXIII to XXIV. The most important visual errors result from having too small a palette:

1. **False contouring** appears in smooth areas represented with too few intermediate tones or shades. It is caused by the highly correlated nature of the error which results from these large regions being approximated with a single colour. The sharp and irregular boundaries of the artefact objects with constant colour are particularly easily perceived in large areas of little content variation within the original image.

2. **Colour shifts** appear in areas of pixels with similar approximation error. If the mean error within such a region is not equal to zero, the shift is perceived thanks to the averaging properties of the human eye, as well as its high sensitivity to hue variations.

3. **Vanishing colours** are a problem in the case of images containing some very small objects of a unique colour. Due to the very small influence they have on the error measures usually applied within the palette generation algorithm (discussed in section 13.3.4), it is unlikely that they appear in the resulting palette. In consequence, the sparse/infrequent colours get very distorted by the approximation using quite different colours from the palette.

The visual annoyance of the colour approximation errors depends heavily on the following properties of the system:

- The number of palette elements.

- The choice of colours in the palette which determines the maximum (E_{max}) and total squared (TSE) error of the approximation, respectively:

$$E_{max} = \max_{i,j} \left| \underline{X}_{i,j} - \underline{C}_n \right|$$

$$TSE = \sum_{i,j} \left| \underline{X}_{i,j} - \underline{C}_n \right|^2$$

where $n = \hat{X}_{i,j}$.

- The applied classification technique, which affects the mean error (ME) and the perceived artefacts which are related to the local squared error (LSE):

$$ME = \frac{1}{N_1 N_2} \sum_{i,j} \left(\underline{X}_{i,j} - \underline{C}_n \right)$$

$$LSE(i,j) = \sum_{k=i-\frac{M}{2}}^{i+\frac{M}{2}} \sum_{l=j-\frac{M}{2}}^{j+\frac{M}{2}} \left| \underline{X}_{k,l} - \underline{C}_n \right|^2$$

(where M determines the size of some local neighbourhood).

There are two alternative classes of colour palette. A **standard palette** is a comprehensive set of colours obtained as combinations of the three uniformly quantized colour components.

This kind of palette is sometimes provided and forced by the display system for all displayed images, and is called in computer terminology a **system palette** or a **shared colourmap**. The advantage of using a system palette is the possibility of simultaneously displaying several images, albeit with compromised quality. The drawback of this approach is that the actual colours of the image are represented very roughly, so the visual and objective quality is low. In order to virtually increase the available gamut of tones and shades some multilevel halftoning techniques are usually applied in such cases (see section 13.3.5).

A **custom palette** is an optimized palette designed individually for the particular image, to minimize the approximation error, defined in some arbitrary terms.

The advantage of using a custom palette is the possibility to design it as well as possible using one of the well developed techniques. The obvious drawback is the limitation of most display hardware devices which are able to use only one colour palette simultaneously. The latter is called in computer terminology a **private colourmap**. In such devices only one image can be displayed correctly using its individual palette among several others presented on the screen. Some efforts have been done towards the development of an efficient technique for combining palettes of two or more images (Iverson and Riskin, 1993) in order to make the limitation less annoying. The algorithms perform pair-wise nearest neighbour matching between elements of two or more palettes for quickly combining the palettes of several images into one palette allowing their simultaneous display with only slight distortion.

The classification procedure is applied at the stage of pixel mapping, where the original values of the colour components are substituted by the indices of one of the palette elements. Its purpose is to find the best match for the given original colour in terms of some predefined quality criteria. There are several approaches to the classification problem. In the simplest case, the error function:

$$\varepsilon(\underline{X}_1, \underline{X}_2) = \sqrt{(R_1 - R_2)^2 + (G_1 - G_2)^2 + (B_1 - B_2)^2} = \|\underline{X}_1, \underline{X}_2\|_2$$

is applied to equation (13.8) and minimized for each pixel independently. In other words, for each pixel, the nearest colour (in terms of Euclidean distance in the colour space) is chosen as the approximation of its original colour. This leads to minimum objective distortion in terms of total squared error. A more precise colour difference formula (discussed in section 3.4) may also be applied here.

Further investigations (Gentile, Walowit and Allebach, 1990; Tremeau, Calonnier and Laget, 1994; Chaddha, Tan and Meng, 1994; Kim, Lee, Lee and Ha, 1996) proved that the pixel-wise Euclidean error measure does not necessarily correspond to the perceived visual quality loss. Therefore, some better (in terms of visual image quality) pixel mapping algorithms have been proposed in order to minimize colour artefacts. Usually, they take into account the spatial activity in a local region of perceptually adjacent pixels. For example, in the work of Liu and Chang (1994), morphological operations were employed to detect and dither the false contours (see also section 13.3.5).

13.3.4   Colour quantization

The objective of **colour quantization** (Jain and Pratt, 1972) is to perform a conversion between true colour and palettized image formats. The process of colour quantization consists basically of two steps: **palette design**, in which the palette colours are specified; and **pixel mapping**, in which each input pixel in the original image is assigned one of the colours from the palette. Since natural images typically contain a large number of distinguishable colours, displaying such images with a limited palette is difficult. Therefore the visual quality of images quantized to a smaller number of colours depends critically on the way the limited set of colours is chosen.

While the pixel mapping stage is involved with the classification problem discussed earlier, the palette design stage may be described from several points of view. Most of them are based on the theoretical fundamentals of vector quantization. In this framework, the colours in the palette can be considered as codes in a codebook, and the pixel mapping as the image pixel coding. The most straightforward approach to colour quantization is related to the application of a **fixed palette** (or **universal palette**, in order to emphasize its independence from the image contents) however images quantized using a **custom palette** (a palette designed adaptively to the image contents) practically always show better visual quality, even with very fast and simple algorithms. This is because the fixed palette has to cover a wide gamut of perceivable colours in order to stay universal, and, in most cases, many of the colours are wasted since they do not appear in the particular image being quantized. Therefore, the image quantized with a fixed set of colours can be thought of as one quantized with a much smaller custom palette. Nevertheless, some effort has been expended towards the proposal of universal palettes for displaying various classes of images (McFall, Mitchell and Pennebaker, 1989; Venable, Stinehour and Roetling, 1990; Kolpatzik and Bouman, 1995).

The task of finding an optimal set of representative colours may be formulated as a large scale clustering problem (Wan, Wong and Prusinkiewicz, 1988; Orchard and Bouman, 1991; Wu, 1992). Given an image $\mathbf{I}$, the goal is to find an optimal partition of the set of M colours in $\mathbf{I}$ into $K < M$ subsets in respect to the predefined error criterion. The centroids of the data clusters thus defined form the codebook entries which are optimal in terms of minimal quantization error. As the error measure, mean squared error is commonly used. It is known that the problem of finding the global minimum of MSE is NP-complete. Consequently, any computationally efficient solution to the colour quantization problem will be sub-optimal (Wu, 1992). Moreover, as previously stated, MSE is by no means the ultimate criterion of visual quality (and neither is any other proposed objective measure).

The existing adaptive techniques to design a colour palette can be divided into three categories:

1. The methods based on iterative optimization start from a given initial set of colours and try to refine the codebook in respect to the global quantization error criterion:

   The **Linde-Buzo-Gray** (LBG) or so called **K-means algorithm** (also known as the **Generalized Lloyd Algorithm** (GLA)) has been proposed for construction of a sub-optimal codebook for application in vector quantization (Linde, Buzo and Gray, 1980) (section 13.2.6). Despite being very intensive computationally, due to its exhaustive checking of every data vector against every codebook element for the closest match, this algorithm has been successfully implemented for final refinement of colour palettes designed using other techniques (Heckbert, 1982; Orchard and Bouman, 1991; Pei and Cheng, 1995).

   A fuzzy extension of the LBG algorithm consists of relaxing the strict classification rule within the centroid recalculation step of the original algorithm. Each data point may be a fuzzy member of several clusters with the membership relation taking a continuous value between 0 and 1. Experiments prove, that, thanks to better dealing with false classification, this fuzziness yields better convergence and, when applied to the palette design problem (Kok, Chan and Leung, 1993), it results in visually better colour quantized images with colour shifts strongly reduced.

   The **genetic C-means clustering** algorithm (Scheunders, 1996) is a hybrid approach, combining the Generalized Lloyd Algorithm together with a Genetic Algorithm (Goldberg, 1989; Davis, 1991) as an attempt to avoid its convergence to a local optimum only. This technique operates on a set of palette colours represented as a string of their components. The iterative palette refinement consists of the three stages typical for the genetic approach, namely regeneration, crossover and mutation followed by Lloyd optimization. Experimental results show that numerous repetitions of such a sequence lead to an optimal (in terms of TSE) set of palette colours, virtually independent of the initial colour set. A similar algorithm has recently been proposed by Freisleben and Shrader (1997).

   Palette design algorithms exploiting the optimization properties of neural networks have been recently proposed by Réndon, Salgado,

Menéndez and Garcia (1997) and Verevka and Buchanan (1995). These methods employ so called self-organizing topological maps introduced by Kohonen (Kohonen, 1982; Kohonen, 1984). In both approaches, an initial palette is created by subsampling the original colour image. Subsequently, a network training process is applied which consists of the iterative adjustment of the winning neuron representing a palette entry. This corresponds to refinement of the palette. Some original simplifications have been proposed for speeding-up this time consuming process. The resulting distribution of the palette colours in the colour space follows almost faithfully the colour distribution of the given input image. The palette obtained is ordered, which is of great importance when further compression of the palettized image is considered (as discussed in section 13.3.6).

2. Heuristic methods, instead of finding local minima by numerous iterations, try to produce an acceptable solution very rapidly. Most of the proposed algorithms are based on statistical analysis of the colour distribution of image pixels within the colour space. All of these methods can be classified as either **divisive** or **agglomerative**. The divisive methods iteratively subdivide the 3-dimensional colour space into cells according to various criteria to determine which cell to partition further and where exactly to place the partitioning surface. The agglomerative methods analyse the set of colours present in the image and apply a **hierarchical clustering** by iteratively merging the data clusters according to various rules.

The **popularity algorithm** (Heckbert, 1982) uses the image colour histogram to select the K most frequently occurring colours. To improve this initial palette, an iterative procedure derived from the LBG algorithm has been proposed. The **Improved Popularity Algorithm** (Braudaway, 1987) artificially reduces the histogram values in a region surrounding each assigned colour, in order to avoid concentrating too many palette colours about one histogram peak.

In the **median cut algorithm** (Heckbert, 1982) the initial palette is constructed by assigning to each colour in the palette an approximately equal number of pixels from the image. The colour space is recursively split into rectangular boxes. At each step, the box with the largest number of pixels is divided into two smaller ones, perpendicularly to the axis with the largest data spread, through the median of the colour distribution within the box. Again, the LBG algorithm is proposed as an option to improve the solution by searching for a local minimum of the TSE. Plate XXIII shows the 'Lena' image quantized to 16 colours using the median cut algorithm. A modification of the median cut algorithm proposed by Joy and Xiang (1993) places the partitioning plane through the algebraic mean of the colour data projected onto the direction of splitting.

The **variance minimization algorithm** adaptively assigns more clusters to the regions with large quantization errors, which corresponds to the data variance (Wan, Wong and Prusinkiewicz, 1988). This approach is based on the rule of maximum reduction of the quantization error at each step of bipartition. The colour space is iteratively divided into rectangular boxes until K clusters are generated. The partition strategy is based on the minimization of the sum of the variances of the data in the resulting subboxes. In the final partition, different boxes may contain quite different numbers of data points, but the contribution of each box to the quantization error is approximately equal. Plate XXIV shows the 'Lena' image quantized to 16 colours using the variance minimization algorithm.

The **binary splitting algorithm** (Orchard and Bouman, 1991; Balasubramanian, Allebach and Bouman, 1994) at each step splits one data cluster in a colour space. The respective cluster is divided along the direction of maximum colour data variation around the cluster centroid by a hyperplane which passes through the centroid. This optimal direction is determined by the principal eigenvector of the covariance matrix of the data within the cluster. The order of splitting the clusters is determined by the magnitude of this vector. Unlike the other splitting techniques, this approach leads to general polytopal regions.

The generalization of the splitting-based algorithms (Wu, 1992) is based on **optimal principal multilevel quantization**, in which the chosen cluster is divided into several smaller ones by a set of planes perpendicular to the direction of the principal eigenvector. The positions of these division planes are optimized by a dynamic programming algorithm, which tends to minimize the resulting error by preserving the statistical moments of the data projected onto the direction of splitting.

The **greedy tree growing** technique (Liu and Chang, 1994; Liu and Chang, 1995) splits the colour space into rectangular boxes. The decision of which region to divide at each step is based on a trial-and-error analysis of the total quantization error reduction resulting from particular division. For simplicity, the splitting is performed along one of the

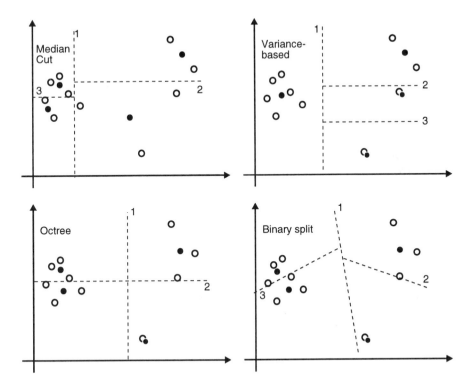

Fig. 13.17 Two-dimensional example of selected partitioning techniques (Median Cut, Variance-Based, Binary Splitting, Octree) used to design a 4-element palette.

axes of the colour space. The plane perpendicular to the axis for which the error reduction is largest is selected as the bipartition plane.

The **octree** algorithm (Gervautz and Purgathofer, 1990; Ashdown, 1994) builds a colour histogram of the image data in the form of a tree data structure, where each node has eight child nodes. Subsequently, similar pixel colours are grouped and replaced with their average. The node merging process starts from the deepest level in the tree hierarchy and is iterated as long as the desired number of leaf nodes is obtained. This strategy leads to minimization of the maximum quantization error. A similar **agglomerative clustering** algorithm (Xiang and Joy, 1994) merges the nodes between different branches of the tree thus minimizing the increase of the quantization error resulting from each reduction of the colour gamut associated with the merging process. Plate XXV shows the 'Lena' image quantized to 16 colours using the octree algorithm.

3. Finally, there is a group of very simple algorithms which can be classified as based on improved scalar quantization rather than on vector quantization. They attempt to exploit within scalar quantization the statistical dependency and dimensionality of image data in a way similar to that use in vector quantization.

   In the **sequential scalar quantization** algorithm (Balasubramanian, Bouman and Allebach, 1994; Balasubramanian, Bouman and Allebach, 1995) the marginal distributions of the colour data are analysed. The scalar colour components are individually quantized in a sequence, with the quantization of each component utilizing conditional information from the quantization of the previous component. The Lloyd-Max algorithm e.g. (Gersho and Gray, 1992) is used as the optimal 1-D quantization strategy. The final palette colours are calculated as the centroids of data clusters from regions of support of the input distribution. The very similar **dependent scalar quantization** algorithm (Pei and Cheng, 1995) applies binary moment-preserving thresholding to consecutive data bands during the scan of the marginal image colour distribution. This is preceded by an automatic bit allocation for each data component.

The algorithms discussed, in their basic forms, produce colour palettes and yield quantized images of various qualities, depending on the optimization criteria applied, their sophistication, and finally, the given colour image. In general, basic and simplified methods (such as the popularity algorithm, the median cut algorithm) result in relatively poor palettes in terms of PSNR and perceived visual quality while the highly-optimized, uncompromised techniques (such as binary splitting, optimal principal multilevel quantization, genetic algorithms, and neural network-based algorithms) offer visually much more pleasant results. Surprisingly, also the very simple sequential scalar quantization algorithm gives very good results. Nevertheless, all colour quantized images suffer from false contouring in smooth regions, if no dithering technique is applied at the pixel mapping stage (see section 13.3.4). Therefore, numerous variations of such techniques have been proposed together with the basic palette design algorithms.

Several modifications of the core palette design procedures have also been proposed in order to take into account the spatial dependencies between neighbouring image pixels (Orchard and Bouman, 1991; Balasubramanian, Allebach and Bouman, 1994; Chaddha, Tan and Meng, 1994; Tremeau,

Calonnier and Laget, 1994; Kim, Lee, Lee and Ha, 1996). Usually, these modifications are related to the use of a subjectively weighted error measure:

$$\text{WTSE} = \sum_{i,j} W(i,j) \left| \underline{X}_{i,j} - \mathbf{C}_n \right|^2$$

The weighting factor $W(i,j)$ is a function of the local spatial activity of the image contents, such as:

- the local luminance gradient smoothed by convolution with a smoothing kernel;

- a maximum value of horizontal and vertical gradient calculated for simplicity only from the green component within an 8×8 pixel block;

- a sum of the absolute deviation of the colour from the mean colour within an 8×8 pixel neighbourhood;

- a maximum of coefficient-weighted difference between input colour and local mean colour within a 4×4 pixel wide neighbourhood.

Finally, it has been proved (Gentile, Allebach and Walowit, 1990; Kolpatzik and Bouman, 1995) that performing colour quantization in a more appropriate colour space, i.e. in a perceptually more uniform space, can substantially improve quantized image quality in comparison with colour quantization in the RGB coordinate system.

Objective comparison of the various colour quantization techniques is difficult due to the various spatio-colorimetric distortions caused by different types of quantization (the numerical discrepancies between original colours and quantized colours), viewing environment, image content, aesthetic considerations and even the viewer experience (Xiang and Joy, 1994). Therefore, only crude comparisons may be performed to some extent as given in Table 13.4. Few researchers present the RMSE ratings (Orchard and Bouman, 1991; Liu and Chang, 1995) or PSNR (Pei and Cheng, 1995; Réndon, Salgado, Menéndez and Garcia, 1997) of their techniques as a function of growing palette size. These errors, however are calculated pixelwise in RGB colour space which is irrelevant for such comparisons. Furthermore, only the basic forms of the presented algorithms can be compared in this way. Because the perceived effects of the halftoning process are transparent to the metric, a mean-squared error criterion is not meaningful for halftoned images (Gentile, Walowit and Allebach, 1990). A much more relevant quality measure can be defined using a perceptually uniform colour space (sections 3.4.1,

Table 13.4 Comparison of performance for some heuristic palette design algorithms for the 24-bit image 'Lena' (see Plate XIV) quantized to $K = 256$ colours ($N_1 = N_2 = 512$).

Algorithm	RMSE	Computational cost	Enables fast pixel mapping?
LBG	8.3	very high	No
popularity algorithm	22.1	very low	No
median cut	8.9	low	No
variance minimization (or variance based)	9.0	low	Yes
binary split	5.9	low	Yes
neural networks	11.1	very high	No
dependent scalar quantization	15.6	very low	Yes

4.7 and 4.8) (Wyszecki and Stiles, 1982; MacAdam, 1974; Miyahara and Yoshida, 1988) and applying some perceptual metric to the colour difference (Tremeau, Calonnier and Laget, 1994; Kolpatzik and Bouman, 1995; Xu and Hauske, 1995).

The practical usefulness of various algorithms strictly depends on the computational cost related to the palette design step as well as the pixel mapping stage. The number of calculations needed for the palette design typically depends on the target palette length K. The common naive implementations involve a colour analysis of each image pixel and therefore the cost is also proportional to the number of pixels $N_1 \times N_2$, whereas a simple fast histogramming step can save many of the calculations thus making it proportional to the number of distinct colours M encountered within the image. Several researchers propose a prequantization of the colour components to as low as 5 bits prior to histogramming in order to speed-up this stage as well as to make it less memory consuming (e.g. (Heckbert, 1982; Balasubramanian, Bouman and Allebach, 1994)). This scalar prequantization leads however to limited accuracy of the initial statistics which the algorithm relies on and its influence on the resulting visual quality of palettized images is significant. As an alternative, image subsampling (usually in intervals determined by prime numbers) has been proposed as yielding reduced data at the beginning of palette generation by Verevka and Buchanan (1995) and Freisleben and Shrader (1997). Several algorithms have been developed with the possibility of fast implementation in mind. Typically, a tree-structured (Frieman, Bentley and Finkel, 1977; Gray, 1984) palette is proposed in order to speed up the time consuming nearest-neighbour search. Despite their slight loss of accuracy, tree-structured search algorithms are often applied, since the search cost is reduced logarithmically.

13.3.5   Multilevel halftoning and colour dithering

Multilevel halftoning can be thought of as a postprocessing stage of the colour quantization procedure after the pixel mapping but it is often very naturally incorporated into this process (Gentile, Walowit and Allebach, 1990; Tremeau and Laget, 1995). Halftoning techniques can induce the perception of the presence of colours that lie between those that are actually present in the image printed or displayed with a limited palette. They rely on the viewer making a local spatial average over patterns of alternating colours. Thus, these techniques sacrifice spatial resolution for tonal resolution, which sometimes leads to visible texture artefacts.

The **error diffusion technique** (Floyd and Steinberg, 1975; Gentile, Walowit and Allebach, 1990) spreads the approximation error at the stage of pixel mapping into its neighbourhood. To implement error diffusion, the image is scanned during the mapping process. After each pixel is mapped, the error between the original and the substituted colour is calculated. This error is distributed to neighbouring pixels which have not yet been mapped and is added to their original values with the weights shown in Fig. 13.18. This mechanism can be thought of as a recursive highpass filtering of the colour quantization error exploiting low sensitivity of the human eye to high frequency noise. Plate XXVI shows the 'Lena' image dithered using Floyd-Steinberg dithering and a 16-colour palette designed using the octree algorithm.

The **ordered dither** technique (Jarvis, Judice and Hinke, 1976; Gentile, Walowit and Allebach, 1990) is performed by adding a pseudorandom two-dimensional periodic dither signal to each colour component of the pixel prior to finding its best match in the colour palette. The spatial structure of the

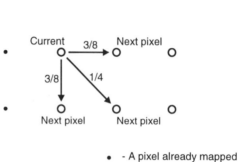

Fig. 13.18   The scheme of propagation of the colour quantization error using error diffusion (Floyd and Steinberg, 1975)

dither signal is chosen to generate patterns of alternating colours that have low visibility. This structure is determined by the dither matrix:

0	8	2	10
12	4	14	6
3	11	1	9
15	7	13	5

which indicates the spatial ordering of the dither signal values.

These two basic approaches suffer from a few well known problems (Akarun, Yardimci and Cetin, 1995). Firstly, accumulation of the approximation error from previous pixels may lead to quite wrong classification of the colour of the current pixel. This manifests itself as a colour impulse, which is very disturbing visually. Therefore one has to take care with round-off errors. Secondly, due to the dithering process, the colours from either side of an edge are smeared towards each other and sharp edges are converted to jagged edges.

Several improvements have been proposed to the basic colour halftoning algorithms. Some researchers propose to adapt the coefficients of the error diffusion filter to the local statistical properties of the image in order to minimize the mean squared error between the average colour of the original and dithered images. Also, only downward rounding rules should be applied. The proposed improvements are also related to the spatial masking properties, whereas the human eye is not sensitive to resolution deficiency of edge areas. This phenomenon suggests that error diffusion needs to be applied only in smooth regions.

- An error diffusion filter with adaptive feedback has been proposed (Orchard and Bouman, 1989; Orchard and Bouman, 1991) in order to damp undesirable error oscillations whenever they exceed a given threshold.

- Optimizing the coefficients of the error diffusion filter for noise shaping (Kolpatzik and Bouman, 1992) together with some additional dithering results in visibly improved halftoned images, especially when a palette covering a large colour gamut is applied.

- A neural network based optimization of the halftoning filter has been proposed (Ling and Just, 1991) in terms of a frequency domain error minimization. Since the energy of the diffusion error is concentrated at high frequencies only it is much less annoying for the observer thus making halftoned pictures of superior quality over other non-optimized techniques.

Experimental results show that for colour palettes of sizes 32 to 64 the quality of halftoned pictures is often acceptable. For 8 to 16 colours, however, the reduction of the colour gamut is usually too great (Chau, Wong, Yang and Wan, 1992). In such cases, it is not possible to find a colour extreme enough to produce a desired tone with the aid of spatial averaging properties of the human eye. It has been proved that the best palettes for halftoning purposes are not necessarily those which produce minimum quantization error when no halftoning is applied. Rather they should contain a wide range of colours spanned over the whole gamut of the original image. To prevent the quantizer from saturation of accumulated diffused error, the colours of the original image should be completely contained in the convex hull of the palette colours. Some original methods for designing an improved palette for application in halftoning have been proposed (Chau, Wong, Yang and Wan, 1992; Kolpatzik and Bouman, 1995). These algorithms extend the bordering colours of an image-dependent palette towards the boundaries of the colour gamut of the original image.

There is much demand from the printing industry to find the best algorithms for halftoning colour images printed with raster devices operating with only 4 basic colour dyes (cyan, magenta, yellow and black). This is why there is much research in this field which takes into account various fine aspects of the human perception of colour.

13.3.6   Further compression of colour quantized images

Palettized or indexed colour images were the dominant type of image file format used on the information highway and in public image databases such as CompuServe archives and World Wide Web (WWW) resources. The most common GIF87 image format has even been extended to support progressive transmission, so called **transparent background**, and basic animation facilities in order to provide more attractive WWW pages. These extensions were incorporated into the GIF89 version format shortly before its common use dropped rapidly due to the decision of IBM and Unisys Corporation to enforce their patent rights for the LZW compression algorithm. Public demand for a royalty-free replacement resulted in a new **Portable Network Graphics** (PNG) format (Crocker, 1995; PNG, 1996). It supports indexed-colour, greyscale, and true colour images with lossless compression based on the original method proposed by Ziv and Lempel (1977). This simple algorithm uses a sliding window of fixed size that is logically passed over the data stream. Adaptive Huffman code is applied to compress symbols within the window. In addition, the compression technique implemented in PNG is able to find the best orientation of image scanning to exploit repetitions of long data sequences in rows or columns, in opposition to the row-only orientation of GIF. The compression ratios offered range from 1:2 for natural scene images to 1:8 for lineart graphics, depending on the complexity of the image. First of all, however, the colour-quantized image intended for further compression must not be halftoned. Since halftoning techniques introduce high-frequency components into the image, they strongly reduce its smoothness and correlation between neighbouring pixels, thus making the compression task extremely difficult.

More sophisticated lossless techniques have been proposed in order to achieve higher compression ratios by the use of two-dimensional prediction together with entropy coding:

Bit-plane coding (Gormish, 1995) has been proposed on the basis of the existing JBIG compression standard (IEC, 1993) and the idea of representation of the indexed colour images using binary planes. Separation of the image into individual bit planes and colour planes have both been tested prior to binary predictive coding using JBIG. 30% to 40% improvements in bitrates compared to the standard Lempel-Ziv compression are possible. The loss-less JPEG coder chooses for each pixel the best one of 8 various predictors which exploit the correlation of the pixel with neighbouring pixels. Either Huffman coding or arithmetic coding can be used as the entropy coding of the prediction error. This technique significantly outperforms other lossless prediction-based compression schemes. A comparative study of various techniques operating on different classes of images can be found in Yovanof and Sullivan (1992).

All the predictive, statistical-based and dictionary-based techniques just discussed suffer from relatively low compression ratios in comparison to more efficient lossy techniques. Although the application of lossless transform coding has not been extensively studied yet, it seems that, as in the case of compression of non-quantized colour images, only lossy techniques can bring substantial improvement here.

An extension of the LZW algorithm has been recently proposed by Chiang and Po (1997). By introduction of a distortion threshold into the string matching process, a generalization of the original LZW algorithm to the lossy class of techniques has been obtained. The adaptive thresholding designed according to the properties of the human visual system yields minimum visible distortions within given constraints. Typically, 25% to 50% improvement is offered in compression efficiency over the PSNR range of 33 to 40 dB with a good quality of decompressed images. The great usefulness of this technique is related to its compatibility with the standard LZW decompression. Thus, original GIF decompression software may be used for decoding of the produced bitstreams.

More powerful general image compression techniques can be adopted for coding of colour quantized images. In general, two approaches are possible here:

1. A three-channel class of techniques involves colour dequantization, i.e. each pixel is represented by the three quantized colour components from the palette. Such an image is compressed using one of the techniques developed for colour images, treating each channel either separately or simultaneously. After decompression, in order to restore its indexed representation and make it suitable for display with the limited palette, the reconstructed colours should be requantized, because they are no longer the elements of the original palette. A mapping to the old palette may be performed as well as to a new palette designed for the reconstructed image. Both approaches introduce additional errors which accumulate with the first quantization errors and coding artefacts. Due

to the mapping stage, decompression is computationally intensive and makes the technique unattractive in practical applications.

2. Instead of treating the colour quantized image as a normal colour image, the problem of its compression can be reduced to one dimension – the compression may be performed in the one-dimensional space of the palette indices. In this case, the indices representing palette colours for consecutive pixels form a pseudo-greyscale image which is compressed using some lossy technique. Because the mapping from 3-dimensional colour space to 1-dimensional number space is highly non-linear, small coding errors within the pseudo-greyscale image can introduce significant colour artefacts into the reconstructed colour image and *vice versa*. The main advantage of this approach is that it guarantees that the reconstructed image will contain strictly the same set of colours and there is no need to perform requantization.

Experimental results (Zaccarin and Liu, 1991) show that the indices of palettized images are highly correlated with the adjacent indices if the colour palette is appropriately arranged as illustrated by Fig. 13.19, i.e. if consecutive indices are assigned to visually similar colours and distant indices represent colours that are visually different. In such a case, possibly few high frequency artefacts are introduced into the pseudo-greyscale image within areas where its colour counterpart is visually smooth. Therefore the ideal ordering will satisfy the relation:

$$|i - j| < |n - m| \Leftrightarrow |\underline{C}_i - \underline{C}_j| < |\underline{C}_n - \underline{C}_m| \qquad \forall i, j, n, m \in \{1 \dots K\}$$

It can be proved easily that this kind of ordering is only possible in the case of palette colours distributed along a straight line in the colour space, therefore for most natural images it does not exist. Several heuristic algorithms have been proposed for ordering of the palette colours together with various compression techniques. Experiments with different palette ordering algorithms show, however, that compression of the pseudo-greyscale image must deal with the problems resulting from the non-linear transformation related to the mapping from 3-dimensional colour space to 1-dimensional index space. In particular, the pseudo-greyscale image possesses some undesirable properties. The impulses and abrupt changes within its content apparent in Fig. 13.19 result from the fact that the smooth change of colour components within the original image may not appear along the chain of ordered palette entries. While the former can be easily removed using median or vector median filters without significant degradation of the image, the latter strongly

Fig. 13.19 A pseudo-greyscale image of a colour image 'Lena' (see Plate XIV) quantized to 256 colours. Left: palette ordered randomly (as created by the palette generation algorithm). Right: palette ordered along the luminance component of their colours.

reduces the efficiency of the coding methods which usually benefit from the smoothness of the image. On the other hand, blurring introduced by various compression techniques may lead to severe colour artefacts, because a value intermediate between two colour indices does not necessarily correspond to an intermediate colour. Therefore some colour correction scheme is usually proposed with each of the proposed techniques. The possibilities are as follows.

A transform-based compression method proposed by Zaccarin and Liu (1991) and (1993) exploits the fact that colour variation is usually less important locally. Therefore the image is divided into square blocks of 8×8 pixels and for each block its local subset of colours is ordered according to the luminance values. Such ordering is quite appropriate since usually the luminance axis is the one with the greatest spread of colour distribution. Orderings optimized using the **Travelling Salesperson Problem** algorithms (Lawler, 1985) and using the farthest insertion rule have also been considered. The blocks of indices are coded using a DCT approach similar to that used in the JPEG standard. After decompression the pseudo-greyscale image is distorted, so the actual colour is chosen from the local subset of palette colours as a colour which gives minimum luminance distortion. The local lists of allowed colours within each block are stacked and losslessly encoded using JBIG compression. Bitrates around 1 to 1.5 bpp are reported as yielding good image quality, comparable with the standard 3-channel JPEG coding.

A similar hybrid compression technique, also based on the DCT and JBIG scheme (Chen, Petersen and Bender, 1993) divides the palette colours into groups of visually similar colours by means of K-dimensional clustering in perceptually uniform colour space. The colours within groups are ordered according to their luminance. At the decoding stage, the distorted colour is mapped to its closest member within the group. In contrast to the spatial partitioning this approach leads to intra-group errors. This means that the maximum distortion is constrained within each chrominance group, not within a spatial square block. Similar bitrates are reported in comparisons of this technique to the previously discussed one. A modification of this approach (Tremblay and Zaccarin, 1994) consists of segmentation-based rather than colour-difference-based partitioning of the set of colours and a variable-size block-based correction scheme. Experimental results show that the performance can be improved in this way both in terms of visual quality and MSE ratings. Bitrates of about 1.3 bpp are achievable for typical colour images.

A more sophisticated compression technique based on the transform coding of variable-shape objects has been proposed in (Overloop, Philips, Torfs and Lemahieu, 1997). In this approach the image is segmented using the luminance component and the luminance is coded for each region using weakly-separable transform base functions and a Huffman entropy coder. The reconstruction of colour information from the decoded luminance is possible thanks to the lists of allowed colours which are determined for each region and compressed using JBIG. Experimental results show visually better reconstructed images in comparison to JPEG coding despite slightly worse MSE ratings at moderate to high compression ratios. In particular, annoying blocking artefacts are completely eliminated.

A subband-based coding method has been proposed to compress the pseudo-greyscale image of palette indices (Waldemar and Ramstad, 1994), because of its good performance in traditional compression of monochrome images without causing visible patterns of blocking artefacts (Woods, 1991). A global palette sorting algorithm has been applied here, however, a block-based colour correction scheme, similar to the one used in the previously discussed approaches appeared to be necessary to resolve severe colour errors caused by coding artefacts. An original palette ordering algorithm with the aim of maximizing the pixel to pixel correlation has been tested together with a simple luminance ordering. Experimental results show better visual quality of reconstructed images in comparison to JPEG coding of the pseudo-greyscale image. The MSE ratings are also clearly better at bitrates of 0.8 to

3.3 bpp for test images such as 'Lena'.

13.4   COLOUR VIDEO COMPRESSION

13.4.1   Component-wise video coding

In video coders accepted as industrial standards, the components of digital colour video signals are compressed separately with shared control and motion estimation mechanisms. Most standards use versions of the DCT-based block hybrid coder. Such a coder operates in two basic modes:

Intraframe (I) : The frames from a video sequence are compressed independently from each other. The coding algorithm is similar to that described for DCT-based coding of still images discussed in section 13.2.4.

Interframe (P) : The current frame is predicted on the basis of the previous frame. The DCT-based coder already described encodes the prediction error, i.e. the difference between the actual and predicted frame (Fig. 13.20).

Prediction of the next frame is made more difficult by the fact that objects seen in one frame can appear in a different position in the next frame. To achieve high compression, it is necessary to estimate the motion of objects in order to exploit similarities between blocks of pixels which appear in different positions in successive frames as a result of object motion. Motion estimation is a very expensive step in the technique and only luminance pixels are regarded in the calculations. For a block of 16×16 luminance samples from the current frame the most similar block in the previous frame is searched for. Differences in the coordinates of these blocks define the elements of a so called **motion vector**. A motion vector has two elements (horizontal and vertical) usually numbers in the range about $(-15, 15)$. More advanced techniques estimate motion vectors with half-pixel accuracy even for 8×8 blocks.

The current frame is predicted from the previous frame with its blocks of data in all the colour components shifted according to the motion vectors which have to be transmitted to the decoder as side information.

The video coder tends to generate a bitstream with very variable bitrate. In order to match this bitrate to the transmission channel capacity, the coder parameters are controlled according to the output buffer occupancy. By growing buffer occupancy the coder reduces the bitrate which is related to increased impairments in the decoded images and *vice versa*, the buffer becoming emptier stimulates the coder to increase the bitrate hence improving the quality of the images reconstructed in the decoder.

Bitrate control is performed by:

- switching between interframe and intraframe modes;

- changing the quantization steps for the DCT coefficient quantization;

- avoiding the encoding of some images or portions of them.

The decisions upon these three features are to be undertaken at least for macroblocks which include arrays of 16×16 pixels of luminance and the corresponding two 8×8 blocks of chrominances (for 4:2:0 chrominance subsampling). In order to make the decisions the coder estimates quality and bitrate. Intraframe mode is to be chosen by the detection of fast motion or a rapid change in the scene. Interframe coding is more efficient for slow and moderate motion. A more modern approach to video compression is based on encoding of irregularly shaped objects extracted from the picture by segmentation (Musmann, Hötter and Ostermann, 1989). The objects are encoded by their shape and texture. Also motion compensation is performed for objects, therefore motion parameters are sent as side information for all objects rather than for rectangular macroblocks as in the classic approach. Also the model of motion used in motion estimation and compensation is improved. Instead of motion of flat rigid blocks in the plane, motion of three-dimensional objects in the space is modelled to some extent. Therefore motion-compensated prediction is more exact thus reducing the amount of information needed to encode the residual prediction error signal. The technique is very suitable for encoding video signals which are to be transmitted by very low bitrate channels, i.e. under 64 kbps.

Encoding of irregularly shaped objects has stimulated development of shape-adaptive transform coding (Gilge, Engelhardt and Mehlan, 1989; Sikora and Makai, 1995; Stasiński and Konrad, 1997).

To illustrate the preceding discussion, Plate XXVII shows an original frame from the colour video sequence known as 'Salesman'; a frame from the same sequence coded in the *intraframe* mode of a standard H.263 coder at a target bitrate of 64 kbps; and a 'P' frame (coded in *interframe*, or predictive mode), the latter showing the distortions introduced by the coder when rapid motion occurs in the scene.

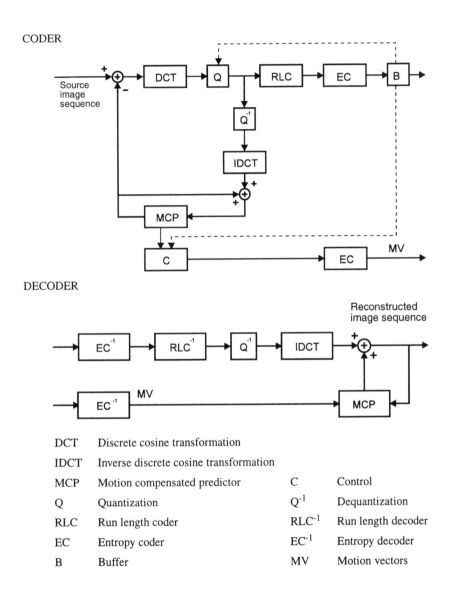

DCT   Discrete cosine transformation

IDCT   Inverse discrete cosine transformation

MCP	Motion compensated predictor	C	Control
Q	Quantization	Q^{-1}	Dequantization
RLC	Run length coder	RLC^{-1}	Run length decoder
EC	Entropy coder	EC^{-1}	Entropy decoder
B	Buffer	MV	Motion vectors

Fig. 13.20 Motion compensated interframe coder and decoder.

Compression standards

A comprehensive overview of standards for video compression is given in Table 13.5. Except for the first (ITU-T J.80) standard stated all other standards include the DCT-based hybrid coder previously described. The exceptional recommendation J.80 describes a high-fidelity intraframe codec based on DPCM. Moreover the coming MPEG-4 standard will handle visual objects as the basic units.

13.4.2   Chrominance vector quantization of video

This section deals with an exemplary technique based on the vector approach to colour instead of separate processing of components. The technique exploits properties of the two-dimensional histograms of chrominance.

Colour images of natural scenes typically contain a small subset of all possible colours. The two-dimensional histograms of chrominance are sparse especially for low resolution images. CIF and QCIF frames contain only 25 632 and 6408 chrominance samples. If all samples exhibit different chrominance values and the components are represented by 8-bit numbers, only 39.1% and 9.8% of all combinations would be used. As the assumption that all pixels in an image exhibit different colours is highly unrealistic, the actual percentage of combinations used is much lower.

Moreover, the chrominance histograms of natural images consist usually of some regions of concentration where the histogram values are relatively high. Outside these small areas, the values of the histogram are zero or close to zero as shown in the example of Fig. 13.21. Also very saturated colours are relatively rare in a typical image of a natural scene. The chrominance histograms are relatively stable across frames, particularly for videophone sequences, as shown by Fig. 13.22. The histogram properties stated above mean that two-dimensional vector quantization in the chrominance plane of YC_bC_r or CIELAB space leads to very efficient representation of the colour data. Experimental results show that for natural scenes the number of codebook entries can be very small (usually about 20–30 for typical sequences in QCIF format) without noticeable degradation of the picture quality. Note that the codebook entries represent chrominance values only and that even with this small number of entries, there is still a possibility to generate many colours in combination with the individual luminance values.

Chrominance quantization is exploited in the proposal for improved colour in high-compression coding of Bartkowiak, Domański and Gerken (1996). In the proposed system represented in Fig. 13.23, instead of two

Table 13.5 Video compression standards.

Bitrate (bps)	International Standard	Application area
High: ≈ 140 M	ITU-T J.80 (formerly CCIR Rec. 721)	Contribution quality TV
Medium: 2–90 M	ITU-T J.81 (formerly CCIR Rec. 723)	34–45 Mbps – Contribution quality TV.
	M-JPEG*	90 Mbps - Contribution quality TV. Applicable to a large range of bitrates.
	ISO/IEC IS 13818-2 = ITU-T H.262 (MPEG-2)	30–50 Mbps – Contribution quality TV. ≈ 15 Mbps – HDTV. 8–10 Mbps – Studio quality TV. 4–6 Mbps – Broadcast quality TV.
Low: 64 k to 1.92 M	ISO/IEC IS 11172-2 (MPEG-1)	About 1.5 Mbps – CD ROM, multimedia
	ITU-T H.120	2.048 or 1.544 Mbps – Videoconferences
	ITU-T H.261	$p \times 64$ kbps, $p = 1, \ldots 30$ mainly 64, 128 or 384 kbps videoconferences on ISDN.
	ISO/IEC IS14496 (MPEG-4)	Many areas of application.
Very low: 8–64 k	ITU-T H.263	Videoconferences, video-telephony, multimedia.
	ISO/IEC IS 14496 (MPEG-4) in development	Videoconferences, video-telephony on PSTN, wireless videophone and videoconferencing, tele-education, teleconsulting, video e-mail, audio-visual data bases, remote surveillance.

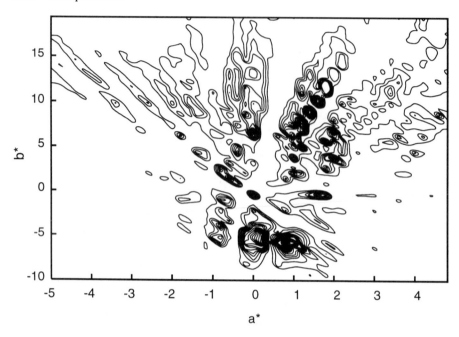

Fig. 13.21 Two-dimensional chrominance histogram for a frame from the video-phone test sequence 'Salesman' drawn as contour lines on the plane of the coordinates a^*b^* of the CIELAB colour space.

independent video components C_b and C_r only one scalar signal of indices to a chrominance codebook is fed into the source coder (a DCT-based coder or an object-based analysis-synthesis coder) independently from the luminance signal. Both encoded signals are transmitted to a receiver through a communication channel. After decoding, the luminance and the scalar chrominance are reconstructed. Then, vector dequantization is performed in order to reconstruct the two chrominance components from the scalar chrominance.

The system includes a two-dimensional vector quantizer with codebooks automatically generated using a fast binary-split algorithm. The codebook entries have to be ordered properly as discussed in section 13.3.6 in order to ensure high coding efficiency. The codebook itself is losslessly encoded and added as side information to the bitstream.

This technique has been shown to be advantageous by high-compression JPEG coding as well as by intraframe very low bitrate coding of video in a H.263 coder.

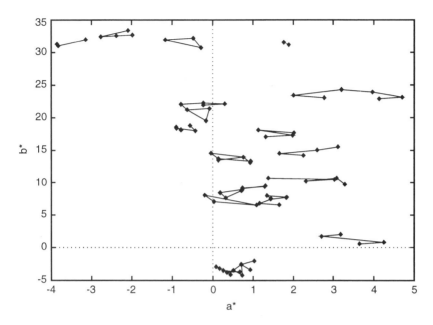

Fig. 13.22 Temporal changes of the most frequent chrominance values from the histogram presented in Fig. 13.21 over 10 consecutive frames.

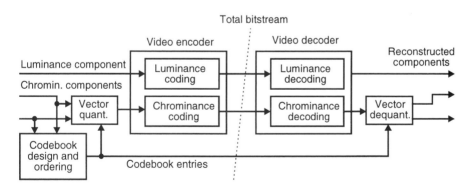

Fig. 13.23 General structure of the proposed system employing chrominance vector quantization.

FURTHER READING

A comprehensive study of different compression techniques was given in Jayant and Noll (1984). Although great progress has been made since publication this book still gives a valuable insight into basic compression techniques. A more recent overview of image compression is to be found in Clarke (1985) and in Bhaskaran and Konstantinides (1995). The last book also describes image and video compression standards. Another book by Tekalp (1995) gives a nice tutorial on video compression techniques. A somewhat more research-oriented overview of video coding can be found in Torres and Kunt (1996). For lossless techniques Held and Marshall (1996) and Gersho and Gray (1992) can be referenced. The last book also contains a comprehensive study of other compression techniques with particular attention to vector quantization.

A comprehensive objective performance comparison of various modern compression algorithms using several standard colour images can be found at (http://links.uwaterloo.ca/bragzone.base.html).

The latest research results can be found in *IEEE Transactions on Image Processing, IEEE Transactions on Circuits and Systems for Video Technology, Signal Processing: Image Communication* as well as in the *Proceedings of the International Conference on Image Processing*, the *Proceedings of the European Signal Processing Conference* and the *Proceedings of the Picture Coding Symposium* and many others.

PART FOUR

APPLICATIONS

14

Colour management for the textile industry

Peter A. Rhodes

Colour management systems are rapidly becoming commonplace in both low-end desktop publishing systems and high-end graphic art systems alike. They typically focus on the imaging path from input device (such as scanner or digital camera) via screen to print. Their goal is quite often simply to achieve a 'pleasing' colour reproduction of an original image: the sky should look blue, the grass green, and flesh tones should appear natural. For the naive user in particular, this is a fundamental expectation. A number of colour management systems are already beginning to realize this goal and their level of performance is already sufficient for many application domains. There are, however, several areas where conventional colour management is still not sufficient.

The textile industry is one such area. The colour of a garment has a great influence on its popularity and hence an impact on its sales. Often, the consumer will choose several coordinated items from the same store and expect them to match one another. Alternatively, clothes may be chosen from a mail order catalogue. If the delivered goods do not live up to expectations – perhaps because they appear different to their printed counterparts – they are likely to be returned. The consequence is that the customer loses confidence and the supplier loses business. It is therefore vital that colour accuracy is maintained both across garments and between garment and printed catalogue.

Increasingly, products such as garments and cars are being created using computer-aided design systems which may or may not include colour management capabilities. Most colour management systems are capable of dealing with a quite limited range of media typically comprising just scanned original, screen and print. While they may be adequate for complex (photographic) scenes, they have great difficulty in directly matching just a few colours which may be seen as relatively large solid areas as is the case with cars or clothing.

In the remainder of this chapter, we will see just why conventional colour management fails to fulfil these needs and what can be done to address the problem. In addition to input from and output to colour imaging devices, consideration is given to the communication of colour data *between* computer systems. The flow and management of colour information in the textile business (which this chapter uses as an example application domain) is discussed in more detail next. To illustrate the practical application of colour management and colour communication to this industry, reference is made throughout this chapter to a system called **ColourTalk** which embodies the principles of WYSIWYG[1] colour management and communication.

14.1   OVERVIEW OF COLOUR FLOW IN THE TEXTILE INDUSTRY

Many industries, including cosmetics, furnishings and clothing, are to some degree dependent upon the consumer's ever-changing tastes. Accurately predicting future trends and reacting quickly to new directions both play an essential role in staying competitive in the marketplace. The textile industry, for the most part, develops garments for two separate seasons: spring/summer and autumn/winter. Designs created for each season are on the whole quite distinct; e.g. different materials are used and different styles and colours are considered appropriate. Despite this, the fundamental design processes are similar. Designers work well in advance of the current season (perhaps three or four seasons ahead of the current one) and because of this are reliant upon accurate forecasts of future market trends.

Colour information is communicated between colour forecaster, or **colourist**, and designer in the form of fashion colour palettes. The colour palette provides a physical means of representing a fixed arrangement of a collection of colours which may be further subdivided into groups of colours such as 'classics', 'neutrals', 'bright' and 'tonal'. Predicting colour trends relies on a number of diverse sources including recurring themes in the media together with any emerging trends observed at street level. Previous years' palettes are often used as a starting point, especially for those shades that have proven popular in the past. Palette development is aided by national organisations such as the British Textile Colour Group. These provide colourists with a

---

[1]WYSIWYG='What you see is what you get'.

forum for the creation of national palettes. Annual fairs such as Première Vision (an international fabric trade fair held in Paris) present an opportunity to consolidate or confirm predictions and to establish new directions.

Colour fidelity is of concern to textile garment design for a variety of reasons. In the case of corporate fashion, there is often a need to match pre-existing company colours. Elsewhere, trends in fashion (including colour) greatly influence sales. For the case where several designers are asked to match their client's colour palette, different garments produced by different suppliers will be required to match one another in order that the consumer may choose from a range of colour-coordinated items from the same store. In selling designs to clients, greater realism is desirable as this lessens the buyer's uncertainty.

The exchange of colour information can take a number of different forms. Prior to actual garment design, the primary medium used is the colour palette. In addition, colour communication can occur at other times and may involve a variety of other media including fabric swatches, yarn or printed samples. Fabric must be dyed towards the end of the process and this typically involves communication with an external dye house. Together with shade information, it is necessary to specify a tolerance describing the acceptable colour variation. There is the potential for loss in accuracy or error each time a colour is transmitted. Every time colours are viewed, if conditions are not properly controlled, there is a possibility that they will be misperceived.

There is thus a need for both high fidelity colour and the accurate, unambiguous communication of colour. Colour fidelity is important not only between screen and garments (including fabric, buttons and zips) but also across any of the other media that may be used for design presentation. Communication between colourists, designers, buyers and mills is an essential part of the business and as these groups become increasingly dispersed across the globe, speed and lead time, in addition to fidelity, must be important considerations. The ability to quickly respond to market directions and demands is highly desirable and *can* be achieved through the judicious application of information technology. In the sections that follow, details are given of how the application of colour management and colour notation systems is able to realize this goal.

14.2 COLOUR MANAGEMENT SYSTEMS

In the past, the job of 'colour management' would have fallen upon a highly skilled operator who would have been tasked with the colour correction of images between, for example, scanner and printer. Apart from the considerable investment required to train operators, the end results are liable to be somewhat subjective and hence different for different operators. As colour imaging devices and the computer systems that drive them have become more powerful and increasingly affordable, it has become desirable to automate the process. This has been further complicated by the introduction of new imaging technologies (e.g. digital cameras and dye sublimation printers) and the requirement for colour to be consistent across a range of different computer systems and application software. Why is colour management necessary?

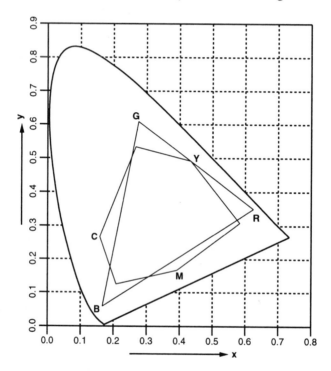

Fig. 14.1   CIE xy chromaticity diagram illustrating gamut mismatch

Different colour output devices frequently have dissimilar device primaries which, together, combine in fundamentally different ways. For colour input devices, there are differences in spectral sensitivity, illumination and measurement geometry. Another key difference is in the range of colours – or **colour gamut** – which can be produced. Figure 14.1 illustrates sample colour gamuts for a printer (area *CMY*) and display (area *RGB*) which are overlaid

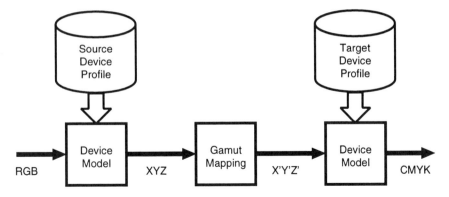

Fig. 14.2 Conventional colour management

on the same chromaticity diagram. It should be obvious that although there is considerable overlap between the two areas, the two gamuts are by no means the same. There are colours that can be viewed on the screen that cannot be faithfully reproduced in print. Conversely, some colours can be printed but not accurately portrayed on screen. The precise degree of mismatch is device-specific and depends on such factors as device primaries, luminance range and media white point.

A **colour management system** (or CMS) is a collection of software and data used by application programs to achieve improved cross-device colour rendering. The data describe the properties of an imaging device in terms of a **reference colour space** and are usually known as a **device profile**. Colour management software is composed of a number of elements including device modelling and gamut mapping. A **device model** or **colour matching** method is an algorithm which, when used in conjunction with the profile, characterizes the behaviour of a peripheral in a device-independent manner. (Several different approaches for deriving such models for CRT displays are given next in section 14.3). Typically, it is desirable to use the same generic device model for a class of imaging devices but then have separate device profiles for each instance of a device within that class. **Gamut mapping** is the means by which mismatches between source and target gamuts are taken into account in order to preserve the overall appearance of an image. There are many strategies to gamut mapping and the particular method chosen may well depend on the nature of the image (e.g. business graphics *versus* reproductions of oil paintings). This subject is covered in more depth in section 15.4. Figure 14.2 depicts the basic operation of a conventional CMS. In this case, the CIE XYZ system is shown fulfilling the role of the reference space although some systems that are in common use today elect to use simpler colour spaces (such as screen RGB) for the sake of efficiency. Another point to note is that the device characterization models need to be bidirectional mappings between device and reference colour spaces.

In addition to supporting colour matching between two or more imaging devices, a number of other useful features are commonly found amongst today's CMS. These include:

- the provision of an on-screen simulation of what an image would appear like when printed to a particular hardcopy device (also known as **soft proofing**);

- the ability to perform manual colour correction of images;

- colour selection via some systematic arrangement of colours; and

- the device-independent exchange of images between computer systems.

Colour selection is achieved using a graphical tool known as a **colour picker**. These vary in their sophistication ranging from basic device dependent colour spaces such as HLS to more elaborate colour notation systems (see section 14.5). The exchange of image data is accomplished by 'tagging' the image files with device profiles so that the receiving system can simply apply this information to the image data just as it would for any other image imported into the system from, for example, a scanner. As well as sharing images themselves, in a networked environment it is economically sound to share expensive colour imaging devices between several heterogeneous computer systems. Since there are countless different CMS currently available, each potentially having a different architecture and providing software with an incompatible applications program interface (API), this could be problematic. Fortunately, there is a standard for the cross-platform exchange of device profiles defined by the International Color Consortium (InterColor) that is being adopted by all the major developers of CMS software.

Colour management technology draws upon at least three main disciplines: colour science, device modelling and human-computer interface design. Colour science supplies the interchange spaces (such as the CIE XYZ system) which, together with suitable device transforms, enable device-independent colour reproduction across different media. The user interface directly determines the system's effectiveness although it is often desirable that the CMS be transparent to the user. Another essential requirement is that colour management be used consistently by all applications.

This dictates that it be implemented as part of the operating system (otherwise, software would need to be specially modified). In some cases, as in the case of Apple Computer's ColorSync™ software (Apple, 1995), the CMS may just be a common framework for colour management and provide only minimal colour matching. Such systems are highly flexible in that their default colour matching methods may be superseded by more elaborate third-party algorithms as and when they become available.

Conventional colour management is currently being used successfully for a wide variety of applications. The use of gamut mapping means that the particular limitations of a hardcopy device need not be of concern and the user can expect pleasing or eye-catching results with the minimum of fuss. There are, however, several limitations to this approach which will be further addressed in section 14.4. These limitations mean that conventional colour management is not generally applicable to all situations, one of these being the textile industry.

14.3   CRT CHARACTERIZATION

The cathode ray tube (CRT) is the most widespread electronic display technology in use today and the shadowmask CRT is the most common variety (illustrated in Fig. 14.3 and in more detail in Fig. 5.12 on page 104). It was developed *circa* 1950 as a further development of the black and white tube. Unlike the monochrome CRT, it uses three electron guns (one each for the red, green and blue primary colours) focused onto the screen. The screen is coated with a regular pattern of red-, green- and blue- emitting phosphor dots. A shadowmask contains a regular set of tiny holes, which correspond to the RGB phosphor dots on the inside face of the screen. By careful arrangement of the holes and positioning of the mask, only those electrons originating from a particular gun will strike the corresponding phosphor dot causing it to emit light. The amount of light emitted is controlled by the applied voltage which, in turn, is governed by the value loaded into the digital to analogue converter (DAC) which makes up the computer's graphic controller. In the CRT display, colour reproduction is dependent upon additive mixing of the light emitted from each of the three types of phosphors. This trichromatic system is similar in many respects to the CIE system of colorimetry and thus it is relatively straightforward to apply the CIE system for the analysis and characterization of such displays. When the luminance measured for each gun is plotted

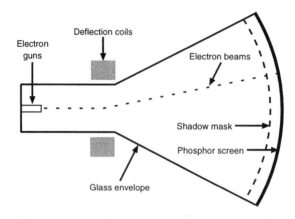

Fig. 14.3 The cathode ray tube. (See also Fig. 5.12 on page 104.)

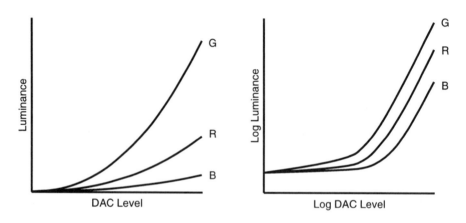

Fig. 14.4 The relationship between DAC values and luminance

against the DAC value used to produce it, the results can often be far from linear as evidenced for the particular monitor shown on the left of Fig. 14.4. Re-plotting both using log scales improves things somewhat as shown on the right of the figure but for many displays, the relationship between DAC value and luminance is not so simple. Departures from a straight line log-log relationship are due to such factors as ambient light or the inability of some instruments to accurately measure darker colours. There now follows a brief description of several characterization models that can be used to define a mapping between RGB and CIE XYZ values. Characterization involves two stages. Firstly, a relationship is determined between the light output and DAC values. This can be accomplished by modelling the curve obtained by the log-log plot mentioned earlier. The exact method used to fit the data depends on the model used. The second stage involves the measurement of the luminances (T_r, T_g and T_b) and chromaticities (x_r, y_r, x_g, y_g, x_b and y_b) for each of the three phosphors (when driven at their respective maximum DAC values) and is common to all models. The relationship (Sproson, 1983) is then:

$$
\begin{bmatrix} X \\ Y \\ Z \end{bmatrix} = \begin{bmatrix} x_r/y_r & x_g/y_g & x_b/y_b \\ 1 & 1 & 1 \\ z_r/y_r & z_g/y_g & z_b/y_b \end{bmatrix} \begin{bmatrix} T_r \\ T_g \\ T_b \end{bmatrix}
$$

By convention, the XYZ values are usually normalized so that Y is set to 100. The predictive performance of any model is very much dependent upon the DAC distribution used. The choice of how many DAC values to measure and their interval is a trade-off between measurement time and model accuracy. A number of studies have been made to compare the performance of various characterization models including those of Post and Calhoun (1989) and Luo, Xin, Rhodes, Scrivener and MacDonald (1991). These studies include additional models not covered here.

14.3.1   Log-Log

This model, also known as **gamma correction**, assumes a straight-line relationship between the log of the output luminance and the log of the applied DAC value (or corresponding gun voltage). The slope of this line is often referred to as the display's **gamma**. Thus, the model can be written as:

$$
\log T = c \log D
$$

where T is the particular gun's luminance and D is the normalized DAC value. This model is easily implemented and may perform adequately enough for some applications. As was seen in Fig. 14.4, however, it can have problems and so its general application is not recommended where high fidelity is required.

14.3.2   Log-Log2

A second-order variant of gamma correction defines the relationship:

$$\log T = c_1 + c_2 \log D + c_3 (\log D)^2$$

While this may give a better fit to the characterization data than the Log-Log model, it is much harder to implement. Determining the coefficients c_1, c_2 and c_3 requires the use of a quasi-Newton optimization technique (Gill and Murray, 1982).

14.3.3   Lin-Lin2

This is a second-order polynomial model that, unlike the other approaches covered here, relates the (non-logarithmic) luminance and DAC value. The relationship is as follows:

$$T = c_1 + c_2 D + c_3 D^2$$

The coefficients can be derived through linear regression which adds to the complexity of the algorithm.

14.3.4   PLCC

The **piecewise linear interpolation** assuming **constant chromaticity co-ordinates** model is very easy to implement and makes the assumption that gun luminance behaves linearly between and outside of each of the measured DAC values. As already noted, performance is highly dependent upon the number of values sampled and their spacing. This is particularly true for this model. In the study made by Luo, Xin, Rhodes, Scrivener and MacDonald (1991), judicious selection of DAC values (in this case, eighteen equal steps per gun) enabled this model to out-perform all other models studied.

14.4   WYSIWYG COLOUR MANAGEMENT

The claim of *what you see is what you get* is made by numerous colour management systems. In section 14.2 we saw how conventional colour management software applies gamut mapping to make best use of a given imaging

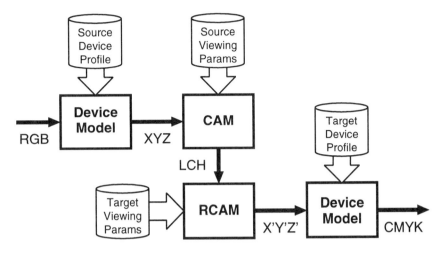

Fig. 14.5 WYSIWYG colour management

device's limited colour gamut. This means that colours are in some way shifted from their ideal (non-gamut mapped) form. Another limitation is that CMS often assume that a colorimetric match is valid across different media. The CIE system of colorimetry has nothing to say about such situations and was originally intended to be used where both media and viewing conditions are identical. For true WYSIWYG fidelity, colours must match one another absolutely across multiple media.

The cornerstone of WYSIWYG colour management is the colour appearance model. Just as a device characterization model enables us to achieve device independency, a colour appearance model provides a means of representing colour in a manner independent of medium and viewing condition. By applying colour appearance and dispensing with gamut mapping altogether, a four-stage transform (Yousif and Luo, 1991) can be constructed as illustrated in Fig. 14.5. It is instructive to contrast this with Fig. 14.2. In addition to the requisite device profile data, the transform requires additional parameters to describe the source and target viewing conditions. Typical viewing conditions that may be considered include media type, white point, illumination and background. All of these factors can have a profound effect on how colours are perceived and need to be taken into account when source and target conditions differ. There are a number of colour appearance models suited to the task which have varying levels of sophistication according to such criteria as complexity (and hence ease of implementation and required processing

Table 14.1 A comparison of colour management philosophies

	Conventional CMS	*WYSIWYG CMS*
Goal	Effective use of gamut to achieve 'pleasing' or 'eye-catching' results	Accurate colour appearance
Matching path	Usually just screen to print	Between all media concerned
Colour reproduction	Generally medium dependent	Medium independent
Out of gamut colours	May be dealt with by gamut compression	Out of gamut colours cannot be accurately reproduced
Typical uses	Photographic imagery, business graphics	Spot colours, mail order catalogues, garment labels and packaging, product design and visualization
Implementation	Device characterization and gamut mapping	Device characterization and colour appearance modelling

power) and number of viewing conditions covered. One of the most comprehensive models currently available is the Hunt colour appearance model (Hunt, 1991a). While this does cover a wide variety of conditions, deriving its inverse (required for the four-stage transform) is non-trivial. Another model, LLAB (Luo, Lo and Kuo, 1996), provides reasonable trade off between complexity and comprehensiveness and is ideal for all but the most demanding of applications. WYSIWYG colour management is most suited to those applications where colours must appear identical across a range of different media. This includes such scenarios as mail order catalogues, product packaging and, of course, the textile industry. Conventional colour management strives to make best use of the available colour gamut and hence produce appealing images. The problem of gamut mismatch is one that is unlikely to be solved in the near future and inevitably there will always be colours which are out of the range reproducible by a particular imaging device. This is a problem for any CMS irrespective of its underlying philosophy. Table 14.1 provides a comparison between conventional and WYSIWYG colour management.

For those colours that can be directly reproduced by a given device, WYSIWYG colour management works very well. It should already be apparent, however, that the principal shortcoming of WYSIWYG colour management is in its dealing with out of gamut colours. Essentially, the process is all or nothing. As a result, the type of images for which this approach is most suitable are those containing a relatively small number of colours. Photographic images portraying complex scenes are probably still better catered for by conventional colour management. In the long term, as device gamuts increase with improved technology, WYSIWYG may become more attractive to other areas. In the meantime, one solution might be to have hybrid CMS that are capable of using either technique according to gamut limitations or user preference.

14.5   COLOUR NOTATION

Colourists designing future fashion palettes, in common with many other industries, need to create new colours. One reference point for this task might be a printed colour atlas of some sort or, where a CAD system is employed, a computer-based colour picker. Alternatively, existing colours may need to be interactively adjusted on screen by the user in some systematic way. In section 14.2 it was suggested that colour selection was a desirable component of any CMS and in this section we outline how such systems might be implemented. To this end, the ColourTalk system (described later in section 14.7) is used as an example.

A **colour notation system** is a logical scheme for ordering and specifying the world of colour according to three attributes that constitute the co-ordinates of the colour system. (As mentioned in Chapter 4, these three attributes form a colour space.) Such systems encompass **colour order systems**. These consist of material standards that represent a physically realizable subset of a notation system and are arranged systematically in some form of colour atlas. (As with CMS earlier, colour gamut is of concern to users of colour specifiers[2].) There are many ways of describing and ordering colour and various colour notation and order systems in common use. These can be applied either on an industry-wide basis, as a national standard or simply due to personal preference or familiarity. The use of colour order systems has both advantages and disadvantages as given in Table 14.2. Overall, the benefits out-weigh the limitations and even these may be overcome by implementing computer-based notation systems. By describing colour order

---

[2]A colour specifier is a (usually) physical standard used to describe a limited number of colours.

Table 14.2   The pros and cons of colour order systems

Advantages	Disadvantages
Portable	Samples wear or fade with use
Useful for colour specification	Limited number of samples
Useful for colour communication	Many different systems in use
Easy to use	Relatively expensive

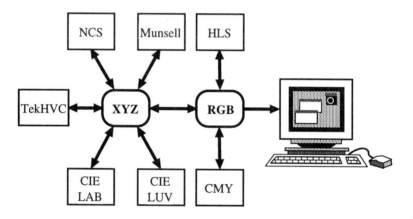

Fig. 14.6   Interrelating colour order systems

systems mathematically, it is possible to use interpolation techniques to produce many more shades than would be practical or economical in a physical
atlas. Computer colours do not fade or wear out over time (provided that the
display calibration is maintained). In addition, because the colours are represented as numbers, they can be replicated quickly and without additional
expense and subsequently communicated electronically in a matter of seconds. In the ColourTalk system, various colour pickers are implemented as
mappings to and from the CIE XYZ colourspace (Rhodes, Luo and Scrivener,
1990) as shown in Fig. 14.6. This is analogous to the common reference space
used by CMS and described in section 14.2. As well as enabling the colours
themselves to be viewed on screen (*via* the display's device characterization
model), the bi-directional mappings permit colours to be exchanged between
incompatible notation systems, thereby breaking down the 'colour language'

barrier. Although the exact form of the mapping is dependent upon the nature of the notation system, mappings can be classified as follows:

- **Device dependent systems** such as those derived directly from RGB such as HLS or HSV (Foley and van Dam, 1982, pages 593-623). These systems can be related to CIE XYZ values *via* the display's characterization model thus overcoming their device dependency. However, such systems are also perceptually non-uniform meaning that equal steps in their coordinates produce colours which do not have visually equal colour differences.

- **Mathematical systems** that are transformations from CIE tristimulus values, e.g. CIELAB or CIELUV.

- Systems based on **aim point databases** that define (usually, again, in terms of tristimulus values) a series of fixed points in the notation system's colour space. Despite representing only a limited number of shades, techniques such as trilinear interpolation can be used to find intermediate colours. This representation is suitable for describing colour order systems whose colour patches can be measured and used as aim points.

ColourTalk implements a total of nine notation systems taken from all of the preceding three categories. Since device dependent colour spaces are already in widespread use in computer graphics systems (despite the limitations previously noted), they will not be covered here. Chapter 4 discusses these colour spaces.

14.5.1 CIELAB and CIELUV

Both systems are defined in CIE (1986) and in Chapter 3 and Chapter 4 and are transformations of the CIE XYZ system which aim to describe a uniform colour space. Both systems share a common lightness attribute, L^*, with the remaining two attributes expressed as either Cartesian (a^*, b^* or u^*, v^*) or polar (hue and chroma) coordinates. CIELUV, because of the additive nature of its chromatic attributes, is useful for those applications involving the mixture of light (such as video) whereas CIELAB is used by various surface colour industries including textiles. Plate XXVIII illustrates ColourTalk's implementation of the CIELAB system. CIELUV, because of its similarity, essentially follows the same pattern.

The CIELAB colour picker is divided up into three distinct areas. The top left area represents a cross-section perpendicular to the L^* axis. In this plane, the neutral centre intersects with the lightness axis. There is an increase in chroma (or colourfulness) as distance from the centre increases. Around the centre, hue is arranged in a rainbow scale. A small cursor is used to indicate the location of the currently selected colour. Clicking in this area changes hue and chroma accordingly. The right hand area of the window presents a plane of constant hue with lightness increasing vertically from black and chroma increasing horizontally from neutral. Any of the colour patches can be selected by a click of the mouse. This, together with the top left area, may be used for coarse colour selection.

The irregular boundary defined by colours in these two areas is limited by the display's gamut. Furthermore, CIELAB or CIELUV can be used in this way as tools to visualize device gamuts. Colours do exist in the colour space beyond those displayed and the system permits them to be selected even though they cannot be faithfully reproduced on the screen. However, because ColourTalk is an example of a WYSIWYG system, no attempt is made to gamut-compress the display. To do so would severely distort the displayed colour space.

The lower left part of the display is used for fine selection. This area presents two micro-space planes: lightness-hue and chroma-hue. Variable zoom is controlled for each attribute using the three scrollbars. The target colour is displayed in the centre of a 5×5 grid with neighbouring colours differing according to the corresponding zoom controls. The window's over-all background can be set via a pop-up menu. The choice of backgrounds is limited to just black, white and grey (the default). This is not only to avoid undesirable colour contrast effects but also because this is the CIE recommendation.

14.5.2   The Munsell colour system

This is the oldest colour order system still in everyday use, being originally devised in 1905 by the artist Albert Munsell and subsequently published in 1915 as the Atlas of the Munsell Color System. Since then, the system has undergone revision to improve sample spacing culminating in the Munsell Renotation System (Newhall, Nickerson and Judd, 1943). The guiding principle is that of equality of visual spacing between each of the three attributes: Munsell Hue, Value (lightness) and Chroma (colourfulness). Munsell Hue is scaled according to five primary colours: red (5R), yellow (5Y), green (5G),

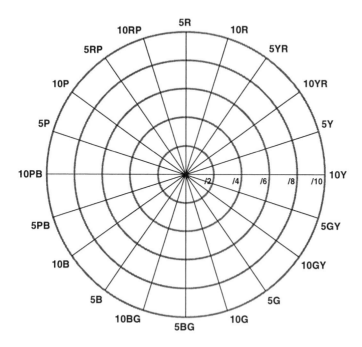

Fig. 14.7 The Munsell hue scale

blue (5B) and purple (5P). A chromatic hue is described according to its re-semblance to one or two adjacent hues. The hue scale, depicted in Fig. 14.7, is probably the most difficult concept to grasp when using the system. For example, is 2.5BG closer to blue or green? (Answer: green.)

The system is available in a variety of forms including the Munsell Book of Color and the three dimensional Munsell Tree. Munsell notation is popular amongst artists and is a standard in Japan and the USA. The atlas aim points have been published (Kelly, Gibson and Nickerson, 1943) and these were used as the basis for the Munsell colour picker shown in Plate XXIX. The arrangement of the display is similar to that given previously for CIELAB.

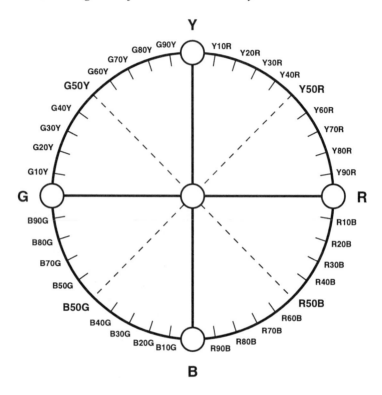

Fig. 14.8 The Natural Colour System hue scale

14.5.3   The Natural Colour System

The Natural Colour System (Hard and Sivik, 1981), or NCS, has its origins in
the work of the German psychologist Ewald Hering (Hering, 1964) in 1875
in which he proposed the existence of four unique hues: red, yellow, green
and blue. This was somewhat contrary to the trichromatic nature of colour
vision advocated elsewhere. Using this system, a chromatic colour could
therefore be represented as a combination of (up to) two of these hues. Hues
such as red and green or blue and yellow cannot be perceived together in a
single colour stimulus and are known as **opponent** hues. NCS further extends
this concept to describe a colour in terms of its redness (r), yellowness (y),
greenness (g), blueness (b), whiteness (w) and blackness (s) using a percentage scale. For example, G80Y is 20% green and 80% yellow. This scale
is defined such that $r+y+g+b+w+s = 100$. Another scale, chromatic-

ness (*c*), is trivially defined as the sum $r + y + g + b$. The arrangement of hues in the system is illustrated in Fig. 14.8. NCS is a standard colour order system throughout Scandinavia and other European countries. In the UK it has been used by Imperial Chemical Industries (ICI) to describe their **Colour Dimensions**™ paint range. It is available in atlas form and its aim points have been published by the Scandinavian Standards Institute (SIS, 1982). A computer-based implementation of the system is shown in Plate XXX.

14.5.4   TekHVC

Tektronix produced one of the first commercially available CMS called **Tek-Color**™, which was distributed with their range of colour printers (Tektronix, 1990). One component of the system was TekHVC™ (Taylor, Murch and McManus, 1989). This is a uniform colour space derived from CIELUV using a simple transform. Its attributes are Hue, Value and Chroma (not to be confused with the device dependent HVC system mentioned earlier). The implementor's guide, in addition to mathematically defining the system, provides guidelines as to the characteristics of the user interface. The ColourTalk realization is depicted in Plate XXXI. This provides a hue adjustment scrollbar together with a hue leaf that permits modification of the target colour's value and chroma. When used as part of TekColor, the printer gamut is often shown overlaying the screen colour's gamut as an aid to making colour adjustment. TekColor is also to be found as part of the X Window CMS known as *Xcms*.

14.5.5   Other sample-based systems

In many different surface colour industries, it is still common practice to employ other colour order systems based on arbitrary physical samples. A number of commercial products based on printing ink formulations have been in widespread use by the graphic arts for many years. In the paint and textile domains, suppliers frequently produce colour cards for customers that illustrate their product ranges. Because such systems are composed of relatively few samples, they are often reasonably cheap to produce. However the limited number of samples, the arbitrary nature of their selection, and the 'non-scientific' arrangement of shades does limit their efficacy. Furthermore, most systems are essentially one-dimensional with each colour being described by either a name (e.g. 'magnolia') or a number. This can make it difficult to locate a particular shade and, more seriously, rules out the possibility of using interpolation since there are no in-between colours. Instead, the best that can be done is to find the closest matching shade (i.e. that exhibiting the least colour difference) to a given target as illustrated in Plate XXXII. As can be seen, this can sometimes yield unacceptably large colour differences.

14.6   COLOUR QUALITY CONTROL

So far, we have seen how colourists select shades to create colour palettes. Colours may originate from measured physical samples or be created entirely on a computer screen. Before going into production, however, there are two principal aspects of quality control that require special consideration: colour tolerance and colour constancy. While these are very important to the textile domain, they are by no means specific to it.

One of the most fundamental questions one might ask is whether a given colour matches another sufficiently well. Colour difference formulae were introduced in Chapter 3 and these are used to quantify the closeness of a pair of colours. A number of different formulae are in everyday use within the textile industry including CMC, CIELAB and, to a lesser extent, M&S and CIE94 (McLaren, 1987). In addition to quantifying a match numerically, it is often prudent to assess this property visually. An example of this is given in Plate XXXIII which shows two samples viewed side by side together with their overall CMC(2:1) colour difference, ΔE, further broken down into individual ΔL (lightness), ΔC (chroma) and ΔH (hue) components. Note that the samples are presented against a neutral background and have a thin dividing line separating them. These conditions mimic the typical viewing conditions that would otherwise be used to judge physical fabric or thread samples in a matching cabinet.

Judging colour difference is really the end of the colouration process. A much better approach is to specify a colour tolerance before the shades are to be reproduced. Again, using the CMC formula, Plate XXXIV illustrates a tool used in the ColourTalk system for determining such constraints. (The colour difference used here is deliberately large so that its effect can be clearly seen. Much smaller values, typically less than 2 CMC ΔE units would be used in practice.) Using this tool, either an overall ΔE or separate tolerances for ΔL, ΔC and ΔH can be assigned *via* slider controls. Individual differences might need to be specified if a particular application required a tighter control of, say, hue. To help visualize the effect of such values, three cross-sections through CMC colour difference space are shown. These depict the follow-ing combinations of colour difference: lightness *versus* chroma (left), hue *versus* chroma (middle) and lightness *versus* hue (right). Each of these cross-sections has the target colour at its centre. Eight neighbouring colours that vary according to the specified colour difference surround this. Note that these illustrate the maximum colour difference from the standard. In prac-tice, batch colours should fall somewhere between the target and one of these extremes, although they could equally well vary in opposite directions from the target, which would therefore result in much larger mutual colour differ-ences.

Unlike many other industries, when a shade is to be produced, textile workers necessarily specify colour in terms of its spectral reflectance. Re-flectance values are directly input when existing physical colours are mea-sured using a spectrophotometer. For shades created on the screen and which are described colorimetrically, a corresponding reflectance curve is produced by recipe prediction *via* a colour physics system. Such systems can predict the final reflectance curve based on the physical properties of real colorants with reasonable certainty (McDonald, 1987a). Once obtained, a reflectance curve enables the determination of a shade's colour constancy, i.e. the extent to which it appears to vary under different light sources (see section 2.5.1). One can either examine this effect visually or quantify it numerically (Kuo and Luo, 1996). Obviously, the goal is to maximize colour constancy and this may be a deciding factor in selecting which combination of dyes is to be used.

A related phenomenon, **metamerism**, discussed in section 2.3.1 and in section 3.3.3 is also important to the textile industry. Often, a garment is made up from different textile materials, each of which (because of the physical properties of their substrates) require different dye recipes to attain a colour match. Since they use different dye recipes their spectral reflectance is also likely to be different. Consequently, the different materials may match one another quite well under, for example, daylight but appear somewhat different when seen under tungsten lighting. It is highly desirable to minimize this effect and, as with colour constancy, this can be visualized as shown in Plate XXXV. Both metamerism and colour constancy are computed from the samples' reflectance data together with the spectral power distribution for each of the light sources under consideration. This yields a different set of tristimulus values for each viewing condition corresponding to a non-adapted state. To visualize the colours after adaptation requires the use of a chromatic adaptation transform or colour appearance model.

There are several essential considerations when using the display to make assessments of colour difference (Berns, 1991). Again, colour gamut is one of them since such assessment can only be made through the use of a true WYSIWYG system. Another is the fundamental limitations intrinsic to CRT technology. Lack of screen uniformity and quantization errors in the digital-to-analogue conversion process both produce undesirable effects. Restricting colour assessment to the centre of the screen (where luminance is usually at its highest) can help to minimize the effect of non-uniformity. Using high precision 10- or 12-bit DAC devices will reduce quantization error. Note that these are both limitations of current technology and not of the basic approach. Finally, the accuracy of the device characterization model is also critical. For ColourTalk, this limited the system's use to colour differences greater than 0.5 CMC(1:1) ΔE units approximately. Large colour differences are also problematic but for a different reason: the CMC colour difference formula was not designed to work with large differences.

14.7   THE COLOURTALK SYSTEM

The ColourTalk system (Rhodes and Luo, 1996) has been mentioned on numerous occasions throughout this chapter as an example of a computer system providing WYSIWYG colour management for textile applications. We now detail the system's principal features and summarize those covered previously. The system comprises a high-performance workstation with a high-resolution 24-bit colour display, a viewing cabinet (for visually assessing physical samples), a spectrophotometer (to measure physical samples) and a colour printer. The software is written in the C language and consists of a number of independent tools between which the user may exchange colour information. The X Window System™ is used for graphical display purposes and the applications themselves run under the Unix™ operating system.

14.7.1   The display

A Barco Calibrator™ (Barco, 1990) is used as the system's display. In addition to day to day use, colours are often viewed over a much longer period of time. Whatever the time scale, users expect their colours to be the same. To address this requirement, the Barco monitor incorporates a light sensor that, under microprocessor control, is capable of correcting for both short-term and long-term colour drifts. Performing a routine calibration at the start of each day ensures colour stability. After the calibration has been completed, the display's colour accuracy can be verified visually by comparing a physical test chart against a screen simulation. The system uses the PLCC characterization model described in section 14.3.

14.7.2 Palette creation

The colour palette is the main form used to represent colours. In the Colour-Talk system, these are represented as a fixed arrangement of six rows of four colours presented against a neutral background. Each colour can optionally have a name associated with it. Within the palette window, shown in Plate XXXVI, shades may be manipulated using the same cut and paste paradigm which should be instantly familiar to users of modern word processing systems. Cut and paste is used throughout the system enabling individual colours to be exchanged between each of the various applications.

In creating colours to be used in the palette, the user has a number of options. The simplest method is to directly measure an existing physical sample using the attached spectrophotometer. Where this is not possible – possibly because the sample is too small or non-uniform to be measured – the shade can be matched visually using the colour adjuster tool. This permits a colour's lightness, chroma and hue to be modified until the desired match is obtained between the screen colour and the sample viewed in the matching cabinet. Alternatively, this tool can also be used to modify existing colours in some way. Plate XXXVII illustrates an example of this whereby several colours are all adjusted at once by the same relative amount. Yet another mechanism for colour input is by selection from one of ColourTalk's colour pickers.

14.7.3 Colour pickers

A total of eight different colour notation systems (section 14.5) are provided. These are Munsell, NCS, CIELAB, CIELUV, TekHVC and together with the device dependent HLS, RGB and CMY colour spaces. Each of these is presented as an electronic colour atlas that can be navigated for purposes of colour selection. By copying a colour from one notation system's window and pasting it into a different notation system's window, it is possible to achieve transparent interrelation between colour notations. Experience has shown that the various colour pickers also make excellent educational tools that can be used to introduce basic colour science concepts.

14.7.4   Visualization

Various colour phenomena can be visualized by the system. These include colour difference, metamerism and colour constancy (section 14.6). Because colour palettes only present colours in isolation, it is important to the colourist to understand how the shades will interact when used together. To this end, simple colour patches can be resized and viewed against each other using the colour viewer tool. This permits effects such as simultaneous colour contrast and the effect of sample size to be clearly seen.

14.7.5   Colour output

Completed colour palettes can be output to various media. For hardcopy purposes, a colour thermal wax printer has been characterized (Luo, Rhodes, Xin and Scrivener, 1992) and, using WYSIWYG colour management, was able to reproduce (in gamut) colours to an average accuracy of around 1.1 CMC(1:1) ΔE units. This is considered to be adequate for the purposes of colour proofing. In addition, the system was linked to an external colour physics system and to other colourant formulation software enabling the reproduction of colours across such media as packaging inks, textile and paint.

SUMMARY

This chapter has discussed the application of colour image processing techniques to the textile industry. Throughout, reference has been made to a computer-based system called ColourTalk (section 14.7). This is a WYSIWYG colour management system (section 14.4) used by the textile industry and incorporating various colour notation systems (section 14.5) together with facilities for colour quality control (section 14.6). The focus of the system is the creation and manipulation of colour palettes (section 14.1). High fidelity cross-media colour is achieved through the application of device characterization (section 14.3) and colour appearance modelling. The work leading to its development was carried out as part of a three-year research project (Luo,

Rhodes, Xin and Scrivener, 1992) involving Coats Viyella, Crosfield Electronics and Loughborough University and sponsored by the UK Government under its Information Engineering Advanced Technology Programme. Further information about ColourTalk and related developments can be found at the Internet address: `http://ziggy.derby.ac.uk/colour/`.

15

Colour management for the graphic arts

Ján Morovic

Imagine yourself in a crowd where everybody speaks a different language and there are words in each language which are untranslatable. Now try to communicate. What you need is a dictionary, or even better - an interpreter. Substitute colours for words and colour imaging devices for languages and what you get is the current graphic arts environment. The aim of this chapter is to look at the idea of colour management (the interpreter) in the context of the graphic arts and give an introduction to the issues involved in its implementation.

It would be valid to ask why the graphic arts need to be treated separately from the textile industry and the surface colour industries in general which were discussed in Chapter 14. The answer lies in the fact that what is reproduced in the graphic arts are **complex images** (e.g. photographic images and computer-generated images) whose reproduction needs to be treated differently from the reproduction of individual colours. This is because the reproduction of a complex image is judged in its entirety, rather than as a collection of the reproductions of its constituent **individual colours**. In our opening analogy this entails a shift of emphasis from the dictionary to the interpreter. The importance for the surface colour industries lies with the dictionary (characterization + colour appearance) because the accuracy of translation is judged relative to individual words (colours). In the graphic arts the dictionary still needs to be accurate – and the same principles apply as were discussed in Chapter 14 – but the stress is on the interpreter (gamut mapping) as what needs to be achieved is a communication of overall meaning rather than an exact translation of individual words.

Having said this, one must not forget that the reproduction of some colours in the graphic arts needs to follow the principles which apply in the surface colour industries. This is the case for colours which in themselves carry significant information; e.g. colours which are part of a company's corporate identity, colours used in maps, or colours used in many forms of packaging. In essence, the difference in approach is determined by whether what is reproduced is seen as an individual colour or as a complex image. This chapter will therefore focus on the reproduction of complex images, as individual colours have already been dealt with in Chapter 14.

15.1   OVERVIEW OF THE GRAPHIC ARTS ENVIRONMENT

To understand why colour management in the graphic arts has become of fundamental importance in recent years one only needs to look at the rapid development of colour imaging in the last few decades. The range of colour reproduction technologies which were developed and the need to transfer colours between them have made a systemic approach necessary. In the past, colour reproduction was entirely in the domain of professional service providers who used a limited number of colour imaging devices in a closed system operated by highly trained professionals. With digital technologies like colour printers, scanners and digital cameras becoming more easily affordable, the general public has also become involved in colour reproduction. Even though digital imaging technologies are used in the consumer market, it is primarily the professional sphere where they are employed at first. Both these applications require a faithful reproduction of colour across the different imaging devices available at any given moment and it is the role of colour management systems to facilitate it. To illustrate the diversity of the current environment, Fig. 15.1 shows the range of colour imaging devices frequently used at present.

Even though Fig. 15.1 already contains a significant number of **device types**, the list is by no means complete. A more detailed diagram could be constructed and each of the above types could further be extended, e.g. the printer would include digital printing technologies (such as ink jet, laser, dye sublimation and thermal wax) as well as traditional printing methods (such as offset lithography, gravure, flexography and screen printing).

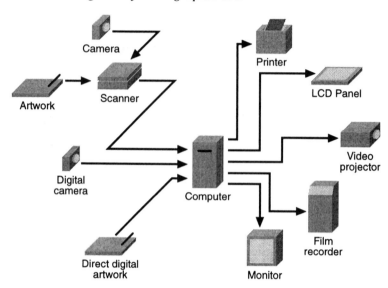

Fig. 15.1 An overview of colour imaging device types

15.2   COLOUR MANAGEMENT SYSTEMS OVERVIEW

A large number of different colour imaging devices are used in the present graphic arts environment and the task of a colour management system is to transfer information between them. The aim of colour management can be expressed by the acronym WYSIWYG[1], however, as the origins of this acronym are in rendering fonts on monitors (at least as far as the digital world is concerned), its relevance to colour reproduction for the graphic arts needs to be specified less ambiguously. WYSIWYG is most often interpreted as describing a situation where a colour seen on one device is represented on another device in a perceptually indistinguishable way. This interpretation of the term is used primarily for colour reproduction in the surface colour industries (e.g. textile and paint manufacturing) as described in Chapter 14. However, in the graphic arts, or any application where complex images are reproduced, this interpretation is not applicable. First of all, it is impossible to achieve as there are usually some colours in the source colour gamut which cannot be reproduced on the target device due to physical limitations. Secondly, it is more important to reproduce the overall appearance of a given

---

[1] WYSIWYG='What you see is what you get'.

image than the appearance of its constituent colours.

In terms of sources of difference between colour imaging devices (which make colour management necessary), the same considerations apply as are discussed in Chapter 14 on the textile industry. As a reminder, these would include differences in colour reproduction method, colorants/sensors, the colour-carrying medium itself and the viewing conditions. Also the components of a colour management system (device characterization, colour appearance and gamut mapping) are the same in essence and are used in the context of the five-stage transform in the graphic arts as well. As has already been said, the difference is in emphasis and the following sections should be seen as a description of what is different in the graphic arts as compared to the surface colour industries. The reader is therefore advised to consult Chapter 14 for the fundamentals of colour management in general.

15.3   CHARACTERIZATION AND CALIBRATION OF SYSTEM COMPONENTS

Before individual characterization and calibration techniques are discussed, it is important to have a clear idea of what is meant by these two terms. The definitions shown below are those given by Johnson (1996) and will be used throughout this chapter:

Calibration   — the adjustment of a device or process so that it gives repeatable data.

Characterization   — defines the relationship between the device-dependent colour space and a device-independent colour space (usually the CIE system of colour measurement).

The need for these two techniques is intrinsic when colours are transferred between two media. As each medium or device uses different colorants, its way of describing colours is device-dependent. Therefore to obtain a match between two **device-dependent descriptions** of colour, it is necessary to transform both into a system which is **device-independent**.

One could argue that it is possible to transform directly between device-dependent colour spaces (which is true), but such a characterization would have to be done for each pair of devices, rather than just once for each individual device. Furthermore, such a transformation would not facilitate the inclusion of gamut mapping, as this needs to be carried out in a perceptually uniform environment. The need for calibration should also be clear, as characterization will only hold true when a device is in the state for which it was characterized.

The following sections will describe common techniques for characterizing the colour reproduction devices most frequently used in the graphic arts. As the characterization of computer displays is described in Chapter 14, what will be described here are methods which can be used with scanners and printers. Note that all colour difference values given in the following sections are specific to a particular device/medium combination and results for other devices or media can be significantly different.

15.3.1   Generic characterization models

Even though the methods for characterizing a particular device depend to a significant extent on the given device, there are two approaches – full characterization, and higher order matrix transformations – which can be used for any device. In this section colorant amounts (cyan, magenta and yellow) will be used as the device-dependent coordinates and CIE XYZ tristimulus values as the device-independent coordinates. Nonetheless, these two techniques can be used with other combinations of device-dependent and independent colour spaces as well. This section on characterization and calibration is based mainly on a set of papers by Johnson, Luo, Lo, Xin and Rhodes (1997a) and (1997b).

Full characterization

The colour reproduction characteristics of any device can be determined by producing all (or a large subset) of the colours it can reproduce and measuring their CIE XYZ tristimulus values. A lookup table (LUT) can then be set up and used for obtaining the device-dependent coordinates for a given set of device-independent coordinates. The following procedure for full characterization is generally used in printing and is based on Johnson, Luo, Lo, Xin and Rhodes (1997b):

1. Produce a series of colour patches with increasing concentration for each colorant at equal steps.

2. Measure the samples and calculate the lightness (CIE L^*) of each patch.

3. Plot lightness *versus* colorant level and if the relationship is signifi-cantly nonlinear, define a LUT between quantization level and light-ness.

4. Select a number of colorant levels, n, at equal lightness intervals. This set should include the minimum and maximum amounts of colorant.

5. Print test patches containing all combinations of the above determined steps for each colorant (resulting in n^3 patches).

6. Measure each colour patch and generate a three dimensional LUT. The CMY coordinates can then be obtained for each of the n^3 XYZ colours directly and for intermediate colours they can be obtained by trilinear interpolation in the LUT using the formula given below.

$$
\begin{aligned}
N = \ & (1-D_c)(1-D_m)(1-D_y)N_1 \ + \ D_c(1-D_m)(1-D_y)N_2 \ + \\
& (1-D_c)D_m(1-D_y)N_3 \ + \ D_cD_m(1-D_y)N_4 \ + \\
& (1-D_c)(1-D_m)D_yN_5 \ + \ D_c(1-D_m)D_yN_6 \ + \\
& (1-D_c)D_mD_yN_7 \ + \ D_cD_mD_yN_8
\end{aligned}
$$

where $N \in \{X,Y,Z\}$ and D_c, D_m, D_y are the distances from the origin of the selected sub-cube in terms of device-dependent coordinates and N_i are the tristimulus values of the vertices of that sub-cube (this method of interpola-tion is illustrated in Fig. 15.2). Note that other interpolation methods, like tetrahedral interpolation or PRISM interpolation can be used as well. A very good overview of interpolation methods can be found in Kasson, Nin, Plouffe and Hafner (1995). Evidently the above method can also be used for calcu-lating D_c, D_m and D_y from XYZ. Increasing the number of steps (n) used in the LUT will increase the accuracy of the model only up to a certain point, which depends on the device being characterized. In the case of certain ink jet printers, for example, the accuracy changes only slightly when more than 9 steps are used along each colorant axis. In contrast, graphic arts scanners used by commercial service providers can often use as many as 16 steps in their LUTs. In the majority of cases full characterization gives the highest accuracy. However, it also involves a large number of measurements and is therefore considerably time consuming. This is especially disadvantageous if the variables of the device change often as is the case for most digital printing devices.

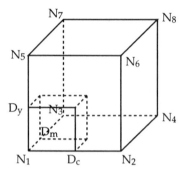

Fig. 15.2 Trilinear interpolation within a sub-cube of the n^3 LUT

Higher-order matrix transformations

This approach is also known as the **masking equations** method which was originally used for characterizing printing devices. However, it can equally well be applied to other devices such as scanners or CRTs. The method first suggested by Yule (1987, pages 205-232) assumes that the relationship between colorant amounts and the resulting tristimulus values can be modelled using the matrix transformation $\mathbf{D} = \mathbf{AS}$ where $\mathbf{D} = [D_r D_g D_b]^T$ and $\mathbf{S} = [cmy]^T$. Here D_n with $n \in \{r,g,b\}$ are **colorimetric densities** and are calculated as follows:

$$D_r = \log(X_0/X), D_g = \log(Y_0/Y), D_b = \log(Z_0/Z)$$

$\mathbf{A}$ is a 3×3 matrix obtained using the least squares method, (c,m,y) are the colorant amounts for the combination of which the XYZ coordinates are to be calculated and (X_0, Y_0, Z_0) is the reference white. The choice of **reference white** is very important and needs to be made carefully. If the output of a device is viewed in isolation, the reference white will usually be the white obtainable by the given medium (e.g. the substrate for printing, or R=G=B=100% for a monitor), but if it will be viewed alongside other media, a common reference white needs to be found (this could be the colour of a perfect diffuser lit by the current illuminant). The colour chosen as the reference white should be the colour to which the observer's visual system is adapted when viewing a given scene.

The **least squares method** used for calculating the matrix coefficients is based on the idea that the matrix gives the best predictions when the sum of the squares of the differences between predicted and original $D_r D_g D_b$ values is minimized. The minimum value can be found by calculating the point at which the first derivative equals zero, and even though points for which the second derivative is positive are also minima, the functions used in least squares systems tend to have minima as their extremes (for more details see Björck (1996)). This makes it sufficient to use the first derivative only and matrix **A** can be found by solving the linear system $[\mathbf{U}|\mathbf{V}]$ where:

$$\mathbf{U} = \sum_{i=1}^{n} \mathbf{S}_i \mathbf{S}_i^{\mathrm{T}} \quad \text{and} \quad \mathbf{V} = \sum_{i=1}^{n} \mathbf{S}_i \mathbf{D}_i^{\mathrm{T}}$$

Here $[\mathbf{U}|\mathbf{V}]$ is the augmented matrix which is to be solved, $\mathbf{S}_i$ is the ith CMY matrix, $\mathbf{D}_i$ is the ith $D_r D_g D_b$ matrix and n is the number of colours used for calculating **A**. Clearly the inverse transformation (from $D_r D_g D_b$ to CMY) can be achieved analogously, i.e. by swapping the two matrices (**S** and **D**) before performing the least squares calculation.

This simple linear method does not give good results for many devices, as for example the assumption of additivity and proportionality of colorant densities does not hold for many printers. However, when higher order and cross-product terms are introduced, as has been suggested by Clapper (1961) and Yule (1938), the precision of this method becomes comparable to full characterization. To include these new terms in the model, it is only necessary to change matrix **S** to the following form:

$$\mathbf{S} = \begin{bmatrix} c \, m \, y \, c^2 \, m^2 \, y^2 \, cm \, cy \, my \, c^3 \, m^3 \, y^3 \, c^2 m \, c^2 y \, m^2 c \, m^2 y \, y^2 c \, y^2 m \end{bmatrix}^T$$

Again we use the least squares method as shown above to calculate the transformation matrix. The precision of this method depends on the number of colours (n) used for obtaining the coefficients of the matrix and it has been found that if the colours are selected carefully (in the same way as for full characterization) there is little benefit to be gained by using more than 64 colours for characterizing printers (Johnson, Luo, Lo, Xin and Rhodes, 1997b). This number will, however, vary with the device being characterized.

A potential problem with using higher order masking equations is the possibility that local minima and maxima present in the equation will create artefacts in some parts of the colour space. Also, before any transformations are carried out, it is useful to normalize and scale the data in both colour spaces (the device-dependent and the device-independent) using the coordinates of black and white achievable by the device in question.

15.3.2   Scanner characterization

A significant number of images used in the graphic arts are acquired *via* scanners. Even though different scanning technologies exist, it is not necessary to distinguish between them for the purpose of characterization or calibration. However, as the response of a scanner is usually not linearly related to the standard observer's colour matching functions, it is necessary to carry out characterization for every substrate-colorant combination used. As the physical properties of the scanner are fixed, it is not possible to calibrate the scanner apart from stabilizing it and taking into account the intensity of the illuminant used in the scanner (most scanners perform this calibration on power-up or before a scan). For flatbed scanners which do not keep their light source switched on throughout their operation it is important to determine the distance from the origin beyond which the light source has a stable output. For example it has been found by Kang (1992) that, for the scanner he was using, the light output was stable only in the last two thirds of the scanning distance. Most of the characterization techniques which will be described here are based on a paper by Johnson (1996).

An additional improvement to any of the techniques mentioned below can be achieved if the scanner responses are first normalized so as to give equal amounts of RGB for neutral colours throughout the entire lightness range (Kang and Anderson, 1992).

Linear transformation

If the spectral sensitivities are linearly related to standard observer matching functions and there is a linear relationship between luminance and digital values, then it is possible to convert between scanner RGB and XYZ using a simple 3×3 matrix whose members can be determined either analytically (i.e. using three colours - e.g. R_{max}, G_{max}, B_{max}) or using the least squares method on a larger number of sampled colours. When the relationship between luminance and digital values is nonlinear, linear transformation can still be used, if the digital values are first linearized in terms of luminance.

The results of this approach had an average ΔE^*_{ab} error of prediction of between 2.54 and 5.97 (depending on substrate) for the Sharp JX-450 scanner used by Kang (1992).

Higher order matrix transformation

If the previous method yields unacceptable results, it is possible to use higher order terms in the transformation matrix whose members are calculated using the least squares method in the same way as shown earlier. Average ΔE^*_{ab}

errors for a 3×6 second order matrix were between 2.51 and 4.26 (Kang, 1992).

It is important to note that the higher the order of the transformation, the less generalizable the results (i.e. the errors will be significantly larger – approximately three times in this case – for colours which have not been used for the least squares fit).

The above two methods, together with full characterization, are likely to be successful for the majority of scanning devices. More specialized methods are available for dealing with scanners having special characteristics; e.g. transparency scanners with narrow-band spectral sensitivities can be characterized using dye modelling algorithms (Johnson, 1996).

15.3.3 Printer characterization

Unfortunately there is no simple relationship between colorant amounts and tristimulus values nor is there a simple additive relationship between the stimuli caused by individual colorants. This makes the whole process of characterizing a printing device fairly problematic and no single method proposed to date can be applied to a particular device with *a priori* confidence in its success. Due to the nature of most printing processes, there arise some additional problems (Johnson, Luo, Lo, Xin and Rhodes, 1997b) which include trapping: (the density of a second colorant will differ from that of the first colorant as they are printed onto different substrates — the first one is printed onto paper, whereas the second one is printed onto a layer of colorant); back transfer; multiple internal reflections; first surface reflection; and sideways scattering of light. Other factors which also influence the relationship between colorant amounts and tristimulus values are the halftone structure of the printed dots (i.e. screen ruling and frequency of amplitude modulated halftone screens or the pixel size of frequency modulated screens) and the spectral absorption characteristics of the colorants.

As most printers use four colorants (i.e. cyan, magenta, yellow and black), there is no longer only a single possible combination for obtaining a given colour stimulus (except for stimuli which can be reproduced by one or two colorants in a CMY system). This necessitates an algorithm for determining what combination is to be used for a particular colour. It is advisable to deal first with obtaining CMY values for a desired colour and, only after that, to calculate the amount of black colorant.

The following section deals with the transformation between CMY and XYZ which characterizes a given printing device.

Transformation models between CMY and XYZ
The models for performing this task can be divided into three groups: full characterization, masking equations and Neugebauer equations. The first two have already been discussed so it is only the Neugebauer equations which remain to be dealt with.

Classical Neugebauer equations   As its name implies, this model was originally proposed by Neugebauer (1937) and many modifications to it have since been devised. It is based on the assumption that the colour of a unit area is determined by the addition of the tristimulus values of the different combinations of colorants present in that area. For a three-colour system, there are eight possible combinations, whose fractional dot areas can be expressed as follows:

$$f_1 = (1-c)(1-m)(1-y) \quad \text{(clear substrate)}$$
$$f_2 = c(1-m)(1-y) \quad \text{(cyan)}$$
$$f_3 = m(1-c)(1-y) \quad \text{(magenta)}$$
$$f_4 = y(1-c)(1-m) \quad \text{(yellow)}$$
$$f_5 = my(1-c) \quad \text{(overprint of } m \text{ and } y\text{)}$$
$$f_6 = cy(1-m) \quad \text{(overprint of } c \text{ and } y\text{)}$$
$$f_7 = cm(1-y) \quad \text{(overprint of } c \text{ and } m\text{)}$$
$$f_8 = cmy \quad \text{(overprint of } c, m \text{ and } y\text{)}$$

Here c, m and y are the percentage dot areas of the three colorants which can be calculated from density using the Murray–Davies equation (Murray, 1936):

$$m = \frac{1 - 10^{-D_t}}{1 - 10^{-D_s}}$$

Here m is a percentage dot area, D_t is its density and D_s is the density of the solid (100%) colorant.

Each of these percentage dot areas has an associated set of tristimulus values, which is obtained by measuring the colorant combination it represents (e.g. $X_4Y_4Z_4$ are the tristimulus values of a solid patch printed with colorant y). The Neugebauer equations can now be expressed as:

$$\begin{bmatrix} X \\ Y \\ Z \end{bmatrix} = \sum_{i=1}^{8} \left(f_i \begin{bmatrix} X_i \\ Y_i \\ Z_i \end{bmatrix} \right)$$

Note that the above equation is identical to the trilinear interpolation equation used in full characterization, which makes the Neugebauer equations a $2 \times 2 \times 2$ LUT with trilinear interpolation.

The accuracy of the classical Neugebauer equations, as stated above, can be further improved and a range of modifications have been proposed in the past (Laihanen, 1987; Heuberger, Jing and Persiev, 1992).

So far, the models have only dealt with obtaining XYZ coordinates from colorant amounts, however in practice there is a need also for the inverse of this transformation. Where an inverse of the model cannot be obtained analytically, the transformation can be achieved using iterative numerical methods, e.g. using partial differential equations.

CMYK to XYZ transformation – black printer algorithms

As has already been mentioned, there may be no unique colorant combination to produce a given colour when using a black ink in addition to CMY, therefore different algorithms are required for the CMYK to XYZ and the XYZ to CMYK transformation[2]. The problem which needs to be addressed in this section is that of additivity failure, i.e. the density of several inks superimposed is not the sum of their individual densities (Yule, 1987).

To calculate XYZ coordinates for a given set of CMYK dot percentages, the following method can be used (Lo, 1995):

1. Calculate density values from percentage dot areas for all four inks D_c, D_m, D_y and $(D_r D_g D_b)_k$ *via* a LUT for the given printing device.

2. Calculate the colorimetric density of the combination of CMY inks $(D_r D_g D_b)_{3c}$ using masking equations, or other CMY to XYZ models with the individual ink densities as the inputs.

3. Use $(D_r D_g D_b)_{3c}$ and $(D_r D_g D_b)_k$ as inputs to a black printer algorithm to give $(D_r D_g D_b)_{4c}$, which can then be converted to XYZ using the equations for calculating colorimetric density.

A good overview of black printer algorithms can be found in Lo and Luo (1994) and in Johnson, Luo, Lo, Xin and Rhodes (1997b), where the following algorithm is also described.

Simple additivity failure model   As with the CMY to XYZ transformation, the black printer algorithm is based on knowing the tristimulus values of certain

---

[2]K is an abbreviation of 'key' and represents the black colorant.

colours. The colours needed in this case are combinations of neutrals obtained from CMY and the corresponding black. In Johnson, Luo, Lo, Xin and Rhodes (1997b) ten levels of each – CMY neutrals and black-only neutrals – were combined with each other and the black levels were also printed on their own (i.e. resulting in 110 colours). When the four-colour density is then plotted against the density of the black on its own, the lines for individual CMY levels should be parallel, however, due to additivity failure, they all approximately converge to a point on the 45° $y = x$ line. Knowing the coordinates of this point, it is possible to determine the colorimetric density of all four colorants superimposed, given that we know $(D_rD_gD_b)_{3c}$ and $(D_rD_gD_b)_k$. The equation for the red colorimetric density is shown below (green and blue are calculated analogously):

$$D_{r(4c)} = D_{r(3c)} + D_{r(k)} - \frac{D_{r(3c)}D_{r(k)}}{k_r}$$

Here k_r is the coordinate of the convergence point. The error of prediction of this model was 1.31 $\Delta E_{CMC(1:1)}$ units for the set of 110 colours and 3.51 $\Delta E_{CMC(1:1)}$ for a set of 31 test colours, which were not used to derive the model.

Three further models were developed by the same authors and can be found in the same paper.

XYZ to CMYK transformation

The goal of this transformation is to give the $(D_rD_gD_b)_k$ values from a set of $(D_rD_gD_b)_{4c}$ values, as these can be obtained from the XYZ coordinates of the desired colour using the equations for colorimetric density. $(D_rD_gD_b)_{3c}$ can then be obtained using the inverse of a black printer algorithm and be converted to CMY dot percentages using the inverse of the CMY to XYZ models.

The two main techniques for obtaining $(D_rD_gD_b)_k$ used in the graphic arts are UCR (under colour removal), which replaces only near-neutral low luminance colours by black, and GCR (grey component replacement), which affects almost all colours, by replacing their grey component with black. The grey component of any CMY combination depends on the colorant with the lowest colorimetric density. This density, along with corresponding densities of the other two inks (to give a neutral) can be replaced by black. A colour having 40%, 50% and 60% for C, M and Y respectively, can therefore be replaced by a colour with approximately 40% K, 10% M and 20% Y(approximately, because due to deficiencies of the colorants, equal amounts of CMY do not give a neutral colour). Note that replacing all of the grey component (100% GCR) does not give good results, due to impurity of pigments and deficiency of additivity failure (Lo, 1995). Furthermore, some research has shown that using GCR between 50% and 100% gives better results when compared to other GCR levels and UCR (Fisch, 1994). The procedure for obtaining the black densities using GCR has been described by Lo (1995) as follows:

1. Compute $(D_r D_g D_b)_{4c}$ from XYZ using the equations for colorimetric density.

2. If $(D_r D_g D_b)_{4c} > (D_r D_g D_b)_{k(max)}$ then $(D_r D_g D_b)_k = (D_r D_g D_b)_{k(max)}$, otherwise find the ink with the smallest density and if this density is above a specified threshold value, then determine $(D_r D_g D_b)_k$ as a proportion of it, otherwise set $(D_r D_g D_b)_k$ to zero.

3. Use the inverse of a black printer algorithm to calculate $(D_r D_g D_b)_{3c}$ from $(D_r D_g D_b)_{4c}$ and $(D_r D_g D_b)_k$.

4. Calculate $D_c D_m D_y$ from $(D_r D_g D_b)_{3c}$ using reverse of a CMY to XYZ transformation model.

5. Transform densities to dot percentages using a LUT.

Even though the models described above were developed for printers where halftoning is used to give the illusion of different levels of the colorant, they can be applied to all types of printing devices. However, for printing devices which are capable of altering the density of the colorant and are therefore not forced to use halftoning (e.g. dye diffusion printers) it is possible to use other characterization approaches as well. A characterization model specifically for dye diffusion printers has been developed by Berns (1993) and is based on the Kubelka-Munk theory. As this algorithm is a fairly complex one and applicable only to a very specific type of device, readers are advised to consult the original paper.

Printer calibration

For the majority of digital printers, calibration is not applicable, as it is not possible to alter the variables influencing colour reproduction. Therefore, characterization needs to be carried out when any of these variables change (e.g. colorant or substrate) which makes it particularly important for this characterization to require a small number of measurements.

Calibration as such is mainly applicable to conventional presses where the user has control over parameters like printing speed, viscosity of inks and pressure between blanket and impression cylinders and where it is possible to set target values for solid density and dot gain to which the system needs to be calibrated.

15.4   GAMUT MAPPING

The task of gamut mapping in the most generic sense of the phrase is to assign colours obtainable on a target device to colours from a source device. In this sense gamut mapping is present in every colour reproduction system, even though it is in most cases carried out implicitly. However, gamut mapping can also be seen as an extension of colour appearance modelling, which attempts to match the appearance of complex images, rather than the appearance of individual colours. As has already been emphasized in the introduction to this section, it is this second meaning which will be used here.

As gamut mapping is an area which is still undergoing significant development, what will be presented here is an overview of the subject (Luo and Morovic, 1996). The recommendations made in this chapter were also influenced by the findings of a psychophysical evaluation of the most prominent algorithms, carried out by Morovic and Luo (1997).

15.4.1   Fundamentals of gamut mapping

To understand the various algorithms which are discussed here, it is necessary to have a clear understanding of what the function of gamut mapping is; the stage of the colour reproduction process at which it occurs; and what its main components are.

Function
The gamut of colours reproducible on any given device can be represented as a solid in a colour space (e.g. a space defined by the CIE or a space defined by a colour appearance model). The function of gamut mapping is then to describe a way of representing colours from the original colour gamut in the colour gamut of the reproduction. The aim of this mapping is to ensure that the appearance of a reproduction is as close to the original as possible. It is important to note here that what needs to be maintained is a similarity in overall appearance rather than the appearance of individual colours as the latter is in most cases impossible. To illustrate the importance of gamut mapping, Fig. 15.3 shows the approximate gamuts of a CRT monitor and a printer. The need for a match in overall appearance introduces a further problem of

quantifying the quality of a given gamut mapping algorithm. As the use of colour difference formulae is not suitable and there is currently no model for quantifying complex images, the only solution is the use of psychophysical techniques.

Gamut mapping as part of colour reproduction

Gamut mapping is only one step in the colour reproduction process. Even colour reproduction techniques which do not explicitly have a gamut mapping stage (e.g. colour reproduction in the surface colour industries) do include gamut mapping. This implicit mapping results in a match of colours where it is physically possible and the clipping of colours onto the gamut boundary where it is not (i.e. for out-of-gamut colours). The results of a gamut mapping algorithm also depend on the other stages of the colour reproduction process and the choice of environment (e.g. colour space or appearance model) within which the mapping is carried out.

Components of gamut mapping

As the function of gamut mapping is to achieve an appearance match, it needs to map perceptual attributes — lightness, chroma and hue or brightness, colourfulness and hue. For this to be possible it is necessary to have quantitative information about original and destination gamuts. Even though the match is needed in appearance attributes, most algorithms proposed so far use uniform colour spaces (e.g. CIELAB or CIELUV) and the mappings are in the dimensions of these. In particular, the representation of hue (usually hue angle) in these spaces does not correlate well with perceived hue. Thus undesirable hue shifts can occur, especially in the blue region (Stone and Wallace, 1991).

Choice of gamuts

There are two possibilities for the choice of gamuts to map between. Firstly, it is possible to map between the original device's gamut and the gamut of the reproduction device (**image independent** mapping). This is the more common approach among those actually implemented in commercial products, as it enables the manufacturer to include the gamut mapping algorithm in the colour reproducing device itself. Secondly, it is also possible to map between the original image gamut and the reproduction gamut (**image dependent** mapping). Here the advantage is that the mapping makes the smallest possible distortion to the overall image. If the whole device gamut of the original is considered, some colours are modified excessively so as to accommodate colours which are not even present in the given image.

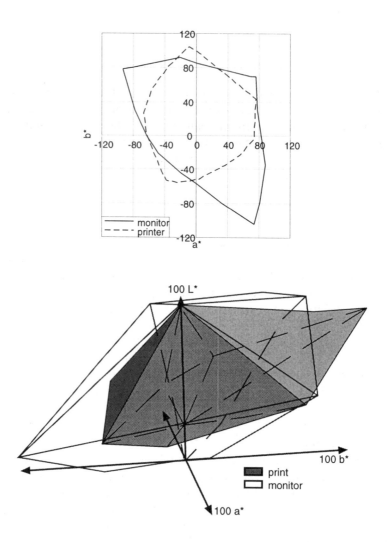

Fig. 15.3 Approximate representations of the gamut of a CRT monitor and a printer in the CIELAB colour space.

The following sections will describe the algorithms proposed for mapping lightness, chroma and hue.

15.4.2   Hue mapping

Because hue is the appearance attribute which we can distinguish with the highest precision and the best agreement between individuals, the majority of authors propose to maintain it unchanged. The only difficulty here is that, even though all but one algorithm maintains hue, they only maintain the hue angle in CIELAB which is an imperfect representation of perceived hue. This can result in a change of perceived hue in spite of colours being moved in planes of constant hue angle.

Based on experimental results a suggestion has been made by Johnson, Luo, Lo, Xin and Rhodes (1997b) to alter hue as well. In the experiment on which this recommendation was based, the ANSI IT8.7/1 colour transmission target for input scanner calibration (McDowell, 1993; ANSI, 1993b) was given to a number of colour reproduction companies in the UK and the USA to be reproduced. The results have revealed the gamut mappings used by scanner operators and they suggest an alteration of the original primary and secondary hues (included in the IT8.7/1 target) approximately half-way towards the primary and secondary hues of the reproduction.

15.4.3   Lightness mapping

The majority of algorithms proposed so far use linear compression of lightness; optionally lightness can be compressed using a soft-clipping function (Stone and Wallace, 1991) similar to the one shown in Fig. 15.4 (the axes are not labelled as the same method can be used for chroma compression as well). This function has the characteristic of not altering the majority of the range and then having a smooth transition to the gamut boundary towards the extreme of the range. It is also similar to tonal reproduction curves used traditionally in the graphic arts and is intrinsically image dependent.

15.4.4   Chroma mapping

As lightness is the attribute whose mapping is seen as being the most important by a large proportion of researchers, it has been explored to a greater

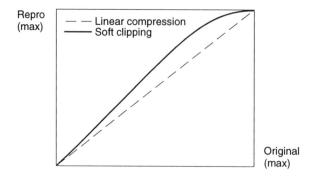

Fig. 15.4 Soft-clipping function used for lightness or chroma compression.

extent. In most studies of lightness mapping, the mapping of chroma is a simple linear compression in planes of constant hue and along lines of constant lightness (Laihanen, 1987) as shown in Fig. 15.5(a). What has also been proposed is a nonlinear compression along this line using a soft-clipping function. These chroma mappings are usually performed after lightness compression. The compression can be determined either for every line independently, or uniformly for the whole colour space. If the latter approach is taken, a compression ratio of half-way between unity and the lightness compression ratio has been found to give the best results.

15.4.5   Simultaneous lightness and chroma mapping

A more successful approach to gamut mapping is taken by those algorithms which map lightness and chroma simultaneously, as this gives smoother transitions between adjacent regions in a colour space. These algorithms are also more likely to preserve more of the colourfulness of an image, which has been found to be of great importance to those who view the images (Morovic and Luo, 1997).

The most prominent of these is a compression which maps colours towards the centre of the lightness axis (i.e. $L^* = 50$) in planes of constant hue (Laihanen, 1987) as illustrated in Fig. 15.5(b).

Again, with this technique, the rate of compression can be determined individually along every line or it can be set to be equal for the entire colour space.

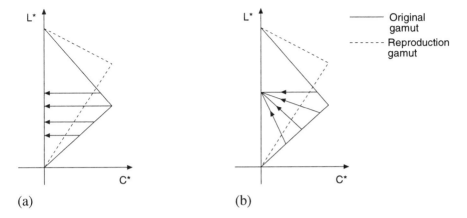

(a)   (b)

Fig. 15.5 (a) Chroma mapping along lines of constant lightness in planes of constant hue; (b) Spherical mapping in planes of constant hue angle.

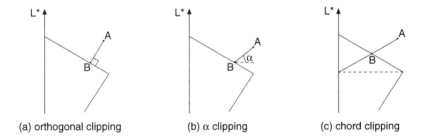

Fig. 15.6 Gamut clipping techniques.

Instead of compression, it is also possible to map only colours from outside the reproduction gamut. This results in the mapping (clipping) of out-of-gamut colours onto the reproduction gamut boundary. The clipping can be done along lines which are at a given angle (α) relative to the C^* axis (Fig. 15.6(b); by finding the closest point on the gamut (Fig. 15.6(a); or by mapping towards a point on the L^* axis (e.g. towards the L^* of the gamut cusp (Fig. 15.6(c)) in a given plane of constant hue. A very promising technique also belonging to this group has been proposed by Johnson, Luo, Lo, Xin and Rhodes (1997b) and it is based on the experiment mentioned in section

15.4.2. The algorithm consists of the following steps:

1. Map white and black points of the two media onto each other and scale the lightnesses between them linearly.

2. Perform additional compression of L^* and C^* depending on the characteristics of the two gamut boundaries in the following way:

   (a) find cusp at the hue angle of each primary and secondary colour (i.e. red, green, blue, cyan, magenta and yellow in the IT8.7/1 target).

   (b) if one gamut completely encloses the other and the L^*s of the cusps are similar for both gamuts, map C^* along lines towards the lightness of the output gamut's cusp on the L^* axis as shown in Fig. 15.7(a), otherwise, if the cusps have different lightness, map along lines going towards the point on the L^* axis where the line going through the two cusps intersects it as shown in Fig. 15.7(b), or, if the gamuts overlap only partially, map towards a point on the C^* axis which is determined by the cusps of the two gamuts as shown in Fig. 15.7(c).

   (c) Compression along a given line is then determined by:

   $$x\frac{(c-b)}{(c-a)}$$

   where x is the distance along the line from the L^* axis, c is the point on the outer gamut, b is the point on the inner gamut and a is the point on the L^* axis as shown in Fig. 15.7. Note that this simple linear compression could be replaced by a nonlinear one (e.g. soft-clipping).

3. Determine the hue shift for the six primary and secondary colours between the two gamuts and translate the original hues half way towards those of the reproduction.

4. For intermediate hues, use proportions of the corrections applied to the two closest primary and secondary hues. If the boundaries are severely concave, more hues need to be chosen and then interpolated between.

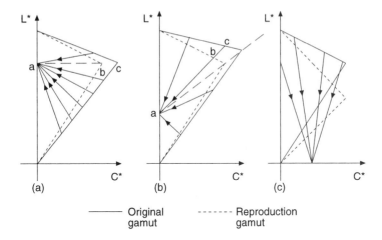

Fig. 15.7 Chroma compression depending on relative relationship between gamut boundaries.

15.4.6   Gamut mapping principles

As has been said in the introduction to this section on page 346, the solutions presented here to the gamut mapping problem are still at an early stage of development. The following suggestions have therefore been included as a guide for those who need to implement gamut mapping, as they represent the recommendations of many researchers who have written on this subject and the conclusions of recent experimental work (Morovic and Luo, 1997).

1. *Make changes to as few colours as possible and make changes as small as possible.*

2. *Use a perceptually uniform environment.* The need for a uniform environment in which the gamut mapping is to be implemented is important. The nonlinearity of iso-hue loci in colour spaces is recognised as a particular problem as chroma compression in such planes results in a perceivable hue change.

3. *Maintain perceived hue, rather than hue angle.*

4. *Apply different compression in different parts of colour space.* This need for nonuniform compression was identified in the earliest papers,

as uniform compression results in excessive compression in most parts of the colour space and is also strongly image dependent.

5. *Maintain as much chroma as possible without sacrificing a significant amount of detail.* Psychophysical experiments suggest the significance of chroma in gamut mapping. This has often been ignored in papers on gamut mapping which consider lightness to be the most important attribute. However the best results will be achieved when a compromise between lightness and chroma is found. To do this, the simultaneous gamut mapping algorithms seem most suitable.

15.5   CURRENT COLOUR MANAGEMENT SYSTEMS

Having discussed the components of colour management systems in general, a more meaningful look can be taken at commercial colour management systems (CMS) available at present. However, instead of describing individual CMSs, information about which is readily available from their publishers, a look will be taken at the colour management framework proposed by the **International Color Consortium** (ICC). The ICC has been founded by a number of companies active in the colour reproduction market with the aim of establishing 'an open, vendor-neutral, cross-platform colour management system architecture and components' (ICC, 1996). The need to work on such an open standard arose from the situation in the colour management market, where there existed a wide range of mutually incompatible solutions usually designed for specific applications. What the ICC has done so far is to propose a colour management architecture standard (ICC, 1996) including a standard for device profiles and colour management modules. The core of the proposed system is the so called **profile connection space** (PCS) using CIE colorimetry (either XYZ or CIELAB) and having the characteristics of an idealised reflection print with unlimited colour gamut. Colour transformations between different devices are accomplished by first transforming colours into the PCS and then transforming them into the colour space of the output device. Clearly for computational purposes this step can be (and often is) implemented as a single transformation. Both transformations are based on information about the input and output devices stored in device profiles. The format of the device profile is set by the ICC and the current version (3.3) includes information about device characteristics like tristimulus values of colorants and media white, tonal reproduction curves of colorants or LUTs for transforming between the device-dependent colour space and the

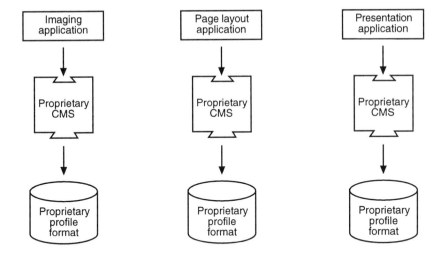

Fig. 15.8 The traditional approach to colour management.

PCS. The actual transformations, based on device profiles, are carried out by **colour management modules** (CMMs) whose architecture is also specified by the ICC. As a minimum, CMMs carry out device characterization for input devices and device characterization with gamut mapping for output devices. Due to the modular nature of the system, it is possible to include any gamut mapping algorithm as well as to use colour appearance models in the process.

Figure 15.9 shows an overview of Apple's ColorSync™ (Apple, 1995), which offers a system-level colour management framework. Note that ColorSync is the colour management system on which the ICC has based its standardization effort. More detailed information about the ICC colour management framework can be found in the standard itself (ICC, 1996) and in MacDonald (1996) where some of the deficiencies of this approach are also discussed.

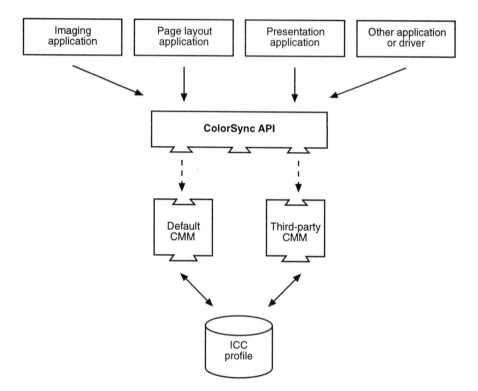

Fig. 15.9 Overview of Apple's ColorSync™ system-level colour management system.

CONCLUSIONS

What has been attempted in this chapter is to give an overview of what is involved in trying to match the appearance of complex images between different media. Many of the components of the colour reproduction system proposed as a solution are already in a mature state of development, however, the element most critical to complex images – gamut mapping – is not. In spite of this, the colour management system presented in this chapter can already be used to achieve satisfactory results.

For further reading on the topics covered in this chapter see Johnson, Luo, Lo, Xin and Rhodes (1997a), Johnson, Luo, Lo, Xin and Rhodes (1997b), Johnson (1996), Johnson and Scott-Taggart (1993), Laihanen (1987), Mac-Donald (1996) and Morovic and Luo (1997).

16

Medical imaging case study: the rôle of colour image processing in the metrics of wound healing

Bryan F. Jones and Peter Plassmann

Pseudo- and true-colour are both widely used in medical imaging. The motivation for using colour in medical imaging can be classified mainly in terms of:

- improving the efficiency of computation;

- segmenting images;

- visualizing data;

- diagnosing diseases by the absolute measurement of colour.

In this case study, MAVIS, which is an instrument for measuring the physical dimensions and colour of wounds uses pseudo and true colour in the four areas identified above. MAVIS (Measurement of Area and Volume Instrument System) uses structured light to construct a 3D model of a wound whose area and volume can then be measured. The projected pattern is colour-coded to improve the efficiency of extracting and labelling the stripes. In spite of the fact that wounds have irregular boundaries and a surface that has a wide variation in reflectivity and colour, wounds have been successfully segmented in some circumstances by analysing the colour of the image. A 3D reconstruction is produced in pseudo colour for the clinician to examine in detail. A true colour image of the wound is used by clinicians to compare the changes in the wound over time and to predict the status of healing of the wound. MAVIS acts as a decision support system for clinicians in the management of

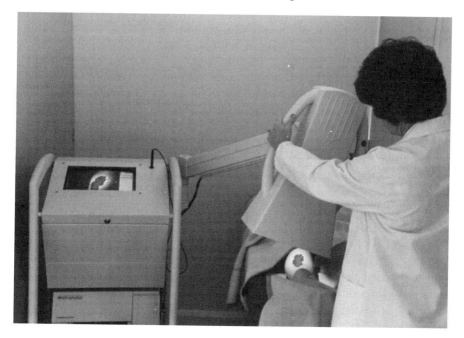

Fig. 16.1 The MAVIS instrument in use.

the treatment of wounds. Figure 16.1 shows MAVIS in use, imaging a wound.

16.1   WOUND METRICS – THE BACKGROUND AND MOTIVATION

When a healthy person cuts or wounds himself (i.e. punctures the protective layer of skin), healing starts first with the formation of a blood clot followed by a scab. The damage is repaired by agents that are released within the wounded subcutaneous tissue and, after a couple of weeks, the scab is sloughed off to reveal new tissue and the wound is forgotten.

In contrast, some people may experience the reverse process: a simple bump on the leg, which for most people would result in no more than slight bruising, may cause the skin to split and die and a leg ulcer to form. Immobile patients, who lie for long periods in the same position, may reduce the blood flow to parts of the skin to a level which allows pressure sores to develop. In each case, the skin is breached and the affected area becomes literally a crater and is susceptible to infection. Infected wounds are often extremely painful and an appropriate antibiotic must be found to counter the infection otherwise the limb may eventually have to be amputated. The generic term **wound** includes ulcers, pressure sores and others such as surgical or malicious wounds. Wounds are a major risk for those suffering from some diseases such as diabetes and for those who take certain medications, such as steroids, over long periods. Pressure sores and ulcers are caused by poor blood perfusion arising from a constricted vein or, less commonly and more seriously, an artery. It is in their nature that these wounds often heal very slowly over a period of months or even years. The cost of wound healing is therefore high both in terms of suffering to the patient and of resources needed to manage the patient. The cost to the British National Health Service is more than £ 200 million each year.

The process of wound healing is not fully understood and there are numerous medications in common use for treating chronic or infected wounds, typically based on impregnated dressings which may also apply pressure to the limb. In addition, more exotic treatments such as the use of maggots are under investigation. During the 1914-18 World War, military Medical Officers were surprised that those soldiers whose wounds were infested by maggots healed more quickly than others. A method of treating wounds using maggots has been developed recently in several centres in the UK, such as Bridgend and Oxford, with encouraging results in combating infection. Because the rate of healing is slow, it is important for a clinician to determine as quickly as possible whether the current treatment is working and that the lesion is healing. In such circumstances, reliable and precise metrics are essential to decide within a week or two whether a wound is decreasing in area and volume. The volume is a more important indicator in the early stages of healing since wounds tend to heal from the base before the area decreases; in fact, the area may increase slightly at first even though the wound is healing.

One current method of area measurement is to place a transparent, sterile, acetate sheet on the wound, and to trace the perimeter. The acetate sheet is then placed on graph paper and the area is determined by counting squares.

Wound volume is measured by several methods. One is to fill the wound with a measured volume of saline solution from a syringe. A source of error is the absorption of the saline by the wound interior. A second is to fill the wound with the alginate paste that is used by dentists to make casts of gums for the fitting of false teeth. When the paste has solidified, it is removed from the wound and weighed and the volume is calculated from the alginate density. A major source of errors is the tendency of the alginate casts to lose moisture; correct storage and prompt measurement are essential to obtain results that are sufficiently precise to assess the progress of healing.

All of these metrics suffer from the major disadvantage that they make physical contact with the wound, so risking the introduction of infection and the infliction of pain on the patient. Furthermore all the methods are slow so that the wound cools during measurement. Some clinicians believe that cooling lengthens the healing process. The objective of the MAVIS project is to devise an instrument that measures the colour, area and volume of a wound and that

- does not touch the patient;

- has a precision of about 5%;

- is easy to use and generates results rapidly;

- is portable and can be used in clinics and in hospital wards.

Structured light has been used in image processing and computer vision to identify objects. This approach has been adapted successfully to wound metrics so that the above objectives are achieved.

16.2   PRINCIPLE OF STRUCTURED LIGHT

A depth map of a surface can be calculated by projecting a pattern of illumination onto it and measuring the distortion of the pattern. As the height of the surface changes, the pattern appears to shift laterally when viewed from above. In Fig. 16.2, P projects a collimated beam of light and the camera, C, captures the image. The angle between the optical axes of the projector and camera is θ. Points A and D are at the arbitrary reference height; the co-ordinates of point A are (x_A, y_A), and so on for points B, C, D and P. The surface that starts and ends at the reference height falls to point B in the middle, and the point of intersection between the light beam and the surface appears to move sideways from A to B from the standpoint of the camera. The lateral shift, $(x_A - x_B)$, is related to the difference in height, $(y_D - y_B)$, by a simple trigonometrical relationship:

$$y_D - y_B = \frac{x_A - x_B}{\tan \theta}$$

The quantity $(x_A - x_B)$ is measured by tracing the lateral movement of the stripe in the image, and θ is known from the calibration of the instrument.

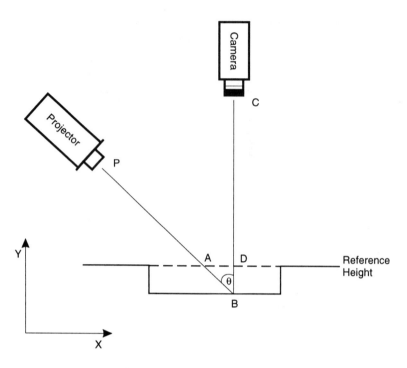

Fig. 16.2 Principle of structured light used in MAVIS.

The depth map, $(y_D - y_B)|_x$, is calculated along the length of the stripe relative to the arbitrary reference height. The analysis may be extended easily to a three-dimensional triangulation (Jones and Plassmann, 1995).

Structured light has been used in computer vision to recognize an object by the characteristic way it distorts the projected stripes. The technique is relatively easy to implement because of the regular shape of most man-made artefacts and the uniformity of the surface reflectance. In contrast, wounds have irregular shapes and the surface may vary in colour from that of healthy skin through blood red to the yellow of slough (dead cells exuded from the wound) and the black of necrotic (dead) tissue. The result is that the reflected stripes in the image are discontinuous because the reflectivity is very low in the dark areas of the lesion and very high when a layer of liquid such as blood gives rise to a specular reflection. A specular reflection is a virtual image of

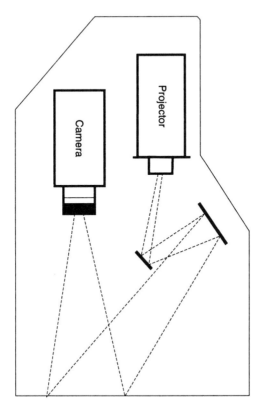

Fig. 16.3 Schematic view of MAVIS.

the light source, and its high luminance masks the lower reflected luminance that is of interest. The resulting gaps in the stripes must be filled, in order to trace a continuous stripe, and to calculate the height of each point relative to the reference level. The gap-closing algorithm used in MAVIS is based on a polynomial interpolation and only operates if the gap parallel to the stripes is smaller than the perpendicular separation, otherwise a new image must be captured.

The instrument (Fig. 16.3) comprises a projector with a 24 V 250 W tungsten halogen bulb housed in a portable box. The dimensions of the box have been kept to a minimum by the use of mirrors which are oriented so that the stripes are projected at an angle of 45° to the optical axis of the camera. During operation, the instrument is held so that the camera views the wound surface more or less normally (that is, the camera axis is perpendicular to the plane of the wound). The housing is completely enclosed and it has a smooth surface for easy disinfection. Its base is a rectangular frame which defines the optimum distance and field of view when it is held close to the skin. Two images of the wound are captured: the first is illuminated by unfiltered light from the tungsten halogen bulb, and the second is illuminated by a slide of alternating red and blue stripes with a single green reference stripe in the centre. The film and colours of the stripes are chosen to match the reflectivity of skin as measured by Anderson and Parish (1981) and Kuppenheim and Heer (1952), to maximize the camera output in each channel, and to minimize the channel crosstalk. The best match is obtained with ordinary Ektachrome™ film in combination with an infrared cut filter. (The spectrum produced by the tungsten halogen bulb contains a high proportion of infrared radiation which would affect the red channel, and also to some extent the green and blue channels without the infrared cut filter.)

The system is calibrated for colour at the colour temperature of the tungsten lamp (~3100 K) rather than daylight (5200 K). A thin chip of magnesium oxide is used as a white standard; this is placed at the base of the instrument, and the colour balancing controls are adjusted to give equal outputs from all three channels. This process must be repeated from time to time because the colour temperature of the lamp changes as it ages.

The output of the camera is fed into a colour framegrabber card housed in an IBM-compatible personal computer. The use of colour-coded structured light improves the efficiency of computation by facilitating the labelling of stripes and the closing of gaps arising from regions of poor reflectivity or specular reflections.

The two images are captured within 0.45 s so that the patient does not move between the images and thus there are no problems of registration. Most of the time is taken up by changing the slides mechanically (0.1 s) and allowing for the mechanical vibrations in the system to be damped (0.2 s). The rest of the time is used to synchronize the camera and framegrabber and to acquire the image.

The system is calibrated geometrically to determine the angle between the optical axis of the camera and the direction of the projected collimated beam. The system housing must be robust enough to prevent changes in the relative positions of the camera and projector during handling and transportation. This is crucial to maintain the precision of the instrument which measures the depth map of the skin surface by triangulation.

16.3   IMPLEMENTATION AND PRECISION OF MAVIS

An overview of the MAVIS software is given in Fig. 16.4 in which the three columns represent the input data, the process and the output data from each process. MAVIS captures two images. One of the advantages of MAVIS is that the first video image is captured under normal lighting conditions and is analysed to classify the wound and stored for the clinician's future reference. Both of these uses are described later. The second image is captured under illumination of alternating red and blue stripes with a central green reference stripe. Typical luminances in the RGB channels for images captured with unfiltered illumination and with stripes are shown in Fig. 16.5. Crosstalk between the channels is evident, especially in the green channel where the green reference stripe is clearly visible along with smaller peaks corresponding mainly to the blue stripes. The high frequency oscillations correspond to the stripes and the low frequency variations follow the changes in reflectivity of the skin. The green reference stripe is the only point at which the luminance of the image with stripes exceeds that of the image without stripes. The centre of the green stripe is defined to be at its centroid. The difference between the blue and red planes is also plotted in Fig. 16.5 along with the mean of the local peaks and troughs. Taking the difference between the blue and red planes enhances the red and blue stripes by reducing the crosstalk between the channels; the peaks correspond to blue stripes and the troughs to red. The peaks and troughs are identified by confirming that peaks and troughs alternate at a minimum separation of 15 pixels and a maximum separation of 30 pixels.

The edges of the stripes are defined to be the points of intersection of the mean and the (blue - red) curve; these intersections occur at y values of a and b. The centroid, C, is given by

$$C = \frac{\sum\limits_{y=a}^{b} yg(y)}{\sum\limits_{y=a}^{b} g(y)}$$

where $g(y)$ is the grey level at position y. Using the dynamic threshold maintains the true centre of the green stripe even when changes in the skin reflectivity affect the width of the stripe.

The instrument is calibrated geometrically to obtain the accurate relative positions of the camera and projector and the focal plane. An error of ± 1 mm

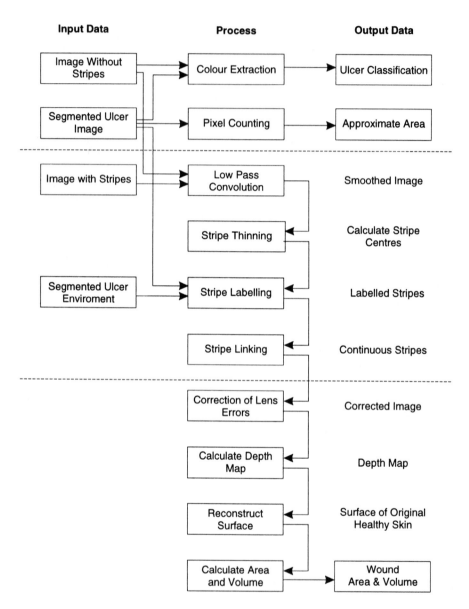

Fig. 16.4 Overview of the MAVIS system software.

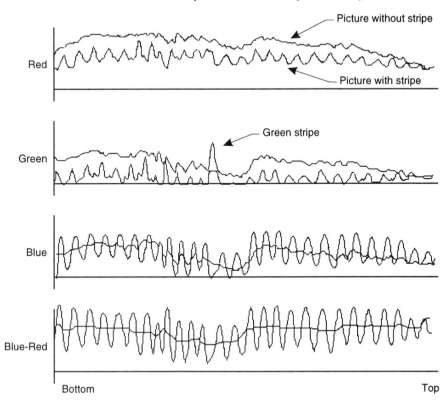

Fig. 16.5 Luminances in the RGB channels for unfiltered illumination and for striped illumination.

in each of these positions leads to an imprecision of 3% in the total area and volume. The next part of the calibration defines a labelled reference grid in the focal plane of MAVIS; this is achieved by projecting the stripes onto a flat plate at its base.

The perimeter of the wound is traced manually on screen using the computer mouse. The surface grid is then subdivided into a mesh of triangles whose areas are integrated numerically to give the wound area.

The volume of a wound is the volume enclosed between the original healthy skin surface and the base of the existing wound. The original healthy skin has been destroyed and must be reconstructed in the first part of the calculation. The operator defines a region surrounding the wound that can be used to reconstruct the original surface by interpolation. The operator is helped by overlaying the extracted stripes which indicate the presence of swellings which are common close to the perimeter of a wound and may form a negative volume alongside the positive volume of the wound. The depth map relative to the focal plane of MAVIS is calculated for the region of interest surrounding the wound using triangulation at the stripe centres. The depth map for each stripe on the healthy skin surrounding the wound is used to form a cubic spline interpolation for the missing wounded skin. The volume of the wound is integrated numerically by forming the product of the area of each triangle that was defined in the area calculation and the average height between the wound base and the interpolated original surface.

The precision of the area measurement is influenced by three main factors:

1. the skill of the operator in following the perimeter of the wound;

2. the quantization of the wound area by the grid; and

3. the distance of the surface to the instrument.

A series of wound model areas were measured at different distances and the precision was within $\pm 0.5\%$ of the total area. The experiment was repeated under more realistic conditions where an operator manually traced rectangular chips of known area. The resulting imprecision was less than 5% apart from small areas (less than $400\,\text{mm}^2$) at distances greater than 40 mm from the instrument.

The precision of volume measurement depends mainly on four factors:

1. the skill of the operator in defining the boundary of the area to be reconstructed;

2. the quantization of the measurement grid;

3. the position calculations for the surface points in the measurement grid; and

4. the ability of the reconstruction algorithm to produce a realistic original skin surface.

In determining the wound volume, MAVIS calculates the position of the base of the wound by triangulation and the position of the former healthy skin by interpolation. The interpolation procedure was assessed by 'reconstructing' a healthy skin surface that was undamaged. There should be no difference between the 'reconstructed' and the measured surfaces. The inaccuracy was less than 2%, but increased as the area to be reconstructed increased, to more than 2000 mm², especially when the surrounding healthy tissue was nearly flat. The undulations in the cubic spline interpolation in these conditions increase the percentage error. The overall imprecision of measuring fifteen wound models was better than 5% apart from wounds with small volumes (less than 1000 mm³) and large areas (more than 300 mm²).

Two limitations of MAVIS are: that wounds with area to volume ratios of less than 0.1 mm⁻¹ cannot be measured because the wound edges are too steep; and that wounds bigger than 7000 mm² occupy so much area that there is insufficient healthy skin visible in the instrument's field of view to reconstruct the original healthy skin surface.

16.4   ASSESSMENT OF THE STATUS OF HEALING

Some clinicians believe that the true colour of a wound is indicative of the status of healing and may even be used to forecast a time to heal. MAVIS incorporates some true-colour measurements to give the clinician an indication of whether a wound is responding to treatment.

Afromowitz, van Liew and Heimbach (1987) investigated the healing of burns which have similar optical properties to wounds; they illuminated the wounds with light-emitting diodes at wavelengths of 640 nm, 550 nm and 880 nm and used simple photodetectors rather than digitized video to measure the reflected radiation. They proposed that the ratio of reflected red, R, to infrared, I, radiation and the ratio of green, G, to infrared radiation could be used to predict the probability, P, of healing within 30 days using the formula:

$$P = \frac{e^k}{1 + e^k}$$

where

$$k = -7.22 - 6.11\frac{G}{I} + 9.22\frac{R}{I}$$

Even though the measurements could be reproduced by a tungsten halogen source and a CCD camera, it is doubtful whether this approach could be transferred directly to wounds whose healing pattern differs significantly from burns.

Arnquist, Hellgren and Vincent (1988) attempted to classify healing leg ulcers by digitally processing colour photographs. They indexed the wound by sixteen parameters from four classes based on wound colour:

Class 1 — black necrotic (dead) tissue;

Class 2 — yellow necrosis and fibrin;

Class 3 — red granulation tissue (indicative of healing);

Class 4 — pink epithelialization tissue (indicating that healing is almost complete).

They monitored the areas of red, yellow and black surface within the ulcer during healing and found over a week of successful treatment that typically the yellow area almost halved and that the black area became very small; this indicated a reduction in the amount of necrotic tissue. In contrast, the red area fell by only 10% indicating a relative increase in granulation and epithelialization.

Herbin, Bon, Venot, Jeanlouis, Dubertret, Dubertret and Strauch (1993) assessed the kinetics of healing using true-colour image processing. They investigated the healing of 8 mm diameter blisters raised by mild suction on the forearms of healthy Caucasian male volunteers. The wounds were tracked over 12 days using colour slide photography, and the slides were digitized with a three-CCD camera. They asserted that wound area is one of the most important indices of healing kinetics and consequently automated the area measurement. They successfully segmented the wound using the colour parameters r, g, b, H and S (section 4.1 and section 4.6), though the blister wounds in their study were delineated much more easily than typical leg ulcers found in wound healing clinics. By day 5 when the redness is greatest due to epithelialization, the parameters r and g had the lowest coefficients of variation and were selected for colour thresholding. They experienced some problems with specular reflections and with disconnected areas in the wound which were closed using standard morphological operators. The parameter g had the biggest value and had an intra-day variability that was lower than the inter-day variability; it was therefore chosen as the colour index of healing.

Hansen, Sparrow, Kokate, Leland and Iaizzo (1997) investigated the correlation between colour and trauma index of wounds inflicted on pigs using hot and cold metal discs applied with different pressures for different times. The motivation for the work was to assess wound severity and to assess the likelihood of a pressure sore developing in immobile patients. Although they concluded that the tissue hue could be used to distinguish between wounds of different severity, they were unable to build a linear model relating H, S and I to wound severity. In any case, it is not clear that a porcine model would transfer to humans with any validity.

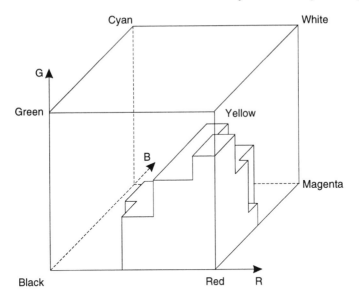

Fig. 16.6 An RGB cube showing a volume which encloses red granulation tissue.

Following the approaches of Afromowitz, van Liew and Heimbach (1987) and Arnquist, Hellgren and Vincent (1988), MAVIS measures the ratio of the black and yellow areas which indicates the presence of necroses (dead tissue) and the red area which shows the presence of granulation tissue formed during healing. The wound is outlined using the computer mouse to trace the perimeter. The colour of each wound pixel is then checked after averaging by convolution with a 3×3 filter. The pixel value in each colour plane is averaged using the following binomial smoothing [1] mask, h:

$$ h = \frac{1}{16} \begin{array}{|c|c|c|} \hline 1 & 2 & 1 \\ \hline 2 & 4 & 2 \\ \hline 1 & 2 & 1 \\ \hline \end{array} $$

The pixels contained within the wound are then classified as either *red, yellow, black* or *unclassified*. One method for expressing colour is the RGB colour cube where the colour is represented as a point in a three dimensional coordinate system. Fig. 16.6 shows the RGB colour cube and the volume that

[1] Binomial smoothing is discussed in detail in section 8.1.2 on page 151.

encloses all the red granulation tissue. This reference volume was defined after manually analysing a large number of wound images captured under illumination from a tungsten halogen lamp which has a continuous spectrum that peaks in the infrared. Other volumes representing yellow and black pixels were similarly defined; pixels that fell outside these volumes were unclassified.

It is interesting to note that the impression of 'red' is caused by the balance between the primary colours rather than the absolute intensity of the red value at a specific pixel. This can be seen in the scan of Fig. 16.5 which shows that in the central area of the picture where the ulcer colour appears to be a saturated red, the luminance in the red channel is less than that in the surrounding healthy, white skin. Here the red is balanced by the relatively strong blue and green luminances. The blue and green values decrease near the centre of the ulcer whereas the red luminance remains high giving the appearance of a saturated red colour.

The use of colour to predict the status of healing is controversial amongst clinicians and the MAVIS instrument therefore simply displays the ratios of the areas of black and yellow pixels and the area of red pixels within the wound.

Further investigation is under way into assessing the significance of colour and thermal imaging in predicting the presence of infection before the suspicions of a clinician would normally be aroused. Harding and Melhuish of the Wound Healing Research Unit in Cardiff, Wales, and Hoppe of the University of Glamorgan, Wales, have found some correlation of wound colour with the presence of infection. Ring and Elvins of the Royal National Hospital for Rheumatic Diseases in Bath, England have investigated the use of infrared imaging in the 8–$13\,\mu$m region to detect infection. Infrared images of wounds themselves disclose little because subcutaneous tissue has a much lower emissivity than skin. However, a low temperature at one part of the wound perimeter is a portent that the wound is about to grow in that direction. Preliminary results are encouraging for the use of colour and thermal imaging.

16.5   AUTOMATIC SEGMENTATION OF THE WOUND

A group of experienced clinicians will disagree about the position of the wound edge and this gives rise to an inconsistency in the measurement of wound dimensions. An automated segmentation that applies rules consistently will provide a better assessment of the progress of wound healing.

We aim to enhance MAVIS by automating the segmentation of the wound in the image, and **colour difference** has been used with some success. The current manual method of tracing the wound perimeter introduces errors in the following ways:

1. The accuracy of the instrument is decreased because the wound edge is often not well defined and its location is a matter of opinion. The skill and level of physical control of the operator determines how faithfully the perimeter can be traced. An algorithm that has been developed using a fixed set of rules agreed by a group of acknowledged experts will improve the accuracy of the instrument.

2. The precision of the instrument is decreased by the difficulty of retracing exactly the original boundary. An algorithm will trace the perimeter consistently leading to an improvement in the precision of the instrument.

Jones (1994) has described in detail the use of colour to segment the wound automatically. A number of pictures captured by the MAVIS instrument show that most of the wounds are in a square central region taking up 5% of the total image area. Unfortunately this is not guaranteed; some operators capture images with the wound some distance off centre. Similarly, there are few rules governing the shape of the wound. Surgical incisions are created deliberately to cover the area of an underlying problem and are aligned in a direction that promotes post-operative healing. In contrast, pressure sores, ulcers and accidental wounds can take any shape. A wound is a part of the body which is not covered by skin. Healthy skin is a highly elastic organ which deforms under stresses caused by changes in the position of limbs, for example. Where the skin is torn, the forces caused by muscular contractions and by gravity combine to form unpredictable sizes and shapes.

As described in the previous section, colour images of wounds provide a good deal of information about wounds; this is apparent even to the unskilled eye to which a black and white image is innocuous compared to a colour image which may show problems arising from infection, necrotic (dead) tissue and so on. The problem of segmenting a wound is a complex one. An image typically includes background, healthy tissue, and wound as may be expected. However it may also include epithelialization and granulation tissue which are good signs, and slough and necrotic tissue which are generally indicators that healing is not making good progress. When present, epithelialization does help to delineate skin from wound, but the question arises as to whether the epithelialization should be classified as wound or healthy tissue. Berriss and Sangwine (1997) have used colour to identify successfully areas of slough, necrotic tissue and epithelialization within wounds.

The colour of healthy tissue, epithelialization and wound have been analysed using the RGB intensities and using hue, saturation and grey level intensity, but unfortunately the RGB intensity histograms do not give a clear segmentation. A well-known drawback of using the RGB planes for analysing colour is that the luminances in the three planes are highly correlated even though the wound may appear to have a strongly saturated red colour. The skin may well have a higher luminance in the red plane than the wound. As may be expected, the wound does have lower luminances proportionately in the green and blue planes than skin; the blue intensity is lower than the green because of its lower intensity in the quartz halogen lamp used in MAVIS. Furthermore, the image of healthy skin may be over-exposed because it has a higher reflectivity than the wound. Epithelialization is darker than healthy skin and wound and is therefore more easily identified in the RGB planes.

The Intensity, Hue and Saturation (IHS) colour space has been used with greater success in segmenting wounds. Individual pixels are classified as either wound or skin by using the IHS space and separating them into clusters using the Jeffries-Matsushita (JM) distance (Niblack, 1986) as the discriminating function. The JM distance between two clusters i and j in a space n is defined as:

$$JM_{ij} = \sqrt{\int_v \left(\sqrt{p(v/i)} - \sqrt{p(v/j)} \right)^2 dv}$$

where $p(v/i)$ is the probability distribution of the cluster i in space v. A value of 0 for JM_{ij} indicates that the clusters i and j are inseparable whereas a value of $\sqrt{2}$ means that the clusters do not overlap. The value of JM was calculated in several wound images where the wound was delineated manually for v corresponding to the three spaces of hue, intensity and saturation or a combination of one or more of them. Whilst hue did not prove to be a useful discriminant, saturation and intensity were more effective though the classification error varied from one image to another. Fewest errors occurred in images where the wound colour was highly saturated. The overlap arises for two main reasons:

1. The condition of the skin surrounding the wound can be very poor with flaking or crusted skin, and red blotches caused by the underlying problems of bad circulation.

2. Dead cells are exuded from the wound forming slough whose colour is yellowish white and often close to that of healthy skin.

A more robust segmentation was obtained when the image was preprocessed with a standard median or low-pass Gaussian filter with a 3×3 kernel followed by morphological closing to fill holes in the wound object.

16.6 VISUALIZATION AND STORAGE OF DATA

Pseudo-colour has a long history of use in medical imaging, and, for example, its use to emphasize small temperature variations and temperature contours in thermal imaging is well established. MAVIS reconstructs a graphical representation of the wound in 3D where the wounded volume identified by the instrument is shown in a different colour from the surrounding tissue. The wound may then be rotated, scaled and viewed by the clinician from different vantage points.

MAVIS stores the data relating to patients in a database which is protected by password to ensure the security of the information. These data include compressed colour images of the wound in addition to values of volume, area, perimeter, and maximum depth. In addition to everyday use in wound healing clinics, the instrument will be used for research into establishing the effectiveness of different methods of treatment and is at present part of a major clinical trial to investigate the use of radio frequencies to encourage wound healing.

CONCLUSION

Colour-coded structured light has been used successfully to measure the physical dimensions and colour of leg ulcers and pressure sores. MAVIS uses colour to improve the efficiency of the segmentation, closing, following and labelling of the stripes extracted from the structured light image. Colour differences in the IHS colour space are used to segment the wound under favourable conditions where the wound has a saturated colour. An advantage of MAVIS over laser scanners is that a colour image is obtained and true-colour measurements are used to predict the status of healing and to identify when the wound is making poor progress; future developments may make the detection of infection possible either by colour measurement or by thermal imaging. Pseudo-colour is used in the visualization of a wound for the clinician.

ACKNOWLEDGEMENTS

We are happy to acknowledge the work on the colour and segmentation of wounds by T. D. Jones and A. Hoppe of the Medical Computing Research Unit at the University of Glamorgan, Wales.

We are indebted to Professor E. F. J. Ring of the Royal National Hospital for Rheumatic Diseases, Bath, England, and External Professor of the University of Glamorgan for introducing us to the problems of wound metrics and for many invaluable discussions and access to clinics.

We are indebted to Dr K. D. Harding of the Wound Healing Research Unit in the University of Wales College of Medicine at Cardiff for many invaluable discussions and access to clinics.

17

Industrial colour inspection case studies

Christine Connolly

The purpose of this chapter is to show the application of colour image processing techniques to a collection of illustrative industrial applications. The aim is to show the advantages of colour image processing over other colour measurement methods, and illustrate how its various techniques may be combined and traded off in practice to form practical solutions. The interesting technical issues are highlighted. The examples are presented in order of increasing technical complexity.

Sorting or recognizing the presence of distinct colours is a fairly simple type of colour image processing application which has been regularly reported in the trade journals for several years now. Examples include checking the colour of pills, sorting plastic bottles (Massen, 1994), sorting limestone from red pegmatite and dark rocks in a limestone mine (Parnanen and Lemstrom, 1995), checking adequate baking of bread and biscuits (Locht, Mikkelsen and Thomsen, 1996), and fruit harvesting. Since the colours involved in a given application are distinct and well-separated from each other, this type of application requries quite low colour resolution. It is simply necessary to distinguish between the possible colours involved. It is usually possible to use the raw RGB data from the camera as a means of classifying colours at this low resolution; but sometimes simple colour transformations such as HSI space are employed (Poston, 1996). Since the processing in these applications is undemanding, it is fast and thus 100% inspection is usually possible.

In contrast, the examples presented in this chapter are not simple and required state-of-the-art colour image acquisition.

17.1   INSPECTION OF PRINTED CARD

In a typical printing works, flat sheets of card, about a metre square, are printed at the rate of several per second. They are later cut up and folded to form packaging sleeves for frozen foods, boxes of chocolates, *etc.* The accuracy of the printed colours is very important, since they are used as a cue for brand recognition, and in some cases the colours themselves may be registered trademarks. In cases where food is portrayed in the printed design, a small change in its colour can make all the difference between an appetising appearance and an off-putting appearance. When the packaging design is developed, the colours are specified precisely, and the printer is expected to reproduce the colours to a tolerance of a few CIELAB[1] units.

Spectrophotometers, colorimeters or densitometers are used in the development of new designs. However during production, colour checking tends to be done by eye. The reason for this is response time: in a multicoloured product, many positions on the card need to be inspected, and contact instruments are too slow in this situation. Where several cards per second are being printed, any delay means the production of waste, amounting to a significant expense. To inspect a card by eye, a sample is taken from the production line every few minutes. It is placed on an inspection table with overhead fluorescent lights, and inspected carefully, placing the target card as close beside it as possible, so as to judge the colours accurately. If any areas of the sample card are found to be off-shade, adjustments are made to the ink flow in that area, and the result is carefully assessed when the next sample is taken a few minutes later. It is not desirable to inspect the cards more frequently, since the ink flow adjustments take a few minutes to make their effect apparent. However, it is important that the result of the inspection is available as soon as possible.

17.1.1   Technical considerations

A large number of colour areas must be inspected to a sufficiently high precision within a matter of minutes. Each different colour area must be inspected, particularly the 'house colours' — that is the brand name, company logo *etc.* These will always be in the same position on the printed card, so for each product the inspection areas may be defined within the captured image, and

---

[1]Colour differences are discussed in section 3.4 and Chapter 15 discusses this problem in more detail.

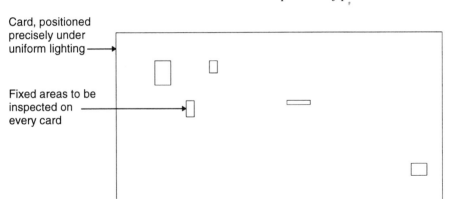

Card, positioned
precisely under
uniform lighting ———▶

Fixed areas to be
inspected on ———————▶
every card

Fig. 17.1 Inspection of fixed areas on printed cards.

saved to disk. Figure 17.1 shows a few inspection areas — each inspection area is rectangular, since this makes efficient use of the pixel layout, but the rectangles are different sizes and shapes, reflecting the underlying design of the print.

The colour precision requirement poses quite a problem. Cameras change their response with setting — iris, gain, shutter time, *etc.* Even if these are held fixed, the colour response of the camera still varies with temperature, humidity, and time. The necessary precision is possible only if the camera is interactively controlled, since otherwise camera drift swamps any colour difference which may be present in the samples. This problem is discussed in detail in Chapter 7, particularly in section 7.1.2, and information is given there on how to obtain repeatable colour measurement from an analogue video camera.

A second factor influencing the precision of colour measurement is the electronic noise in the whole system: the signal-to-noise ratio of the camera; the resolution (number of bits) of the analogue-to-digital converters in the framegrabber; the screening quality of the cables; and the vibrations and ambient electronic noise in an environment where large machinery is drawing heavy electric current nearby. It is essential that the camera and lighting be supplied from a stabilized power supply.

Uniformity of lighting is a problem in this applicaton, since it is very difficult to arrange uniform lighting over a large surface area. A theoretical study on obtaining uniform lighting from straight or circular tube lighting has been published very recently (Davies, 1997), and fibre optics are now advertised with uniformity of 15 greyscale values (out of 256) over an area of almost 700 mm square (Smith, 1997). Lighting is considered in some detail in section 7.1.3.

In this application, colour acceptability decisions must be made which agree with the printer's and the client's perception of colour. It is inappropriate to use RGB or HSI colour spaces, since these are non-uniform. Uniform colour spaces, such as CIELAB, provide a single-figure measure of colour mismatch which can be used very simply and effectively to make pass/fail decisions. So the procedure is:

1. define the 50 areas of the image which will be inspected, and the target colour and colour tolerance of each area;

2. at each image capture, calculate the average RGB value in each area, convert this to CIELAB colour space and make a pass/fail decision.

3. if any area is of unacceptable colour, alert the production staff as to which area is wrong, and the extent of the colour mismatch.

With a Pentium™ 133 MHz processor, and a PCI framegrabber, 50 or so areas can be inspected in less than a second. This compares very favourably with contact colour measuring instruments in its speed and convenience, and in its ability to provide quantitative measurements of colour faults. In contrast, with a densitometer, the measurements have to be taken in each of the 50 different places sequentially, and this takes at least an hour, and is therefore impractical in this production situation.

17.2   INSPECTION OF FAST-MOVING BEVERAGE CANS

Background   In a typical production unit, beverage cans are printed as part of their production process. The metal is drawn into its cylindrical shape, and the cans are coated with varnish on the inside and with white on the outside and then they are colour-printed, before going on to be filled with liquid and having their tops fitted. About 30 cans per second are produced and printed.

Technical Considerations   The cans travel at high speed and in random orientations (they are free to rotate about their cylindrical axis) along the production line. This application has the precision colour-inspection requirements of the printed card problem, plus the high speed movement and the random orientation to contend with.

17.2.1 Image capture

First, a clear, unblurred image must be captured. A beverage can is about 70 mm wide, and travels along the production line at about $3.5\,ms^{-1}$. If the width of the can in the image is 200 pixels, then each pixel represents $(70/200) = 0.35\,mm$ on the can. To keep the can's motion to 0.3 pixels, the image must be captured in a time:

$$0.35 \times 0.3/3500 = 30\,\mu s = 1/33\,333\,s$$

So, a camera with an electronic shutter capable of a 1/30 000 s shutter speed is needed to freeze the motion.

The next problem to solve is the lighting. Since the camera's CCD sensor is only exposed for 1/30 000 s, the lighting must be very bright. The light sensitivity of a camera is stated in its data sheet. For example:

Minimum illumination: 7 lux at **f 1.4** and 18 dB gain.

This figure assumes no shutter, i.e. the CCD sensors are exposed to the light for 1/60 s, before discharge by the CCD readout process. With an exposure time of 1/30000 s, the lighting must be increased by a factor of 30000/60. Furthermore, the camera should, if possible, be used not at its maximum aperture and highest gain, but in its optimum settings of about **f 8** and 0 dB gain. Using 0 dB gain instead of 18 dB requires an 8-fold increase in light intensity. The amount of light reaching the sensor array is inversely proportional to the square of the **f** stop figure, so stopping down the camera from **f 1.4** to **f 8** requires a lighting increase factor of:

$$(8/1.4) \times (8/1.4) = 32.7$$

Thus the intensity of illumination required for this application is:

$$7 \times (30000/60) \times 8 \times 32.7 = 915\,600\,lux$$

This requires a very intense light. Fortunately, some companies specialize in machine vision illumination. For example, Lake Image Systems of Aylesbury, UK, advertize a 150 W quartz halogen fibre-optic illuminator providing 4 Mlux at the fibre-optic insertion plate. A second consideration regarding lighting is the frequency of its power supply. Figure 17.2 shows what happens if the lighting is mains powered. The 1/30000 s shutter samples only a tiny portion of the 50 Hz mains cycle, which may be near its peak, as shown by T1, or near zero crossover T2. So different image acquisitions would

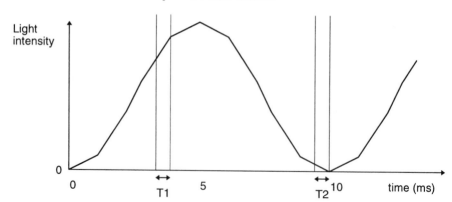

Fig. 17.2 Effect of mains frequency lighting on shuttered image exposure.

give very different image exposures. There are two ways of overcoming this problem. The lighting could be DC powered, in which case the intensity is constant over time. Alternatively, the lighting could be powered by a high frequency supply, so that one or more cycles are completed within the exposure period T1 or T2. (It is not possible to synchronize the image acquisition to the mains cycles, which would be another option in many applications, because the acquisition here must be synchronized with the position of the cans.) For a 1/30000 s shutter, a lighting frequency of 30 kHz or more is required. Ring lights powered at 33 kHz and producing 12 000 lux are obtainable[2]. These provide one possible lighting solution for this application, but the camera has to be used at 18 dB gain and with a fairly wide aperture. Another approach is to use a xenon flash source. This produces light of high intensity, but the flashes are not absolutely repeatable, either in intensity or in spectral balance, so this makes it very difficult to achieve correct colour rendering in the image.

17.2.2 Image processing

Once the image has been captured, by a camera which is tightly controlled, the following two-stage process can then be applied:

1. segment the image to pick out the regions having the expected colours;

[2] An example is the Stocker and Yale Model 13 Steady Lite, supplied in the UK by Special Application Products Ltd.

2. make a more accurate assessment of the colour of each segment.

It is only possible to use this approach because we are dealing with applications where we know what the colours should be, and we have stabilized the camera. Stage 1 picks out the expected colours, and actually amounts to a crude colour-inspection procedure in itself. It is also a means of identifying the positions of each expected colour, so it provides the data which allows stage 2 to be implemented.

For stage 1, it is necessary to convert the whole image from RGB to CIELAB colour space. The image is then thresholded to isolate regions of the various colours which are expected to be present on this particular can. The conversion of the whole image from RGB to CIELAB is not a trivial problem. The transformation, applied to an image of 512×512 24-bit pixels, takes 81 s on a 133 MHz Pentium™ personal computer (PC) if the code is written in the Pascal language. The most time-consuming part is the evaluation of the cube roots in the calculation of L^*, a^* and b^{*} [3]. Programming languages use various techniques for the approximation of non-linear expressions. The techniques differ with regard to speed, memory space and precision. The same program coded in the C language, using the standard exp and log functions in the math.h library, takes 22 s.

It is clear that straightforward computation of the cube root is too slow for real-time inspection applications. Different methods for implementing the cube-root function have been studied (Connolly and Fliess, 1997). The most successful methods were Taylor expansion, Newton-Raphson iteration, and look-up tables, and Table 17.1 summarizes the results. A large look-up table is the most practicable method. Once the positions of the expected colours have been found, the precise inspection of the colours required by stage 2 can proceed along similar lines to the inspection method described in the printed card application.

---

[3] See section 4.7.1 for the equations required to convert RGB to CIELAB

Table 17.1 Comparison of methods investigated for computing cube roots.

Criterion	Taylor expansion	Newton-Raphson	Look-up table
Speed	slow	very slow	fastest
Time to process 512×512 pixels (s)	5.5		1.4
Accuracy	very bad	bad	medium
Maximum error in CIELAB units	176		2.0 for 2000-entries 0.6 for 8000-entries
Storage expense	small	small	Up to 32 kB

References

Abel, J. S., Bhaskaran, V. and Lee, H. J. (1992). Color image coding using an orthogonal decomposition, *Image Processing Algorithms and Techniques III*, Vol. 165, Society of Photo-Optical Instrumentation Engineers.

Adams, E. Q. (1942). X-Z planes in the 1931 I.C.I. (CIE) system of colorimetry, *Journal of the Optical Society of America* **32**: 168.

Afromowitz, M. A., van Liew, G. S. and Heimbach, D. M. (1987). Clinical evaluation of burn injuries using an optical reflectance technique, *IEEE Transactions on Biomedical Engineering* **34**(2): 114–127.

Ailisto, H. and Piironen, T. (1987). Evaluation of color representation methods in a practical vision system, *Proc. of 5th SCIA Stockholm*.

Akarun, L., Yardimci, Y. and Cetin, A. E. (1995). New methods for dithering of color images, *Proceedings of the IEEE Workshop on Nonlinear Signal/Image Processing*, Institute of Electrical and Electronics Engineers, Neos Marmas, Greece.

Ali, M., Martin, W. N. and Aggarwal, J. K. (1979). Color-based computer analysis of aerial photographs, *Computer Graphics and Image Processing* **9**: 282–283.

Anderson, R. R. and Parish, J. A. (1981). The optics of human skin, *Journal of Investigative Dermatology* **77**(1): 13–19.

Andreadis, I., Browne, M. A. and Swift, J. A. (1990). Image pixel classification by chromaticity analysis, *Pattern Recognition Letters* **11**(1): 51–58.

ANSI (1993a). *American National Standard: Graphic Technology – Color Reflection Target for Input Scanner Calibration, ANSI IT8.7/2-1993*, 11 West 42nd Street, New York, New York 10036.

ANSI (1993b). *American National Standard: Graphic Technology – Color Transmission Target for Input Scanner Calibration, ANSI IT8.7/1-1993*, 11 West 42nd Street, New York, New York 10036.

Anstis, S. and Cavanagh, P. (1983). A minimum motion technique for judging equiluminance, *in* J. D. Mollon and L. T. Sharpe (eds), *Colour Vision: Physiology and Psychophysics (Proceedings Cambridge Colour Vision Conference 1982)*, Academic Press, London.

Apple (1995). *ColorSync 2.0 The answer to colour matching*, Apple Inc. See `http://colorsync.apple.com/`.

Arai, Y., Agui, T. and Nakajima, M. (1988). A fast DCT-SQ scheme for images, *Transactions of the IEICE* **E-71**: 1095–1097.

Arnquist, J., Hellgren, L. and Vincent, J. (1988). Semiautomatic classification of secondary healing ulcers, *Proceedings of the 9th International Conference on Pattern Recognition*, Institute of Electrical and Electronics Engineers, New York, pp. 459–461.

Artmann, B. (1988). *The Concept of Number: from Quaternions to Monads and Topological Fields*, Ellis Horwood series in mathematics and its applications, Ellis Horwood, Halsted, Chichester. Translation of: Der Zahlbeginff, Gottigen: Vandenhoeck & Rupprecht, 1983. Translated with additional exercises and material by H. B. Griffiths.

Asano, T., Kenwood, G., Mochizuki, J. and Hata, S. (1986). Color image recognition using chrominance signals, *Proceedings of the 8th International Conference on Pattern Recognition, 27–31 October, Paris*, Vol. Ch. 342, pp. 804–807.

Ashdown, I. (1994). Octree color quantization, *Radiosity: A Programmer's Perspective*, John Wiley.

ASTM (1996). *Document E 308–96: Standard practice for computing the colors of objects by using the CIE system*, American Society for Testing and Materials, West Conshohocken, PA.

ASTM (Undated). American society for testing and materials, `http://www.astm.org/`.

Astola, J., Haavisto, P. and Neuvo, Y. (1990). Vector median filters, *Proceedings of the IEEE* **78**(4): 678–689.

Austermeier, H., Hartmann, G. and Hilker, R. (1996). Color-calibration of a robot vision system using self-organizing features maps, *Proceedings of International Conference on ANNs, ICANN'96, Bochum*, Springer-Verlag, pp. 257–262.

Bajon, J., Cattoen, M. and Kim, S. D. (1985). Real-time colorimetric transformations used in robot vision system, *Proceedings of MICAD*, pp. 76–86. In French.

Bajon, J., Cattoen, M. and Liang, L. (1986). Identification of multicoloured objects using a vision module, *Proceedings of the 6th RoViSeC*, pp. 21–30.

Baker, D. C., Hwang, S. S. and Aggarwal, J. K. (1989). Detection and segmentation of man-made objects in outdoor scenes: Concrete bridges, *Journal of Optical Society of America A – Optics and Image Science* **6**(6): 938–950.

Balasubramanian, R., Allebach, J. P. and Bouman, C. A. (1994). Color-image quantization with use of a fast binary splitting technique, *Journal of the Optical Society of America A – Optics Image Science and Vision* **11**(11): 2777–2786.

Balasubramanian, R., Bouman, C. A. and Allebach, J. P. (1994). Sequential scalar quantization of color images, *Journal of Electronic Imaging* **3**(1): 45–59.

Balasubramanian, R., Bouman, C. A. and Allebach, J. P. (1995). Sequential scalar quantization of vectors – an analysis, *IEEE Transactions on Image Processing* **4**(9): 1282–1295.

Barco (1990). *The Calibrator™ explained Part I*, Barco Video and Communications, Belgium.

Barilleaux, J., Hinkle, R. and Wells, S. (1987). Efficient vector quantization for color image encoding, *Proceedings International Conference on Acoustics, Speech and Signal Processing, ICASSP-87*, pp. 740–743.

Barnett, V. (1976). The ordering of multivariate data, *Journal of Royal Statistical Society A* **139**(Part 3): 318–355. Includes discussion of paper on page 343 *et seq.*

Barni, M., Cappellini, V. and Mecocci, A. (1994). Fast vector median filter based on Euclidean norm approximation, *IEEE Signal Processing Letters* **1**(6): 92–94.

Barth, M., Parthasarathy, S. and Wang, J. (1986). A color vision system for microelectronics: Applications to oxide thickness measurements, *Proceedings of IEEE International Conference on Robotics and Automation*, Vol. 2, Institute of Electrical and Electronics Engineers, pp. 1242–1247.

Bartkowiak, M. and Domański, M. (1995). Vector median filters for processing of color images in various color spaces, *Proceedings 5th International Conference on Image Processing and its Applications*, Institution of Electrical Engineers, Heriot-Watt University, Edinburgh, UK, pp. 833–836. IEE Conference Publication 410.

Bartkowiak, M., Domański, M. and Gerken, P. (1996). Vector representation of chrominance for very low bitrate coding of video, *in* G. Ramponi, G. L. Sicuranza, S. Carrato and S. Marsi (eds), *Signal Processing VIII: Theories and Applications*, Vol. II, Trieste, pp. 1351–1354.

Benson, K. B. (1992). *Television Engineering Handbook*, McGraw-Hill, New York, London. Revised by Jerry C. Whitaker.

Berlin, B. and Kay, P. (1991). *Basic Color Terms: Their Universality and Evolution*, University of California Press, Berkeley CA. First published 1969.

Berns, R. S. (1991). Color tolerance feasibility study comparing CRT-generated stimuli with an acrylic-lacquer coating, *Color Research and Application* **16**(4): 232–242.

Berns, R. S. (1993). Spectral modelling of a dye diffusion thermal transfer printer, *Journal of Electronic Imaging* **2**(4): 359–370.

Berns, R. S., Alman, D. H., Reniff, L., Snyder, G. D. and Bolonon-Rosen, M. R. (1991). Visual determination of suprathreshold color difference tolerances using probit analysis, *Color Research and Application* **16**(5): 297–316.

Berriss, W. P. and Sangwine, S. J. (1997). A colour histogram clustering technique for tissue analysis of healing skin wounds, *Proceedings 6th International Conference on Image Processing and its Applications*, Vol. 2, Institution of Electrical Engineers, Trinity College, Dublin, Ireland, pp. 693–697. IEE Conference Publication 443.

Berry, D. T. (1987). Colour recognition using spectral signatures, *Pattern Recognition Letters* **6**: 69–75.

Beucher, S. (1982). Watersheds of functions and picture segmentation, *Proceedings of IEEE International Conference on Acoustics, Speech, and Signal Processing, 3–5 May, Paris*, Institute of Electrical and Electronics Engineers, pp. 1928–1931.

Bezdek, J. C. and Pal, S. K. (eds) (1992). *Fuzzy Models for Pattern Recognition: Methods that Search for Structures in Data*, IEEE, New York. A selected reprint volume, IEEE Neural Networks Council, sponsor.

Bhaskaran, V. and Konstantinides, K. (1995). *Image and Video Compression Standards*, Kluwer.

Björck, A. (1996). *Numerical Methods for Least Squares Problems*, SIAM.

Blahut, R. E. (1985). *Fast algorithms for digital signal processing*, Addison-Wesley, Reading, MA. Reprinted with corrections 1987.

Boucher, P. and Goldberg, M. (1984). Color image compression by adaptive vector quantization, *Proceedings International Conference Acoustics Speech and Signal Processing*, pp. 2.6.1–2.6.4.

Boynton, R. M. (1989). *Human Color Vision*, Holt, Rinehart and Winston, New York.

Braudaway, G. W. (1987). A procedure for optimum choice of a small number of colors from a large color palette for color imaging, *Proceedings Electronic Imaging, San Francisco*, pp. 75–79.

Brigham, E. O. (1988). *The fast Fourier transform and its applications*, Prentice Hall, Englewood Cliffs, NJ.

Brock-Gunn, S. and Ellis, T. (1992). Using colour templates for target identification and tracking, *Proceedings British Machine Vision Conference '92*, British Machine Vision Association, Springer-Verlag, Berlin, pp. 207–216.

Brockelbank, D. and Yang, Y. H. (1989). An experimental investigation in the use of color in computational stereopsis, *IEEE Transactions on Systems, Man and Cybernetics* **19**(6): 1365–1383.

Buchsbaum, G. and Gottschalk, A. (1983). Trichromacy, opponent colors coding and optimum color information-transmission in the retina, *Proceedings of the Royal Society of London Series B – Biological Sciences* **220**(1218): 89–113.

Burns, S. A., Elsner, A. E., Pokorny, J. and Smith, V. C. (1984). The Abney Effect – Chromaticity coordinates of unique and other constant hues, *Vision Research* **24**(5): 479–489.

Canny, J. (1986). A computational approach to edge detection, *IEEE Transactions on Pattern Analysis and Machine Intelligence* **8**(6): 679–698.

Castleman, K. R. (1996). *Digital Image Processing*, Prentice Hall, Englewood Cliffs, NJ.

Celenk, M. (1995). Analysis of color images of natural scenes, *Journal of Electronic Imaging* **4**(4): 382–396.

Cen (1996). *Proceedings of the CIE Expert Symposium '96 Colour Standards for Image Technology*, Vienna. Publication No. x010-1996.

Chaddha, N., Tan, W.-C. and Meng, T. H. Y. (1994). Color quantization of images based on human vision perception, *Proceedings International Conference Acoustics, Speech and Signal Processing*, Vol. 5, pp. 89–92.

Chang, Y.-C. and Reid, J. F. (1996). RGB calibration for color image analysis in machine vision, *IEEE Transactions on Image Processing* **5**(10): 1414–1422.

Charrier, C., Knoblauch, K. and Cherifi, H. (1997). VQ-coded image quality optimized by color space selection, *Proceedings Picture Coding Symposium, PCS–97*, Berlin, pp. 195–200.

Chau, W. K., Wong, S. K. M., Yang, X. D. and Wan, S. J. (1992). On the selection of small color palette for color image quantization, *in* Sullivan, Rabbani and Dawson (1992), pp. 326–333.

Chen, M. and Yan, P. (1989). A multiscaling approach based on morphological filtering, *IEEE Transactions on Pattern Analysis and Machine Intelligence* **11**(7): 694–700.

Chen, W. H. and Smith, C. H. (1977). Adaptive coding of monochrome and color images, *IEEE Transactions on Communication* **25**(11): 1285–1292.

Chen, Y., Petersen, H. and Bender, W. (1993). Lossy compression of palettized images, *Proceedings IEEE International Conference Acoustics Speech Signal Processing ICASSP-93, 27–30 April, Minneapolis, MN*, Vol. 5, pp. 325–328.

Chiang, S. W. and Po, L. M. (1997). Adaptive lossy LZW algorithm for palettised image compression, *Electronics Letters* **33**(10): 852–854.

Cho, N. I. and Lee, S. U. (1991). Fast algorithm and implementation of 2-D discrete cosine transform, *IEEE Transactions on Circuits and Systems* **38**(3): 297–305.

Chu, C. H. and Delp, E. J. (1989). Impulsive noise suppression and background normalization of electrocardiogram signals using morphological operators, *IEEE Transactions on Biomedical Engineering* **36**(2): 262–273.

CIE (1986). *Colorimetry*, second edn, Vienna. Publication No. 15.2–1986.

CIE (1989). *International Lighting Vocabulary*, Vienna. Publication No. 17.4–1989.

CIE (1995). *Industrial Colour-Difference Evaluation*, Vienna. Publication No. 116–1995.

Clapper, F. R. (1961). An empirical determination of half-tone colour reproduction requirements, *TAGA Proceedings*, pp. 31–41.

Clarke, F. J. J., McDonald, R. and Rigg, B. (1984). Modification to the JPC 79 color-difference formula, *Journal of the Society of Dyers and Colourists* **100**(9): 281–282.

Clarke, R. J. (1985). *Transform Coding of Images*, Academic Press, London; Orlando.

Comer, M. L. and Delp, E. J. (1992). An empirical study of morphological operators in color image enhancement, *in* Sullivan, Rabbani and Dawson (1992), pp. 314–325.

Connolly, C. (1996). The relationship between color metrics and the appearance of three-dimensional colored objects, *Color Research and Application* **21**(5): 331–337.

Connolly, C. and Fliess, T. (1997). A study of efficiency and accuracy in the transformation from RGB to CIELAB color space, *IEEE Transactions on Image Processing* **6**(7): 1046–1048.

Connolly, C. and Leung, T. W. W. (1994). Colour inspection system, UK Patent Application 94 16406.8 in the name of University of Huddersfield.

Connolly, C. and Leung, T. W. W. (1995). Industrial colour inspection by video camera, *Proceedings 5th International Conference on Image Processing and its Applications*, Institution of Electrical Engineers, Heriot-Watt University, Edinburgh, UK, pp. 672–676. IEE Conference Publication 410.

Crevier, D. (1993). Computing statistical properties of hue distributions for color image analysis, *in* D. P. Casasent (ed.), *Proceedings Intelligent Robots and Computer Vision XII: Algorithms and Techniques, 7–9 September, Boston, MA*, Vol. 2055, Society of Photo-Optical Instrumentation Engineers, pp. 613–623.

Crocker, L. (1995). PNG: The portable network graphic format, *Dr. Dobb's Journal* **20**(7): 36–44.

Dai, Y. and Nakano, Y. (1996). Face-texture model based on SGLD and its application in face detection in a color scene, *Pattern Recognition* **29**(6): 1007–1017.

Dalton, C. J. (1988). The measurement of the colorimetric fidelity of television cameras, *Journal of the Institution of Electronic and Radio Engineers* **58**(4): 181–186.

Davies, E. R. (1997). Principles and design graphs for obtaining uniform illumination in automated visual inspection, *Proceedings 6th International Conference on Image Processing and its Applications*, Institution of Electrical Engineers, Trinity College, Dublin, Ireland, pp. 161–165. IEE Conference Publication 443.

Davis, G., Danskin, J. and Heasman, R. (1997). Wavelet image compression construction kit, http://www.cs.dartmouth.edu/~gdavis/wavelet/wavelet.html. Publicly available software.

Davis, L. (1991). *Handbook of Genetic Algorithms*, Van Nostrand Reinhold, New York. Also published 1996 International Thomson Computer Press, London; Boston.

De Valois, R. L. and De Valois, K. K. (1993). A multi-stage color model, *Vision Research* **33**(8): 1053–1065.

DeGroot, M. H. (1989). *Probability and Statistics*, second edn, Addison-Wesley, Reading, MA.

Deknuydt, B., Smolders, J., Van Eycken, L. and Oosterlinck, A. (1992). Color space choice for nearly reversible image compression, *in* P. Maragos (ed.), *Proceedings Visual Communications and Image Processing '92, 18–20 November, Boston, MA*, Vol. 1818, Society of Photo-Optical Instrumentation Engineers, pp. 1300–1311.

Della Ventura, A. and Schettini, R. (1992). Computer-aided color coding for data display, *Proceedings of 11th IAPR International Conference on Pattern Recognition*, Vol. III, The Hague, Netherlands, pp. 29–32. Conference C: Image, Speech, and Signal Analysis.

Delp, E. J. and Mitchell, O. R. (1979). Image compression using block truncation coding, *IEEE Transactions on Communications* **27**(9): 1335–1342.

Derefeldt, G. and Swartling, T. (1995). Color concept retrieval by free color naming – identification of up to 30 colors without training, *Displays* **16**(2): 69–77.

Domański, M. (1988). Digital recursive filters for subband coding of images, *KST Bydgoszcz* pp. 254–262. In Polish.

Domański, M. and Stasiński, R. (eds) (1997). *Proceedings 4th International Workshop on Systems, Signals and Image Processing*, Poznań, Poland.

Domański, M. and Świerczyński, R. (1994). Subband coding of images using hierarchical quantization, *Signal Processing VII: Theories and Applications* pp. 1218–1221.

Dougherty, E. R. (ed.) (1993). *Mathematical Morphology in Image Processing*, Marcel-Dekker, New York.

Duda, R. O. and Hart, P. E. (1973). *Pattern Classification and Scene Analysis*, John Wiley, New York.

Dudani, S. A. (1977). The distance-weighted k-nearest neighbor rule, *IEEE Transactions on Systems, Man and Cybernetics* **15**: 630–636.

Egger, O. and Li, W. (1995). Subband coding of images using asymmetrical filter banks, *IEEE Transactions on Image Processing* **4**(4): 478–485.

Ell, T. A. (1992). *Hypercomplex Spectral Transformations*, PhD thesis, University of Minnesota.

Ell, T. A. (1993). Quaternion-Fourier transforms for analysis of two-dimensional linear time-invariant partial-differential systems, *Proceedings of the 32nd IEEE Conference on Decision and Control, San Antonio, Texas, USA, 15 - 17 December 1993*, Vol. 1-4, Institute of Electrical and Electronics Engineers, Control Systems Society, pp. 1830–1841.

Fairchild, M. D. (1996). Refinement of the RLAB color space, *Color Research and Application* **21**(5): 338–346.

Faugeras, O. D. (1979). Digital color image processing within the framework of a human visual model, *IEEE Transactions on Acoustics, Speech and Signal Processing* **27**(4): 380–393.

Feng, Y. and Nasrabadi, N. M. (1988). A new vector quantization technique: Address-vector quantization, *Proceedings IEEE Global Telecomm.*, Institute of Electrical and Electronics Engineers, pp. 755–759.

Feng, Y. and Nasrabadi, N. M. (1989). A dynamic-address vector quantization algorithm based on inter-block and inter-color correlation for color image coding, *Proceedings International Conference Acoustics, Speech and Signal Processing ICASSP-89*, Vol. III, pp. 1755–1758.

Ferri, F. and Vidal, E. (1992). Colour image segmentation and labeling through multiedit-condensing, *Pattern Recognition Letters* **13**(8): 561–568.

Fisch, R. S. (1994). A study of image colorimetric and tonal changes when transferred from one imaging media to another, *in* W. H. Banks (ed.), *Advances in Printing Science and Technology, 22nd Research Conference of the International Association of Research Institutes for the Graphic Arts Industry, 5–8 September 1993, Munich*, Vol. 22, pp. 482–501.

Fleck, M. M., Forsyth, D. A. and Bregler, C. (1996). Finding naked people, *Proceedings of 4th European Conference on Computer Vision*, Cambridge.

Floyd, R. W. and Steinberg, L. (1975). An adaptive algorithm for spatial gray scale, *Proceedings International Symposium Digital Tech.*, Vol. 36.

Foley, J. D. and van Dam, A. (1982). *Fundamentals of Interactive Computer Graphics*, The Systems programming series, Addison-Wesley, Reading, MA. Reprinted 1984 with corrections.

Freisleben, B. and Shrader, A. (1997). Color quantization with a hybrid genetic algorithm, *Proceedings 6th International Conference on Image Processing and its Applications*, Institution of Electrical Engineers, Trinity College, Dublin, Ireland, pp. 86–90. IEE Conference Publication 443.

Frey, H. (1988). *Colour image processing in colour spaces*, PhD thesis, Technical University Munich. In German.

Frey, H. and Palus, H. (1993). Sensor calibration for video-colorimetry, *Proceedings of Workshop on Design Methodologies for Microelectronics and Signal Processing, October 20-23rd 1993*, Gliwice-Cracow, Poland, pp. 109–113.

Frieman, J. H., Bentley, J. L. and Finkel, R. A. (1977). An algorithm for finding best matches in logarithmic expected time, *ACM Transactions on Mathematical Software* 3(3): 209–226.

Fukunaga, K. (1990). *Introduction to Statistical Pattern Recognition*, second edn, Academic Press, Boston.

Gabbouj, M. and Cheickh, F. A. (1996). Vector median-vector directional hybrid filter for color image restoration, *Proceedings of EUSIPCO-96*, pp. 879–881.

Gagliardi, G., Hatch, G. F. and Sarkar, N. (1985). Machine vision applications in the food industry, *Proceedings Vision '85 Conference, 25 - 28 March, Detroit, MI*, pp. 524–538.

Gan, Q., Kotani, K. and Miyahara, M. (1994). Quantizing accuracy for high quality image processing, *Proceedings International Workshop on Image Processing, Budapest*, pp. 69–71.

Gauch, J. and Hsia, C. W. (1992). A comparison of three segmentation algorithms in four color spaces, *Visual Communications and Image Processing '92, Boston*, Vol. 1818, Society of Photo-Optical Instrumentation Engineers, pp. 1168–1181.

Gauch, J. M. (1992). Investigations of image-contrast space defined by variations on histogram equalization, *CVGIP-Graphical Models and Image Processing* **54**(4): 269–280.

Gauch, J. M. (1997). Image segmentation and analysis via multiscale gradient watershed hierarchies, *IEEE Transactions on Image Processing*. In press.

Gentile, R. S., Allebach, J. P. and Walowit, E. (1990). Electronic imaging – quantization of color images based on uniform color spaces, *Journal of Imaging Technology* **16**(1): 11–21.

Gentile, R. S., Walowit, E. and Allebach, J. P. (1990). Quantization and multilevel halftoning of color images for near-original image quality, *Journal of the Optical Society of America A – Optics and Image Science* **7**(6): 1019–1026.

Gersho, A. and Gray, R. M. (1992). *Vector Quantization and Signal Compression*, Kluwer Academic Publishers, Boston.

Gershon, R. and Jepson, A. D. (1989). The computation of color constant descriptors in chromatic images, *Color Research and Application* **14**(6): 325–334.

Gervautz, M. and Purgathofer, W. (1990). A simple method for color quantization: Octree quantization, *in* A. S. Glassner (ed.), *Graphics Gems*, Academic Press, Boston, pp. 287–293.

Gilge, M., Engelhardt, T. and Mehlan, R. (1989). Coding of arbitrarily shaped image segments based on a generalized orthogonal transform, *Signal Processing: Image Communication* **1**: 153–180.

Gill, P. E. and Murray, W. (1982). Quasi-Newton methods for unconstrained optimisation, *Journal Int. Math. Appl.* **9**: 91–108.

Godlove, I. H. (1951). Improved color difference formula with applications to the perceptibility and acceptability of fadings, *Journal of the Optical Society of America* **41**: 760–772.

Goldberg, D. E. (1989). *Genetic Algorithms in Search, Optimization and Machine Learning*, Addison-Wesley, Reading, MA.

Gong, Y. and Sakauchi, M. (1995). Detection of regions matching specified chromatic features, *Computer Vision and Image Understanding* **61**(2): 263–269.

Gong, Y., Zhang, H. and Chua, H. C. (1994). An image database system with content based indexing and retrieval capabilities, *Proceedings of the 3rd International Conference on Automation, Robotics and Computer Vision*, Singapore.

Gonzalez, R. C. and Woods, R. E. (1992). *Digital Image Processing*, third edn, Addison-Wesley, Reading, MA. Reprinted with corrections 1993.

Gordillo, J. L. (1985). Colour representations for a vision machine, *Proc. 2nd International Conf. on Machine Intelligence*, London, pp. 375–385.

Gormish, M. J. (1995). Compression of palettized images by color, *Proceedings International Conference on Image Processing*, Vol. I, Institute of Electrical and Electronics Engineers, Washington, DC, pp. 274–277.

Gouras, P. (ed.) (1991). *The Perception of Colour*, Vol. 6 of *Vision and Visual Dysfunction*, Macmillan, London. Also published by CRC Press, Boca Raton.

Goutsias, J. (1992). Morphological transformations of image sequences: A lattice theory approach, *in* P. D. Gader, E. R. Dougherty and J. C. Serra (eds), *Proceedings Conference on Image Algebra and Morphological Image Processing III*, Vol. 1769, Society of Photo-Optical Instrumentation Engineers, San Diego, CA, pp. 306–317.

Gray, R. (1984). Vector quantization, *IEEE Acoustics Speech and Signal Processing Magazine* **1**(2): 4–29.

Guild, J. (1931). The colorimetric properties of the spectrum, *Philosophical Transactions of the Royal Society of London* **A 230**: 149–187.

Gunzinger, A., Mathis, S. and Guggenbuehl, W. (1990). Real time color classification, *in* V. Cappelini (ed.), *Time-Varying Image Processing and Moving Object Recognition*, Vol. 2, Elsevier, Amsterdam, pp. 82–87.

Hachimura, K. (1995). Extraction of principal colors from Japanese painting, *Proceedings of 9th SCIA*, Uppsala, pp. 457–464.

Hamilton, W. R. (1866). *Elements of Quaternions*, Longmans, Green and Co., London.

Hansen, G. L., Sparrow, E. M., Kokate, J. Y., Leland, K. J. and Iaizzo, P. A. (1997). Wound status evaluation using color image processing, *IEEE Transactions on Medical Imaging* 16(1): 78–86.

Haralick, R. M. (1984). Digital step edges from zero crossings of second directional derivatives, *IEEE Transactions on Pattern Analysis and Machine Intelligence* 6(1): 58–68.

Haralick, R. M. and Shapiro, L. G. (1991). Glossary of computer vision terms, *Pattern Recognition* 24(1): 69–93.

Haralick, R. M., Sternberg, S. R. and Zhuang, X. (1987). Image analysis using mathematical morphology, *IEEE Transactions on Pattern Analysis and Machine Intelligence* 9(4): 532–550.

Hard, A. and Sivik, L. (1981). NCS Natural Color System: A Swedish Standard for color notation, *Color Research and Application* 6(3): 129–138.

Hardie, R. and Arce, G. (1991). Ranking in $\mathbf{R}^p$ and its use in multivariate image estimation, *IEEE Transactions on Circuits and Systems for Video Technology* 1(2): 197–209.

Healey, G. (1992). Segmenting images using normalized color, *IEEE Transactions on Systems, Man and Cybernetics* 22(1): 64–73.

Healey, G. E., Shafer, S. A. and Wolff, L. B. (eds) (1992). *Physics-Based Vision - Principles and Practice*, Jones and Bartlett, Boston.

Heckbert, P. (1982). Color image quantization for frame buffer display, *Computer Graphics* 16(3): 297–307.

Held, G. and Marshall, T. R. (1996). *Data and Image Compression: Tools and Techniques*, fourth edn, John Wiley, Chichester, England; New York.

Herbin, M., Bon, F. X., Venot, A., Jeanlouis, F., Dubertret, M. L., Dubertret, L. and Strauch, G. (1993). Assessment of healing kinetics through true colour image processing, *IEEE Transactions on Medical Imaging* 12(1): 39–43.

Hering, E. (1964). *Outlines of a Theory of the Light Sense*, Harvard University Press, Cambridge MA. Translated by L. M. Hurvich and D. Jameson. Originally published 1920.

Heuberger, K. J., Jing, Z. M. and Persiev, S. (1992). Color transformations from lookup tables, *TAGA Proceedings*.

Horowitz, S. L. and Pavlidis, T. (1976). Picture segmentation by a tree traversal algorithm, *Journal of the Association for Computing Machinery* **23**(2): 368–388.

Huang, J. J. Y. and Schultheiss, P. M. (1963). Block quantisation of correlated Gaussian random variables, *IEEE Transactions on Communication* **11**(3): 289–296.

Huffman, D. A. (1952). A method for the construction of minimum redundancy codes, *Proceedings of the IRE* **40**(9): 1098–1101.

Hummel, R. A. (1975). Histogram modification techniques, *Computer Graphics and Image Processing* **4**: 209–224.

Hunt, R. W. G. (1991a). *Measuring Colour*, second edn, Ellis Horwood.

Hunt, R. W. G. (1991b). Revised colour-appearance model for related and unrelated colours, *Color Research and Application* **16**(3): 146–165.

Hunt, R. W. G. (1994). An improved predictor of colourfulness in a model of colour-vision, *Color Research and Application* **19**(1): 23–26.

Hurvich, L. M. (1981). *Color Vision*, Sinauer Associates Inc., Cambridge MA.

Hurvich, L. M. and Jameson, D. (1955). Some quantitative aspects of opponent colours theory: II Brightness saturation and hue in normal and dichromatic vision, *Journal of the Optical Society of America* **45**: 602–616.

ICC (1996). *ICC Profile Format Specification*, International Color Consortium, http://www.color.org.

IEC (1993). *IEC 11544(1993–12) Information technology – Coded Representation of Picture and Audio Information – Progressive Bi-Level Image Compression*, Geneva, (http://www.iec.ch/).

IJG (Undated). Independent JPEG group software archive, ftp://ftp.netcom.com/pub/tg/tgl/jpeg/README also at http://www.jpeg.org. Publicly available software.

Ikeda, H., Dai, W. and Higaki, Y. (1992a). A study of colorimetric errors caused by quantizing color information, *IEEE Transactions on Instrumentation and Measurement* **41**(6): 845–849.

Ikeda, H., Dai, W. and Higaki, Y. (1992b). A study on colorimetric effects caused by quantizing chromaticity information, *IEEE Instrumentation and Measurement Technology Conference*, Institute of Electrical and Electronics Engineers, pp. 374–378.

ISO/IEC (1994). *ISO/IEC Standard 10918-1:1994 Information Technology – Digital compression and coding of continuous-tone still images: requirements and guidelines*, Geneva, (http://www.iso.ch/, http://www.iec.ch/). See also ISO/IEC 10918-2, -3 and -4.

ITU (1990). *ITU-R Recommendation BT.709-2: Parameter Values for the HDTV Standards for production and International Programme Exchange*, Geneva, (http://www.itu.ch/).

ITU (1994a). *ITU-R Recommendation BT.500-7: Methodology for the Subjective Assessment of the Quality of Television Pictures*, Geneva, (http://www.itu.ch/).

ITU (1994b). *ITU-R Recommendation BT.601-5: Studio Encoding Parameters of Digital Television for standard 4:3 and wide-screen 16:9 aspect ratios*, Geneva, (http://www.itu.ch/).

Iverson, V. S. and Riskin, E. A. (1993). A fast method for combining color palettes of color quantized images, *Proceedings International Conference on Acoustics, Speech and Signal Processing ICASSP-93*, Vol. V, pp. 317–320.

Jacquin, A. E. (1992). Image coding based on a fractal theory of iterated contractive image transformations, *IEEE Transactions on Image Processing* **1**(1): 18.

Jain, A. and Pratt, W. K. (1972). Color image quantization, *Proceedings National Telecommunications Conference*, Institute of Electrical and Electronics Engineers.

Jarvis, J. F., Judice, C. N. and Hinke, W. H. (1976). A survey of techniques for the display of continuous tone pictures on bilevel displays, *Computer Graphics and Image Processing* **5**: 13–40.

Jayant, N. and Noll, P. (1984). *Digital Coding of Waveforms: Principles and applications to speech and video*, Prentice Hall, Englewood Cliffs, NJ.

Johnson, A. J. (1996). Methods for characterising colour scanners and digital cameras, *Displays*.

Johnson, A. J. and Scott-Taggart, M. (1993). *Guidelines for choosing the correct viewing conditions for colour publishing*, Pira International, Leatherhead, Surrey, UK.

Johnson, T., Luo, M. R., Lo, M. C., Xin, J. H. and Rhodes, P. A. (1997a). Aspects of colour management. Part I Characterisation of three-colour imaging devices, *Color Research and Application*.

Johnson, T., Luo, M. R., Lo, M. C., Xin, J. H. and Rhodes, P. A. (1997b). Aspects of colour management, Part II Characterisation of four-colour imaging devices and colour gamut compression, *Color Research and Application*.

Johnston, J. (1980). A filter family designed for use in quadrature mirror filter banks, *Proceedings IEEE International Conference on Acoustics, Speech, and Signal Processing*, Institute of Electrical and Electronics Engineers, pp. 291–294.

Jones, B. F. and Plassmann, P. (1995). An instrument to measure the dimensions of skin wounds, *IEEE Transactions on Biomedical Engineering* **42**(5): 464–470.

Jones, T. D. (1994). Semi-automatic segmentation algorithms for measuring the area of skin wounds, *Computer Studies Technical Report CS–94–3*, University of Glamorgan.

Joy, G. and Xiang, Z. (1993). Center-cut for color image quantization, *Vision Computing* **10**(1): 62–66.

Judd (1949). Response functions for types of vision according to the Mueller theory, *Journal of Research of the National Bureau Standards* **42**(1): 1–16.

JVC (1996). *Imaging - a Technical Reference Booklet*, JVC Professional Products (UK) Ltd.

Kang, H. R. (1992). Color scanner calibration, *Journal of Imaging Science and Technology* **36**(2): 162–170.

Kang, H. R. and Anderson, P. G. (1992). Neural network applications to the color scanner and printer calibrations, *Journal of Electronic Imaging*.

Karakos, D. G. and Trahanias, P. E. (1995). Combining vector median and vector directional filters: The directional-distance filters, *Proceedings of the IEEE Conference on Image Processing, ICIP-95*, Institute of Electrical and Electronics Engineers, pp. 171–174.

Kasson, J. M., Nin, S. I., Plouffe, W. and Hafner, J. L. (1995). Performing color space conversions with 3-dimensional linear interpolation, *Journal of Electronic Imaging* 4(3): 226–250.

Kay, G. and De Jager, G. (1992). A versatile colour system capable of fruit sorting and accurate object classification, *Proceedings of COMSIG'92*, Cape Town, pp. 145–148.

Kehtarnavaz, N., Griswold, N. C. and Kang, D. S. (1993). Stop-sign recognition based on color/shape processing, *Machine Vision and Applications* **6**: 206–208.

Kelly, K. L., Gibson, K. S. and Nickerson, D. (1943). Tristimulus specification of the Munsell book of color from spectrophotometer measurements, *Journal of the Optical Society of America* **33**: 353–376.

Kender, J. R. (1976). Saturation, hue, and normalized color; calculation, digitization effects, and use, *Technical report*, Carnegie-Mellon University, Pittsburgh.

Kim, J.-Y., Shim, J.-C. and Ha, Y.-H. (1992). Color image enhancement based on modified IHS coordinate system, *in* D. P. Casasent (ed.), *Proceedings Intelligent Robots and Computer Vision XI: Algorithms, Techniques, and Active Vision, 16 - 18 November, Boston, MA*, Vol. 1825, Society of Photo-Optical Instrumentation Engineers, pp. 366–377.

Kim, K. M., Lee, C. S., Lee, E. J. and Ha, Y. H. (1996). Color image quantization using weighted distortion measure of HVS color activity, *Proceedings International Conference on Image Processing*, Vol. III, Institute of Electrical and Electronics Engineers, Lausanne, pp. 1035–1039.

Klinker, G. J. (1993). *A Physical Approach to Color Image Understanding*, A. K. Peters, Wellesley MA.

Kodak (1992). *A planning guide for developers, Photo CD Products*, Rochester.

Kohonen, T. (1982). Self-organized formation of topological correct feature maps, *Biological Cybernetics* **43**(1): 59–69.

Kohonen, T. (1984). *Self-Organization and Associative Memory*, Vol. 8 of *Springer Series in Information Sciences*, Springer-Verlag, Berlin; New York.

Kok, C. W., Chan, S. C. and Leung, S. H. (1993). Color quantization by fuzzy quantizer, *in* E. R. Dougherty, J. Astola and H. G. Longbotham (eds), *Proceedings Nonlinear Image Processing IV, 1-3 February, San Jose, CA*, Vol. 1920, Society of Photo-Optical Instrumentation Engineers, pp. 235–242.

Kollhoff, D. and Kempe, H. (1995). Procedures for colorimetric calibration, *in* V. Rehrmann (ed.), *Proceedings of 1st Workshop 'Farbbildverarbeitung', Koblenz*, pp. 5–8. In German.

Kolpatzik, B. W. and Bouman, C. A. (1992). Optimized error diffusion for high quality image display, *Journal of Electronic Imaging* **1**(3): 277–292.

Kolpatzik, B. W. and Bouman, C. A. (1995). Optimized universal palette design for error diffusion, *Journal of Electronic Imaging* **4**(2): 131–143.

Koschan, A. (1993). Dense stereo correspondence using polychromatic block matching, *in* D. Chetverikov and W. D. Kropatsch (eds), *Computer Analysis of Images and Patterns, Proceedings of 5th International Conference*, Springer-Verlag, Berlin, pp. 538–542.

Kuglin, C. D. and Hines, D. C. (1975). The phase correlation image alignment method, *Proceedings IEEE Conference on Cybernetics and Society*, Institute of Electrical and Electronics Engineers, pp. 163–165.

Kuo, W. G. and Luo, M. R. (1996). Methods for quantifying metamerism. Part 2 – Instrumental methods, *Journal of the Society of Dyers and Colourists* **112**(12): 354–360.

Kuppenheim, H. F. and Heer, R. R. (1952). Spectral reflectance of white and negro skin between 440 and 1000 nm, *Journal of Applied Physiology* **4**: 800–806.

Laihanen, P. (1987). Colour reproduction theory based on the principles of colour science, *IARAIGAI Conference Proceedings: Advances in Printing Science and Technology*, Vol. 19, pp. 1–36.

Lawler, E. L. (ed.) (1985). *The travelling salesman problem: a guided tour of combinatorial optimization*, Wily-Interscience Series in Discrete Mathematics and Optimization, John Wiley, Chichester, West Sussex; New York. Reprinted with corrections 1990.

Ledley, S., Buas, M. and Golab, T. (1990). Fundamentals of true-color image processing, *Proceedings of 10th International Conference on Pattern Recognition*, Atlantic City, pp. 791–795.

Lee, B. G. (1984). A new algorithm to compute the discrete cosine transform, *IEEE Transactions on Acoustics, Speech, and Signal Processing* **32**(6): 1243–1245.

Lee, H.-C., Breneman, E. J. and Schulte, C. P. (1990). Modeling light reflection for computer color vision, *IEEE Transactions on Pattern Analysis and Machine Intelligence* **12**(4): 402–409.

Lee, R. L. (1988). Colorimetric calibration of a video digitizing system: Algorithm and applications, *Color Research and Application* **13**(3): 180–186.

Lennie, P. and D'Zmura, M. (1988). Mechanisms of color-vision, *Critical Reviews in Neurobiology* **3**(4): 333–400.

Levine, M. D. (1985). *Vision in Man and Machine*, McGraw-Hill, New York.

Levkowitz, H. and Herman, G. T. (1993). Glhs: A generalized lightness, hue and saturation color model, *CVGIP: Graphical Models and Image Processing* **55**(4): 271–285.

Li, H. D. and Jain, V. K. (1996). Color image coding by vector sub-bands/ECVQ and activity map, *Proceedings International Conference Acoustics, Speech Signal Processing ICASSP-96*, Vol. IV, Atlanta, GA, pp. 2044–2047.

Lim, Y. W. and Lee, S. U. (1990). On the color image segmentation algorithm based on the thresholding and the fuzzy c-means techniques, *Pattern Recognition* **23**(9): 935–952.

Linde, Y., Buzo, A. and Gray, R. M. (1980). An algorithm for vector quantizer design, *IEEE Transactions on Communications* **28**(1): 84–95.

Ling, D. T. and Just, D. (1991). Neural networks for halftoning of color images, *in* M. R. Civanlar, S. K. Mitra and R. J. Moorhead (eds), *Proceedings Image Processing Algorithms and Techniques II*, Vol. 1452, Society of Photo-Optical Instrumentation Engineers, San Jose, CA, pp. 10–20.

Liu, J. Q. and Yang, Y.-H. (1994). Multiresolution color image segmentation, *IEEE Transactions on Pattern Analysis and Machine Intelligence* **16**(7): 689–700.

Liu, T. S. and Chang, L. W. (1994). Greedy tree growing for color image quantization, *Proceedings International Conference Acoustics, Speech and Signal Processing ICASSP-94*, Vol. 5, Adelaide, Australia, pp. 97–100.

Liu, T.-S. and Chang, L.-W. (1995). Fast color image quantization with error diffusion and morphological operations, *Signal Processing* **43**(3): 293–303.

Lo, M. C. (1995). *The LLAB Model for Quantifying Colour Appearance*, PhD thesis, Loughborough University of Technology, England.

Lo, M. C. and Luo, M. R. (1994). Models for characterising 4-primary printing devices, *Journal of Photographic Science* **42**(3): 94–96.

Locht, P., Mikkelsen, P. and Thomsen, K. (1996). Advanced color image analysis for the food industry: It's here - now, *Advanced Imaging* **November**: 12–16.

Luo, M. R., Clarke, A. A., Rhodes, P. A., Schappo, A., Scrivener, S. A. R. and Tait, C. J. (1991a). Quantifying colour appearance, Part I: LUTCHI colour appearance data, *Color Research and Application* **16**(3): 166–180.

Luo, M. R., Clarke, A. A., Rhodes, P. A., Schappo, A., Scrivener, S. A. R. and Tait, C. J. (1991b). Quantifying colour appearance, Part II: Testing colour models performance using LUTCHI colour appearance data, *Color Research and Application* **16**(3): 181–197.

Luo, M. R., Gao, X. W., Rhodes, P. A., Xin, H. J., Clarke, A. A. and Scrivener, S. A. R. (1993a). Quantifying colour appearance, Part III: Supplementary LUTCHI colour appearance data, *Color Research and Application* **18**(2): 98–113.

Luo, M. R., Gao, X. W., Rhodes, P. A., Xin, H. J., Clarke, A. A. and Scrivener, S. A. R. (1993b). Quantifying colour appearance, Part IV: Transmissive media, *Color Research and Application* **18**(3): 191–209.

Luo, M. R. and Hunt, R. W. G. (1998). Testing colour appearance models using corresponding colour and magnitude estimation data sets, *Color Research and Application*. In press.

Luo, M. R., Lo, M. C. and Kuo, W. G. (1996). The LLAB(l:c) colour model, *Color Research and Application* **21**(6): 412–429.

Luo, M. R. and Morovic, J. (1996). Two unsolved issues in colour management - colour appearance and gamut mapping, *Proceedings of 5th International Conference on High Technology: Imaging Science and Technology - Evolution and Promise, Chiba, Japan*.

Luo, M. R., Rhodes, P. A., Xin, J. and Scrivener, S. (1992). Effective colour communication for industry, *Journal of the Society of Dyers and Colourists* **108**(12): 516–520.

Luo, M. R. and Rigg, B. (1986). Chromaticity-discrimination ellipses for surface colours, *Color Research and Application* **11**(1): 25–42.

Luo, M. R. and Rigg, B. (1987a). BFD(l:c) colour-difference formula, Part I Development of the formula, *Journal of the Society of Dyers and Colourists* **103**(2): 86–94.

Luo, M. R. and Rigg, B. (1987b). BFD(l:c) colour-difference formula, Part II Performance of the formula, *Journal of the Society of Dyers and Colourists* **103**(3): 126–132.

Luo, M. R., Xin, J. H., Rhodes, P. A., Scrivener, S. A. R. and MacDonald, L. W. (1991). Studying the performance of high resolution colour displays, *CIE 22nd Session – Division 1*, CIE, Melbourne, pp. 97–100.

MacAdam, D. L. (1974). Uniform color scales, *Journal of the Optical Society of America* **64**: 1691–1702.

MacDonald, L. W. (1996). Developments in colour management systems, *Displays* **16**(4): 203–211.

Malkin, F. and Verrill, J. F. (1983). *Proceedings CIE 20th Session, Amsterdam*.

Mallat, S. G. (1989). A theory for multiresolution signal decomposition: The wavelet representation, *IEEE Transactions on Pattern Analysis and Machine Intelligence* **11**(7): 674–693.

Mallat, S. and Zhong, S. (1992). Characterization of signals from multiscale edges, *IEEE Transactions on Pattern Analysis and Machine Intelligence* **14**(7): 710–732.

Maloney, L. T. and Wandell, B. A. (1986). Color constancy: a method for recovering surface spectral reflectance, *Journal of the Optical Society of America A-Optics and Image Science* **3**(1): 29–33.

Maragos, P. A. (1989). Pattern spectrum and multiscale shape representation, *IEEE Transactions on Pattern Analysis and Machine Intelligence* **11**(7): 701–716.

Maragos, P. A. (1990). Morphological systems for multidimensional signal processing, *Proceedings of the IEEE* **78**(4): 690–709.

Maragos, P. A., Mersereau, R. M. and Schafer, R. W. (1984). Multichannel predictive coding of colour images, *Proceedings International Conference on Acoustics, Speech and Signal Processing*, pp. 29.6.1–29.6.4.

Maragos, P. A. and Schafer, R. W. (1987). Morphological filters–Part I: Their set-theoretic analysis and relations to linear shift-invariant filter, *IEEE Transactions on Acoustics, Speech, and Signal Processing* **35**(8): 1153–1169.

Marcu, G. and Abe, S. (1995). Blue-print document analysis for color classification, *Proceedings of 9th SCIA*, Uppsala, pp. 569–574.

Marr, D. and Hildreth, E. (1980). Theory of edge detection, *Proceedings of the Royal Society of London, Series B* **207**: 187–217.

Marszalec, E. and Pietikäinen, M. (1994). On-line color camera calibration, *Proceedings of 12th IAPR International Conference on Pattern Recognition – Conference A: Computer Vision and Image Processing*, Jerusalem, Israel, pp. 232–237.

Masaki, I. (1988). Real-time multi-spectral visual processor, *in* T. Pavlidis (ed.), *Proceedings IEEE International Conference on Robotics and Automation*, Vol. 3, Philadelphia, PA, pp. 1554–1559.

Massen, R. (1994). Optical sortation of plastics by colour and composition, Paper presented at Joint Meeting on On-Line Optical Inspection Calibration and Validation, National Physical Laboratory, England, 17th February.

Massen, R. C., Bottcher, P. and Leisinger, U. (1988). Real-time grey level and colour image preprocessing for a vision guided biotechnology robot, *in* W. Guttropf (ed.), *Proceedings of the 7th International Conference on Robot Vision and Sensory Controls*, Zurich, Switzerland, pp. 115–122.

McCabe, A., Caelli, T., West, G. and Reeves, A. (1997). Encoding and processing spatio-chromatic image information using complex Fourier transform methods, *Technical Report 97-3*, Curtin University of Technology, Perth, Western Australia.

McCamy, C. S., Marcus, H. and Davidson, J. G. (1976). A color-rendition chart, *Journal of Applied Photographic Engineering* **2**(3): 95–99.

McDonald, R. (1980a). Industrial pass/fail colour matching, Part I: Preparation of visual colour-matching data, *Journal of the Society of Dyers and Colourists* **96**: 372–376.

McDonald, R. (1980b). Industrial pass/fail colour matching, Part II: Methods of fitting tolerance ellipsoids, *Journal of the Society of Dyers and Colourists* **96**: 418–433.

McDonald, R. (1980c). Industrial pass/fail colour matching, Part III: Development of a pass/fail formula for use with instrumental measurement of colour difference, *Journal of the Society of Dyers and Colourists* **96**: 486–497.

McDonald, R. (1987a). Computer match prediction – dyes, in *Colour Physics for Industry* (McDonald, 1987b), pp. 116–185. See also (McDonald, 1997).

McDonald, R. (ed.) (1987b). *Colour Physics for Industry*, Society of Dyers and Colourists, Bradford. See also (McDonald, 1997).

McDonald, R. (ed.) (1997). *Colour Physics for Industry*, second edn, Society of Dyers and Colourists, Bradford.

McDonald, R. and Smith, K. J. (1995). CIE94 – a new colour-difference formula, *Journal of the Society of Dyers and Colourists* **111**: 376–379.

McDowell, D. Q. (1993). Summary of IT8/SC4 color activities, *in* R. J. Motta and H. A. Berberian (eds), *Proceedings Device-Independent Color Imaging and Imaging Systems Integration Conference*, Vol. 1913, Society of Photo-Optical Instrumentation Engineers, San Jose, CA, pp. 229–235.

McFall, J. D., Mitchell, J. L. and Pennebaker, W. B. (1989). Displaying photographic images on computer monitors with limited color resolution, *in* Y. W. Lin and R. Srinivasan (eds), *Proceedings Digital Image Processing Applications*, Vol. 1075, Society of Photo-Optical Instrumentation Engineers, Los Angeles, CA, pp. 179–184.

McLaren, K. (1987). Colour space, colour scales and colour difference, *in* McDonald (1987b), pp. 97–115. See also (McDonald, 1997).

Milvang, O. and Olafsdottir, B. (1993). Discriminating crates from color images, *Proceedings of 8th SCIA*, Tromso.

Mitchel, J. L. and Pennebaker, W. B. (1988). Optimal hardware and software arithmetic coding procedures for the Q-coder, *IBM Journal of Research and Development* **32**(6): 727–736.

Miyahara, M. and Yoshida, Y. (1988). Mathematical transform of (R,G,B) color data to Munsell (H,V,C) color data, *in* T. R. Hsing (ed.), *Proceedings Visual Communications and Image Processing Conference '88*, Vol. 1001, Society of Photo-Optical Instrumentation Engineers, Cambridge, MA, pp. 650–657.

Mollon, J. D. (1989). ' 'tho' she kneeled in that place where they grew ... ' the uses and origins of primate colour vision, *Journal of Experimental Biology* **146**: 21–38.

Monro, D. M. and Nicholls, J. A. (1995). Low bit rate colour fractal video, *Proceedings International Conference on Image Processing*, Institute of Electrical and Electronics Engineers, Washington, DC, pp. C.264–C.267.

Morovic, J. and Luo, M. R. (1997). Cross-media psychophysical evaluation of gamut mapping algorithms, *Proceedings AIC Color '97, Kyoto, Japan*.

Mullen, K. T. (1985). The contrast sensitivity of human colorvision to red-green and blue-yellow chromatic gratings, *Journal of Physiology-London* **359**: 381–400.

Mullen, K. T. and Kingdom, F. A. A. (1991). Colour contrast in form perception, *in* P. Gouras (ed.), *Vision and Visual Dysfunction 6: The Perception of Colour*, Macmillan, London, pp. 198–217.

Murray, A. (1936). Monochrome reproduction in photoengraving, *Journal Franklin Institute* **221**: 721–744.

Musmann, H. G., Hötter, M. and Ostermann, J. (1989). Object-oriented analysis-synthesis coding of moving images, *Signal Processing: Image Communication* **1**: 17–138.

Naiman, A. (1985). Colour spaces and colour contrasts, *Graphics Interface '85*, Montreal, pp. 313–320.

Naka, K.-I. and Rushton, W. A. (1966). S-potentials from colour units in the retina of fish (cyprinidae), *Journal of Physiology* **185**: 587–599.

Nayatani, Y., Sobagaki, H., Takahama, K. and Yano, T. (1997). Field trials of a non-linear color-appearance model, *Color Research and Application* **22**(4): 240–258.

Neugebauer, H. (1937). Die theoretischen grundlagen des mehrfarbenbuch-drucks in zeitschrift für wissenschaftliche photographie, *Photophysik und Photochemie* **36**(4): 73–89.

Nevatia, R. (1977). A color edge detector and its use in scene segmentation, *IEEE Transactions on Systems, Man and Cybernetics* **7**(11): 820–826.

Newhall, S. M., Nickerson, D. and Judd, D. B. (1943). Final report of the OSA subcommittee on the spacing of Munsell colors, *Journal of the Optical Society of America* **33**: 385–418.

Niblack, W. (1986). *An introduction to digital image processing*, Prentice Hall, Englewood Cliffs, NJ.

Nier, M. C. (ed.) (1996). *Standards for Electronic Imaging Technologies, Devices and Systems*, Vol. CR 61 of *Critical reviews of optical science and technology*, SPIE Optical Engineering Press, Bellingham, Washington. Proceedings of a conference held 1-2nd February 1996, San Jose, CA.

Nikolaidis, N. and Pitas, I. (1995). Multichannel L filters based on reduced ordering, *Proceedings of the IEEE Workshop on Nonlinear Signal/Image Processing*, Institute of Electrical and Electronics Engineers, Neos Marmas, Greece, pp. 542–546.

Novak, C. L., Shafer, S. A. and Wilson, R. G. (1990). Obtaining accurate color images for machine vision research, *Proceedings Conference on Perceiving, Measuring and Using Color*, Society of Photo-Optical Instrumentation Engineers.

Oehler, K. L., Riskin, E. A. and Gray, R. M. (1991). Unbalanced tree-growing algorithms for practical image compression, *Proceedings International Conference Acoustics, Speech Signal Processing ICASSP-91*, pp. 2293–2296.

Ohta, Y., Kanade, T. and Sakai, T. (1980). Color information for region segmentation, *Computer Graphics and Image Processing* **13**: 222–241.

Orchard, M. and Bouman, C. (1989). Color image display with a limited palette size, *in* W. A. Pearlman (ed.), *Proceedings Visual Communications and Image Processing Conference IV*, Vol. 1199, Society of Photo-Optical Instrumentation Engineers, Philadelphia, PA, pp. 522–533.

Orchard, M. and Bouman, C. (1991). Color quantization of images, *IEEE Transactions on Signal Processing* **39**(12): 2677–2690.

Overloop, J. V., Philips, W., Torfs, D. and Lemahieu, I. (1997). Segmented image coding of palettized images, *Proceedings International Conference Acoustics Speech Signal Processing, ASSP-97*, Vol. IV, pp. 2937–2939.

Overturf, L. A., Comer, M. L. and Delp, E. J. (1995). Color image coding using morphological pyramid decomposition, *IEEE Transactions on Image Processing* **4**(2): 177–185.

Paeth, A. W. (1995). Distance approximations and bounding polyhedra, *in* A. W. Paeth (ed.), *Graphics Gems*, Vol. V, Academic Press Professional, Chestnut Hill, MA, pp. 78–87.

Palus, H. (1996). IHS colour space: Properties and modifications, *in* K.-H. Franke (ed.), *2. Workshop Farbbildverarbeitung*, Ilmenau, pp. 73–77. In German.

Parnanen, P. and Lemstrom, G. (1995). In the limelight, *Image Processing* **7**: 30–32.

Pasco, R. (1976). *Source Coding Algorithms for Fast Data Compression*, PhD thesis, Stanford University.

Pei, S.-C. and Cheng, C.-M. (1995). Dependent scalar quantization of color images, *IEEE Transactions on Circuits and Systems for Video Technology* **5**(2): 124–139.

Pennebaker, W. B. and Mitchell, J. L. (1993). *JPEG Still Image Data Compression Standard*, Van Nostrand Reinhold, New York.

Perez, F. and Koch, C. (1994). Toward color image segmentation in analog VLSI: Algorithm and hardware, *International Journal of Computer Vision* **12**(1): 17–42.

Perona, P. and Malik, J. (1990). Scale-space and edge detection using anisotropic diffusion, *IEEE Transactions on Pattern Analysis and Machine Intelligence* **12**(7): 629–639.

Pien, H. H. and Gauch, J. M. (1995). A variational approach to multi-sensor fusion of images, *Applied Intelligence* **5**(3): 217–235.

Pitas, I. (1993). *Digital Image Processing Algorithms*, Prentice Hall International Series in Acoustics, Speech, and Signal Processing, Prentice Hall, New York.

Pitas, I. (1995). Multichannel order statistical filtering, *in* C. Toumazou (ed.), *Circuits and Systems Tutorials*, IEEE Press, pp. 41–49.

Pitas, I. and Venetsanopoulos, A. N. (1990). *Nonlinear Digital Filters: Principles and Applications*, Kluwer Academic Publishers, Boston.

Pitas, I. and Venetsanopoulos, A. N. (1992). Order statistics in digital image processing, *Proceedings of the IEEE* **80**(12): 1893–1923.

Pizer, S. M. (1983). An automatic intensity mapping for the display of CT scans and other images, *Proceedings Information Processing and Medical Imaging*, pp. 276–309.

Plataniotis, K. N., Androutsos, D., Sri, V. and Venetsanopoulos, A. N. (1995). A nearest neighbour multichannel filter, *Electronics Letters* **31**(22): 1910–1911.

Plataniotis, K. N., Androutsos, D. and Venetsanopoulos, A. N. (1995). Colour image processing using fuzzy vector directional filters, *Proceedings of the IEEE Workshop on Nonlinear Signal/Image Processing*, Institute of Electrical and Electronics Engineers, Neos Marmas, Greece, pp. 535–538.

Plataniotis, K. N., Androutsos, D. and Venetsanopoulos, A. N. (1996). Fuzzy adaptive filters for multichannel image processing, *Signal Processing* **55**(1): 93–106.

Plataniotis, K. N., Androutsos, D. and Venetsanopoulos, A. N. (1997). Multichannel filters for image processing, *Signal Processing – Image Communications* **9**(2): 143–158.

Plataniotis, K. N., Sri, V., Androutsos, D. and Venetsanopoulos, A. N. (1996). An adaptive nearest neighbor multichannel filter, *IEEE Transactions on Circuits and Systems for Video Technology* **6**(6): 699–703.

PNG (1996). *PNG (Portable Network Graphics) Specification*, W3C Tech. Reports, ftp://ftp.uu.net/graphics/png/. Thomas Boutell (ed.).

Pomierski, T. and Gross, H.-M. (1996). Biological neural architecture for chromatic adaptation resulting in constant color sensations, *Proceedings of ICNN 1996, IEEE International Conference on Neural Networks*, Institute of Electrical and Electronics Engineers, Washington, DC, pp. 734–739.

Post, D. L. and Calhoun, C. S. (1989). An evaluation of methods for producing desired colors on CRT monitors, *Color Research and Application* **14**(4): 172–186.

Poston, R. N. (1996). Hue, saturation and intensity - colour image analysis for histology, *Image Processing* **December**: 4–8.

Pratt, W. K. (1971). Spatial transform coding of colour images, *IEEE Transactions on Communications* **19**(6): 980–992.

Pratt, W. K. (1991). *Digital Image Processing*, second edn, John Wiley, New York.

Priese, L. and Rehrmann, V. (1993). A fast hybrid color segmentation method, *in* S. Poeppl and H. Handels (eds), *Proceedings 15th DAGM-Symposium Mustererkennung*, pp. 297–304.

Pritchard, A. J., Horne, R. E. N. and Sangwine, S. J. (1995). Achieving brightness-insensitive measurements of colour saturation for use in colour object recognition, *Proceedings 5th International Conference on*

Image Processing and its Applications, Institution of Electrical Engineers, Heriot-Watt University, Edinburgh, UK, pp. 791–795. IEE Conference Publication 410.

Proceedings International Conference on Image Processing (1996). Institute of Electrical and Electronics Engineers, Lausanne.

Ramstad, T. A. (1988). IIR filterbank for subband coding of images, *Proceedings IEEE International Symposium on Circuits and Systems*, Helsinki, pp. 827–830.

Rao, K. R. and Yip, P. (1990). *Discrete Cosine Transform: Algorithms, advantages, applications*, Academic Press, Boston.

Réndon, E., Salgado, L., Menéndez, J. M. and Garcia, N. (1997). Adaptive color quantization through self-organization neural networks, *in* Domański and Stasiński (1997), pp. 195–198.

Rhodes, P. A. and Luo, M. R. (1996). A system for WYSIWYG colour communication, *Displays* **16**(4): 213–221.

Rhodes, P. A., Luo, M. R. and Scrivener, S. A. R. (1990). Colour model integration and visualisation, *in* D. Diaper et al. (eds), *Proceedings of IFIP INTERACT 90: Human-Computer Interaction, Cambridge*, IFIP, Elsevier Science Publishers B.V., North-Holland, pp. 725–728.

Rissanen, J. J. (1976). Generalized Kraft inequality and arithmetic coding, *IBM Journal of Research and Development* **20**(3): 198–203.

Roberts, L. G. (1965). Machine perception of three-dimensional solids, *Optical and Electro-Optical Information Processing*, MIT Press, Cambridge, MA, pp. 159–197.

Robertson, P. K. (1988). Visualizing color gamuts: A user interface for the effective use of perceptual color spaces in data displays, *IEEE Computer Graphics and Applications* **8**(5): 50–64.

Rubin, J. M. and Richards, W. A. (1982). Color vision and image intensities: When are changes material?, *Biological Cybernetics* **45**(3): 215–226.

Rubner, J. and Schulten, K. (1989). A regularized approach to colour constancy, *Biological Cybernetics* **61**(1): 29–36.

Said, A. and Pearlman, W. A. (1996). A new fast and efficient image codec based on set partitioning in hierarchical trees, *IEEE Transactions on Circuits and Systems for Video Technology* **6**(3): 243–250.

Sandini, G., Buemi, F., Massa, M. and Zucchini, M. (1990). Visually guided operations in green-houses, *1st International Workshop on Robotics in Agriculture and the Food Industry*, Vol. 1, Avignon, France, pp. 69–83.

Sangwine, S. J. (1996). Fourier transforms of colour images using quaternion, or hypercomplex, numbers, *Electronics Letters* **32**(21): 1979–1980.

Schettini, R. (1993). A segmentation algorithm for color images, *Pattern Recognition Letters* **14**(6): 499–506.

Schettini, R. (1994). Multicolored object recognition and location, *Pattern Recognition Letters* **15**(11): 1089–1097.

Schettini, R., Barolo, B. and Boldrin, E. (1995). Colorimetric calibration of color scanners by backpropagation, *Pattern Recognition Letters* **16**(10): 1051–1056.

Scheunders, P. (1996). A genetic approach towards optimal color image quantization, *Proceedings International Conference on Image Processing*, Vol. III, Institute of Electrical and Electronics Engineers, Lausanne, pp. 1031–4.

Schmid, R. and Truskowski, T. (1993). Colour contrast calculation for image processing, *Elektonik* **22**: 114–118. In German.

Schonfeld, D. and Goutsias, J. (1991). Optimal morphological pattern restoration from noisy binary images, *IEEE Transactions on Pattern Analysis and Machine Intelligence* **13**(1): 14–29.

Senoo, T. and Girod, B. (1992). Vector quantization for entropy coding of image subbands, *IEEE Transactions on Image Processing* **1**(4): 526–532.

Serra, J. (1982). *Image Analysis and Mathematical Morphology*, Vol. 1, Academic Press, London; New York.

Serra, J. (1988). *Image Analysis and Mathematical Morphology: Theoretical Advances*, Vol. 2, Academic Press, London; New York.

Shah, J. (1991). Segmentation by nonlinear diffusion, *IEEE Computer Society Conference on Computer Vision and Pattern Recognition*, Lahaina, HI, pp. 202–207.

Shapiro, J. M. (1993). Embedded image coding using zerotrees of wavelet co-efficients, *IEEE Transactions on Signal Processing* **41**(12): 3445–3462.

Sharma, G. and Trussell, H. J. (1997). Digital color imaging, *IEEE Transactions on Image Processing* **6**(7): 901–932.

Sikora, T. and Makai, B. (1995). Shape-adaptive DCT for generic coding of video, *IEEE Transactions on Circuits and Systems for Video Technology* **5**(1): 59–62.

SIS (1982). *Swedish Standard SS 01 91 03 CIE Tristimulus Values and Chromaticity Coordinates for Colour Samples in SS 01 91 02*, Sweden.

Slater, J. (1991). *Modern Television Systems to HDTV and beyond*, Pitman, London.

Slaughter, D. C. and Harrell, R. C. (1987). Color vision in robotic fruit harvesting, *Transactions of the ASAE* **30**(4): 1144–1148.

Smith, J. R. (1997). Shine a light, *Image Processing* **April**: 20–22.

Smith, M. J. T. and Barnwell, T. P. (1988). Exact reconstruction techniques for tree-structured subband coders, *IEEE Transactions on Acoustics, Speech, and Signal Processing* **34**(3): 434–441.

Smith, V. C. and Pokorny, J. (1975). Spectral sensitivity of the cone photopigments between 400 and 500 nm, *Vision Res.* **15**: 161–171.

Solinsky, J. C. (1985). The use of color in machine edge detection, *Proceedings Vision '85 Conference*, Detroit, MI, pp. 248–266.

Song, J. and Delp, E. J. (1990). The analysis of morphological filters with multiple structuring elements, *Computer Vision, Graphics, and Image Processing* **50**(3): 308–328.

Song, J. and Delp, E. J. (1991). A study of the generalized morphological filter, *Circuits, Systems, and Signal Processing* **11**(1): 229–252.

Sonka, M., Hlavac, V. and Boyle, R. (1993). *Image Processing, Analysis and Machine Vision*, Chapman and Hall, London.

Speranskaya, N. I. (1959). Determination of spectrum colour coordinates for twenty-seven observers, *Optics and Spectroscopy* **7**: 424.

Sproson, W. N. (1978). PAL System I phosphor primaries - the present position, *IEE Proceedings* **125**: 603–605.

Sproson, W. N. (1983). *Colour Science in Television and Display Systems*, Adam Hilger, Bristol.

Stasiński, R. and Konrad, J. (1997). A new approach to generation of shape-adaptive transforms, *in* Domański and Stasiński (1997), pp. 13–16.

Stevenson, R. L. and Arce, G. R. (1987). Morphological filters: Statistics and further syntactic properties, *IEEE Transactions on Circuits and Systems* **34**(11): 1292–1305.

Stiles, W. S. and Burch, J. M. (1959). N. P. L. colour-matching investigation: Final report (1958), *Optica Acta* **6**(1): 1–26.

Stockham, T. G. (1972). Image processing in the context of a visual model, *Proceedings of the IEEE* **60**(7): 828–842.

Stokes, M., Fairchild, M. D. and Berns, R. (1992). Precision requirements for digital color reproduction, *ACM Transactions on Graphics* **11**(4): 406–422.

Stone, M. C. and Wallace, W. E. (1991). Gamut mapping computer generated imagery, *Proceedings Graphics Interface 91 Conference*, Calgary, Alberta, Canada, pp. 32–39.

Strachan, N. J. C., Nesvadba, P. and Allen, A. R. (1990). Calibration of a video camera digitizing system in the CIE L*u*v* colour space, *Pattern Recognition Letters* **11**(11): 771–777.

Sullivan, J. R., Rabbani, M. and Dawson, B. M. (eds) (1992). *Proceedings Image Processing Algorithms and Techniques III*, Vol. 1657, Society of Photo-Optical Instrumentation Engineers, San Jose, CA.

Sun, F. K. and Maragos, P. (1989). Experiments on image compression using morphological pyramids, *in* W. A. Pearlman (ed.), *Proceedings Visual Communications and Image Processing Conference IV*, Vol. 1199, Society of Photo-Optical Instrumentation Engineers, Philadelphia, PA, pp. 1303–1312.

Sun, Y.-N., Wu, C.-S., Lin, X.-Z. and Chou, N.-H. (1993). Color image analysis for liver tissue classification, *Optical Engineering* **32**(7): 1609–1615.

Swain, M. J. (1990). *Color Indexing*, PhD thesis, University of Rochester.

Tan, K. T., Ghanbari, M. and Pearson, D. E. (1997). A video distortion meter, *Picture Coding Symposium PCS'97*, pp. 119–122.

Taylor, J. M., Murch, G. and McManus, P. A. (1989). TekHVC: A uniform perceptual color system for display users, *Proceedings of the SID* **30**(1): 15–21.

Taylor, R. I. and Lewis, P. H. (1992). Colour image segmentation using boundary relaxation, *Proceedings of 11th IAPR International Conference on Pattern Recognition*, Vol. III, The Hague, Netherlands, pp. 721–724.

Tekalp, A. M. (1995). *Digital video processing*, Prentice Hall.

Tektronix (1990). *TekColor Color Management System: System Implementers Manual*, Tektronix Inc.

Tenenbaum, J. M., Garvey, T. D., Weyl, S. and Wolf, H. C. (1974). An interactive facility for scene analysis research, *Technical Report 87*, Stanford Research Institute, AI Centre.

Thornton, A. L. and Sangwine, S. J. (1995). Colour object recognition using complex coding in the frequency domain, *Proceedings 5th International Conference on Image Processing and its Applications*, Institution of Electrical Engineers, Heriot-Watt University, Edinburgh, UK, pp. 820–824. IEE Conference Publication 410.

Thornton, A. L. and Sangwine, S. J. (1996). Colour object recognition using phase correlation of log-polar transformed fourier spectra, *in* B. G. Mertzios and P. Liatsis (eds), *Proceedings 3rd International Workshop on Image and Signal Processing, UMIST, Manchester, UK, 4-7 November*, Elsevier Science B.V., Amsterdam, pp. 615–18.

Thornton, A. L. and Sangwine, S. J. (1997). Log-polar sampling incorporating a novel spatially-variant filter to improve object recognition, *Proceedings 6th International Conference on Image Processing and its Applications*, Institution of Electrical Engineers, Trinity College, Dublin, Ireland, pp. 776–779. IEE Conference Publication 443.

TIFF (1992). *TIFF Revision 6.0, June 3 1992*, 411 First Avenue South, Seattle, WA 98104-2871. Available by anonymous `ftp` from Silicon Graphics at `ftp://ftp.sgi.com/other/tiff/TIFF6.ps`.

Toet, A. (1989). A morphological pyramidal image decomposition, *Pattern Recognition Letters* **9**(4): 255–261.

Toet, A. (1992). Multiscale colour image enhancement, *Pattern Recognition Letters* **13**(3): 167–174.

Tominaga, S. (1987). Expansion of color images using three perceptual attributes, *Pattern Recognition Letters* **6**(1): 77–85.

Tominaga, S. (1988). A color classification algorithm for color images, *in* J. Kittler (ed.), *Pattern Recognition (4th International Conference, Cambridge, UK, March 28-30th: proceedings)*, Springer-Verlag, Berlin; New York, pp. 163–172. BPRA International Conference on Pattern Recognition.

Tominaga, S. (1992). Color classification of natural color images, *Color Research and Application* **17**(4): 230–239.

Tominaga, S. (1994). Dichromatic reflection models for a variety of materials, *Color Research and Application* **19**(4): 277–285.

Torres, L. and Kunt, M. (eds) (1996). *Video Coding: The Second Generation Approach*, Kluwer, Boston, MA.

Trahanias, P. E., Karakos, D. G. and Venetsanopoulos, A. N. (1996). Directional processing of color images: Theory and experimental results, *IEEE Transactions on Image Processing* **5**(6): 868–880.

Trahanias, P. E., Pitas, I. and Venetsanopoulos, A. N. (1994). Color image processing, *in* C. T. Leondes (ed.), *Digital Image Processing: techniques and applications*, Vol. V 67 0090-5267 of *Control and dynamic systems: advances in theory and applications*, Academic Press, San Diego.

Trahanias, P. E. and Venetsanopoulos, A. N. (1993). Vector directional filters: A new class of multichannel image processing filters, *IEEE Transactions on Image Processing* **2**(4): 528–534.

Tremblay, M. P. and Zaccarin, A. (1994). Transmission of the color information using quad-trees and segmentation-based approaches for the compression of color images with limited palette, *Proceedings International Conference on Image Processing*, Vol. III, Institute of Electrical and Electronics Engineers, Austin, TX, pp. 967–971.

Tremeau, A., Calonnier, M. and Laget, B. (1994). Color quantization error in terms of perceived image quality, *Proceedings International Conference Acoustics, Speech and Signal Processing ICASSP-94*, Vol. 5, Adelaide, Australia, pp. 93–96.

Tremeau, A. and Laget, B. (1995). Quantification couleur et analyse d'image, *Traitement du Signal* 12(1): 1–28.

Tremeau, A., Lozano, V. and Lager, B. (1995). How to optimize the use of the L*H*C* color space in color image analysis processes, *Acta Stereol.* 14/2: 223–228.

Tukey, J. M. (1974). Nonlinear (nonsuperimposable) methods for smoothing data, *Proc. Congr. EASCON*, p. 673.

Uchiyama, T. and Arbib, M. A. (1994). Color image segmentation using competitive learning, *IEEE Transactions on Pattern Analysis and Machine Intelligence* 16(12): 1197–1206.

Vaidyanathan, P. P. (1993). *Multirate Systems and Filter Banks*, Prentice Hall, Englewood Cliffs, NJ.

Valavanis, K. P., Zheng, J. and Gauch, J. M. (1991). On impulse noise removal in color images, *Proceedings of IEEE International Conference on Robotics and Automation*, Institute of Electrical and Electronics Engineers, Sacramento, CA, pp. 144–149.

Van Dyck, R. E. and Rajala, S. A. (1994). Subband/VQ coding of color images with perceptually optimal bit allocation, *IEEE Transactions on Circuits and Systems for Video Technology* 4(1): 68–82.

Venable, D., Stinehour, J. and Roetling, P. (1990). Selection and use of small color sets for pictorial display, *SPSE 43rd Annual Meeting, Rochester, New York*, pp. 90–92.

Venetsanopoulos, A. N. and Plataniotis, K. N. (1995). Multichannel image processing, *Proceedings of the IEEE Workshop on Nonlinear Signal/Image Processing*, Institute of Electrical and Electronics Engineers, Neos Marmas, Greece, pp. 2–6.

Verevka, O. and Buchanan, J. (1995). Local K-means algorithm for color image quantization, *Proceedings of Graphics/Vision Interface GI-95*, Quebec, Canada, pp. 128–134.

Vetterli, M. (1984). Multi-dimensional sub-band coding: Some theory and algorithms, *Signal Processing* **6**(2): 97–112.

Vetterli, M. and Kovacevic, J. (1995). *Wavelets and Subband Coding*, Prentice Hall Signal Processing Series, Prentice Hall, Englewood Cliffs, NJ.

Viero, T., Vistamo, K. and Neuvo, Y. (1994). Three dimensional median-related filters for color image sequence filtering, *IEEE Transactions on Circuits and Systems for Video Technology* **4**(2): 129–142.

Vos, J. J., Estevez, O. and Walraven, P. L. (1990). Improved color fundamentals offer a new view on photometric additivity, *Vision Research* **30**(6): 937–943.

Waldemar, P. and Ramstad, T. A. (1994). Subband coding of color images with limited palette size, *Proceedings International Conference Acoustics, Speech and Signal Processing ICASSP-94*, Vol. V, Adelaide, Australia, pp. 353–356.

Wallace, G. K. (1991). The JPEG still picture compression standard, *Communications ACM* **34**(4): 30–44.

Wallace, R. H. (1976). An approach for the space variant restoration and enhancement of images, *Proceedings Symposium on Current Mathematical Problems in Imaging Science*.

Walraven, J., Benzschawel, T. and Rogowitz (1989). Color constancy interpretation f chromatic induction, *Die Farbe* pp. 67–68.

Wan, S., Wong, S. and Prusinkiewicz, P. (1988). An algorithm for multi-dimensional data clustering, *ACM Transactions on Mathematical Software* **14**(2): 153–162.

Wang, C. Y. and Chang, L. W. (1992). Color image coding using variable vector quantization in (R,G,B) domain, *in* P. Maragos (ed.), *Proceedings Visual Communications and Image Processing 92*, Vol. 1818, Society of Photo-Optical Instrumentation Engineers, Boston, MA, pp. 512–523.

Watson, A. B. (1994). Perceptual optimization of DCT color quantization matrices, *Proceedings International Conference on Image Processing*, Vol. I, Institute of Electrical and Electronics Engineers, Austin, TX, pp. 100–104.

Watson, A. B. and Tiana, C. L. M. (1992). Color motion video coded by perceptual components, *SID Annual Meeting Proceedings*.

Welch, T. A. (1984). A technique for high performance data compression, *IEEE Computer* **17**(6): 8–19.

Westerink, P. H., Biemond, J. and Boekee, D. E. (1988). An optimal bit allocation algorithm for subband coding, *Proceedings International Conference on Acoustics Speech and Signal Processing ICASSP '88*, Institute of Electrical and Electronics Engineers, New York, NY, pp. 757–760.

Whitaker, R. and Gerig, G. (1994). Vector valued diffusion, *in* B. M. ter Haar Romeny (ed.), *Geometry-Driven Diffusion in Computer Vision*, Vol. 1 of *Computational Imaging and Vision*, Kluwer Academic, Dordrecht; Boston, pp. 93–134.

Wixson, L. E. and Ballard, D. (1989). Real-time detection of multicolored objects, *in* P. S. Schenker (ed.), *Proceedings Sensor Fusion II: Human and Machine Strategies Conference*, Vol. 1198, Society of Photo-Optical Instrumentation Engineers, Philadelphia, PA.

Woods, J. (ed.) (1991). *Subband Image Coding*, Kluwer Academic Publishers, Boston.

Woods, J. W. and O'Neil, S. D. (1986). Subband coding of images, *IEEE Transactions on Acoustics, Speech, and Signal Processing* **34**(5): 1278–1288.

Wright, W. D. (1928-29). A re-determination of the trichromatic coefficients of the spectral colours, *Transactions Optical Society* **30**: 141.

Wu, X. L. (1992). Color quantization by dynamic programming and principal analysis, *ACM Transactions on Graphics* **11**(4): 348–372.

Wyszecki, G. and Stiles, W. S. (1982). *Color Science: Concepts and Methods, Quantitative Data and Formulae*, second edn, John Wiley, New York.

Xiang, Z. G. and Joy, G. (1994). Color image quantization by agglomerative clustering, *IEEE Computer Graphics and Applications* **14**(3): 44–48.

Xu, W. and Hauske, G. (1995). Perceptually relevant error classification in the context of picture coding, *Proceedings 5th International Conference on Image Processing and its Applications*, Institution of Electrical Engineers, Heriot-Watt University, Edinburgh, UK, pp. 589–593. IEE Conference Publication 410.

Yagi, D., Abe, K. and Nakatani, H. (1992). Segmentation of color aerial photographs using HSV color models, *Proceedings of MVA'92*, Tokyo, pp. 367–370.

Yakimovsky, Y. (1976). Boundary and object detection in real world images, *Journal of the Association for Computing Machinery* **23**(4): 599–618.

Yamaba, K. and Miyake, Y. (1993). Color character recognition method based on human perception, *Optical Engineering* **32**(1): 33–40.

Yang, C. C. and Rodriguez, J. J. (1996). Saturation clipping in the LHS and YIQ color spaces, *in* J. Bares (ed.), *Proceedings Conference on Color Imaging – Device-Independent Color, Color Hard Copy, and Graphic Arts*, Vol. 2658, Society of Photo-Optical Instrumentation Engineers, San Jose, CA, pp. 297–307.

Yang, C. K., Lin, J. C. and Tsai, W. H. (1994). Color image compression by moment-preserving and block truncation coding, *Proceedings International Conference on Image Processing*, Vol. III, Institute of Electrical and Electronics Engineers, Austin, TX, pp. 972–976.

Yousif, W. S. and Luo, M. R. (1991). An image process for achieving WYSIWYG colour, *Proceedings of the 14th World Congress on Computation and Applied Mathematics, IMACS 91, Dublin*, Vol. 4, pp. 1841–1843.

Yovanof, G. S. and Sullivan, J. R. (1992). Lossless predictive coding of color graphics, *in* Sullivan, Rabbani and Dawson (1992), pp. 68–82.

Yule, J. A. C. (1938). The theory of subtractive colour photography. I. the conditions for perfect colour rendering, *Journal of the Optical Society of America* **28**: 419–430.

Yule, J. A. C. (1987). *Principles of Colour Reproduction: applied to photochemical reproduction, color photography, and the ink, paper, and other related industries*, Wiley Series on Photographic science and technology and the graphic arts, John Wiley, New York.

Zaccarin, B. and Liu, B. D. (1991). Transform coding of color images with limited palette size, *Proceedings International Conference Acoustics, Speech, and Signal Processing, ICASSP '91*, Toronto, Canada, pp. 2625–2628.

Zaccarin, B. and Liu, B. D. (1993). A novel approach for coding color quantized images, *IEEE Transactions on Image Processing* **2**(4): 442–453.

Zhang, Y. and Po, L. M. (1995). Fractal color image compression using vector distortion measure, *Proceedings International Conference on Image Processing*, Vol. III, Institute of Electrical and Electronics Engineers, Washington, DC, pp. 276–279.

Zheng, J., Valavanis, K. P. and Gauch, J. M. (1993). Noise removal from color images, *Journal of Intelligent and Robotic Systems* **7**(3): 257–285.

Ziv, J. and Lempel, A. (1977). A universal algorithm for sequential data compression, *IEEE Transactions on Information Theory* **23**(3): 337–343.

Ziv, J. and Lempel, A. (1978). Compression of individual sequences via variable-rate coding, *IEEE Transactions on Information Theory* **24**(5): 53–60.

Index